T0291758

CAMBRIDGE LIBRARY COLLECTION

Books of enduring scholarly value

Physical Sciences

From ancient times, humans have tried to understand the workings of the world around them. The roots of modern physical science go back to the very earliest mechanical devices such as levers and rollers, the mixing of paints and dyes, and the importance of the heavenly bodies in early religious observance and navigation. The physical sciences as we know them today began to emerge as independent academic subjects during the early modern period, in the work of Newton and other 'natural philosophers', and numerous sub-disciplines developed during the centuries that followed. This part of the Cambridge Library Collection is devoted to landmark publications in this area which will be of interest to historians of science concerned with individual scientists, particular discoveries, and advances in scientific method, or with the establishment and development of scientific institutions around the world.

Mathematical and Physical Papers

William Thomson, first Baron Kelvin (1824–1907), is best known for devising the Kelvin scale of absolute temperature and for his work on the first and second laws of thermodynamics, though throughout his 53-year career as a mathematical physicist and engineer at the University of Glasgow he investigated a wide range of scientific questions in areas ranging from geology to transatlantic telegraph cables. The extent of his work is revealed in the six volumes of his *Mathematical and Physical Papers*, published from 1882 until 1911, consisting of articles that appeared in scientific periodicals from 1841 onwards. Volume 2, published in 1884, includes articles from the period 1853–1856, and puts a special emphasis on the issue of the development of electric telegraphy. Also included is Thomson's Bakerian Lecture on the electro-dynamic qualities of metals.

Cambridge University Press has long been a pioneer in the reissuing of out-of-print titles from its own backlist, producing digital reprints of books that are still sought after by scholars and students but could not be reprinted economically using traditional technology. The Cambridge Library Collection extends this activity to a wider range of books which are still of importance to researchers and professionals, either for the source material they contain, or as landmarks in the history of their academic discipline.

Drawing from the world-renowned collections in the Cambridge University Library, and guided by the advice of experts in each subject area, Cambridge University Press is using state-of-the-art scanning machines in its own Printing House to capture the content of each book selected for inclusion. The files are processed to give a consistently clear, crisp image, and the books finished to the high quality standard for which the Press is recognised around the world. The latest print-on-demand technology ensures that the books will remain available indefinitely, and that orders for single or multiple copies can quickly be supplied.

The Cambridge Library Collection will bring back to life books of enduring scholarly value (including out-of-copyright works originally issued by other publishers) across a wide range of disciplines in the humanities and social sciences and in science and technology.

Mathematical and Physical Papers

VOLUME 2

LORD KELVIN

CAMBRIDGE UNIVERSITY PRESS

Cambridge, New York, Melbourne, Madrid, Cape Town,
Singapore, São Paolo, Delhi, Tokyo, Mexico City

Published in the United States of America by Cambridge University Press, New York

www.cambridge.org
Information on this title: www.cambridge.org/9781108028998

© in this compilation Cambridge University Press 2011

This edition first published 1884
This digitally printed version 2011

ISBN 978-1-108-02899-8 Paperback

MATHEMATICAL

AND

PHYSICAL PAPERS.

MATHEMATICAL

AND

PHYSICAL   PAPERS

BY

SIR WILLIAM THOMSON, LL.D., D.C.L., F.R.S.,

PROFESSOR OF NATURAL PHILOSOPHY IN THE UNIVERSITY OF GLASGOW,
AND FELLOW OF ST PETER'S COLLEGE, CAMBRIDGE.

*COLLECTED FROM DIFFERENT SCIENTIFIC PERIODICALS FROM
MAY,* 1841, *TO THE PRESENT TIME.*

VOLUME II.

Cambridge:
AT THE UNIVERSITY PRESS.
1884

Cambridge:

PRINTED BY C. J. CLAY, M.A. & SON,

AT THE UNIVERSITY PRESS.

PREFACE.

THIS second volume contains the Reprint of my papers on Mathematical and Physical subjects, including the titles of all published from April 1853 to February 1856, and the text of all of them, except those which are to be found in my volume of collected papers on Electrostatics and Magnetism.

I have thought it advisable to bring into one series, in Articles LXXII. to LXXXV., nearly all that I had written on the subject of Electric Telegraphs during the period, May 1855 till December 1865: a period of the greatest importance in the history and development of Submarine Telegraphy; as including within it the earliest, and the finally successful, efforts to establish telegraphic communication between Europe and America. The series concludes with a short paper of date March 1873, describing the siphon recorder, which is the instrument at present chiefly used in doing the work of the great submarine cables.

II.   *b*

With respect to the long series of investigations in Electrodynamic Qualities of Metals extending over twenty-two years, from May 1856 till February 1878, the various papers describing them and the results to which they led, will be found collected under Article xci., the concluding Article of the volume.

The present volume will be followed as speedily as possible by others completing the series to the present date.

WILLIAM THOMSON.

University, Glasgow,
Nov. 6, 1884.

CONTENTS.

MATHEMATICAL AND PHYSICAL PAPERS.

ART. LXIV ON THE MUTUAL ATTRACTION AND REPULSION BETWEEN TWO ELECTRIFIED SPHERICAL CONDUCTORS.

[*Phil. Mag.* April and Aug. 1853. " ELECTROSTATICS AND MAGNETISM," Art. VI.]

ART. LXV. REMARQUES SUR LES OSCILLATIONS D'AIGUILLES NON CRISTALISÉES DE FAIBLE POUVOIR INDUCTIF PARAMAGNÉTIQUE OU DIAMAGNÉTIQUE.

[*Comptes Rendus*, XXXVIII. March, 1854. "ELECTROSTATICS AND MAGNETISM,' Art. XXXVI.]

[From *Edinb. Roy. Soc. Trans.* April, 1854; *Phil. Mag.* Dec. 1854; *Comptes Rendus*, XXXIX. Oct. 1854.]

ART. LXVI. ON THE MECHANICAL ENERGIES OF THE SOLAR SYSTEM.

THE mutual actions and motions of the heavenly bodies have long been regarded as the grandest phenomena of mechanical energy in nature. Their light has been seen, and their heat has been felt, without the slightest suspicion that we had thus a direct perception of mechanical energy at all. Even after it has been shown * that the almost inconceivably minute fraction [a one hundred and forty thousand millionth] of the Sun's heat and light reaching the earth is the source of energy from which all the

---

* Herschel's *Astronomy*, edition 1833.—See last ed., § (399).

mechanical actions of organic life, and nearly every motion of inorganic nature at its surface, are derived, the energy of this source has been scarcely thought of as a development of mechanical power.

Little more than ten years ago the true relation of heat to force, in every electric, magnetic, and chemical action, as well as in the ordinary operations of mechanics, was pointed out*; and it is a simple corollary from this that the Sun, within the historical period of human observation, has emitted hundreds of times as much mechanical energy† as that of the motions of all the known planets taken together. The energy, that of light and radiant heat, thus emitted, is dissipated always more and more widely through endless space, and never has been, probably never can be, restored to the Sun, without acts as much beyond the scope of human intelligence as a creation or annihilation of energy, or of matter itself, would be. Hence the question arises, What is the source of mechanical energy, drawn upon by the Sun, in emitting heat, to be dissipated through space? In speculating on the answer, we may consider whether the source in question consists of dynamical energy [kinetic energy], that is, energy of motion ‡, or of "potential energy," (as Mr Rankine has called the energy of force acting between bodies, which will give way to it unless held); or whether it consists partly of dynamical [kinetic] and partly of potential energy.

And again, we may consider whether the source in question, or any part of it, is in the Sun, or exists in surrounding matter, until taken and sent out again by the Sun, or exists as energy

---

* Joule "On the Generation of Heat in the Galvanic Circuit," communicated to the Royal Society of London, Dec. 17, 1840, and published *Phil. Mag.*, Oct. 1841. "On the Heat evolved during the Electrolysis of Water," Literary and Phil. Soc. of Manchester, 1843, Vol. vii., Part 3, Second Series. "On the Calorific Effects of Magneto-Electricity, and the Mechanical Value of Heat," communicated to the British Association, August 1843, and published *Phil. Mag.*, Sept. 1843. "On the Changes of Temperature produced by the Rarefaction and Condensation of Air," communicated to the Royal Soc., June 1844, and published *Phil. Mag.*, May 1845. Joule and Scoresby, "On the Powers of Electromagnetism, Steam, and Horses," *Phil. Mag.*, June 1846. [All these articles are to be found in the collection of Joule's scientific papers, now in print, and published, or, very soon it is hoped, to be published, by the Physical Society of London. W. T. Nov. 9, 1882.]

† Once every 20 years or so.—See Table of Mechanical Energies of the Solar System, appended.

‡ "Actual Energy," as Mr Rankine has called it.

only convertible into heat by mutual actions between the Sun and surrounding matter.

If it be dynamical and entirely in the Sun, it can only be primitive heat; if potential and in the Sun, it can only be energy of chemical forces ready to act. If not in the Sun, it must be due to matter coming to the Sun ; (for it certainly is not a mere communication of motion to solar particles from external energy, as such could only be effected by undulations like sound or radiant heat, and we know that no such anti-radiation can be experienced by a body in the Sun's circumstances); but whether intrinsically in such external matter, or developed by mutual action between this matter and the Sun, and whether dynamical [kinetic] or potential in either case, requires careful consideration, as will be shown in the course of this communication. We see, then, that all the theories which have been yet proposed, as well as every conceivable theory, must be one or other, or a combination of the following three :—

I.   That the Sun is a heated body, losing heat.

II.   That the heat emitted from the Sun is due to chemical action among materials originally belonging to his mass, or that the Sun is a great fire.

III.   That meteors falling into the Sun give rise to the heat which he emits.

In alluding to theories of solar heat in former communications to the Royal Society, I pointed out that the first hypothesis is quite untenable*. In fact, it is demonstrable that, unless the Sun be of matter inconceivably more conductive for heat, and less volatile, than any terrestrial meteoric matter we know, he would become dark in two or three minutes, or days, or months, or years, at his present rate of emission, if he had no source of energy to draw from but primitive heat†. The second has been not only

* [Note of Nov. 9, 1882. I soon was forced to abandon this conclusion, and to definitely adopt prop. I. as the true theory of solar heat and light, that the Sun is merely a vast fluid mass, cooling by radiation into space. See extract from Presidential Address to the British Association, 1871, appended to the present Article.]

† This assertion was founded on the supposition that conduction is the only means by which heat could reach the Sun's surface from the interior, and perhaps requires limitation. For it might be supposed that, as the Sun is no doubt a melted mass, the brightness of his surface is constantly refreshed by incandescent fluid rushing from below to take the place of matter falling upon the surface after

held by the Fire-worshippers, but has probably been conceived
of by all men in all times, and considered as more or less probable
by every philosopher who has ever speculated on the subject.
The third may have occurred at any time to ingenious minds,
and may have occurred and been set aside as not worth considering; but was never brought forward in any definite form,
so far as I am aware*, until Mr Waterston communicated to the
British Association, during its last meeting at Hull, a remarkable
speculation on cosmical dynamics, in which he proposed the
Theory that solar heat is produced by the impact of meteors
falling from extra-planetary space, and striking his surface with
velocities which they have acquired by his attraction. This is a
form of what may be called the Gravitation Theory of Solar Heat,
which is itself included in the general meteoric theory.

The objects of the present communication are to consider the
relative capabilities of the second and third hypothesis to account·
for the phenomena; to examine the relation of the gravitation
theory to the meteoric theory in general; and to determine what
form of the gravitation theory is required to explain solar heat
consistently with other astronomical phenomena.

In the first place, it may be remarked, that in all probability
there must always be meteors falling into the Sun, since the
fact of meteors coming to the earth† proves the existence of such
bodies moving about in space; and even if the motions of these

becoming somewhat cooled and consequently denser—a process which might go on
for many years without any sensible loss of brightness. If we consider, however,
the whole annual emission at the present actual rate, we find, even if the Sun's
thermal capacity were as great as that of an equal mass of water, that his mean
temperature would be lowered by about 3° cent. in two years. We may, I think,
safely conclude that primitive heat within the Sun is not a sufficient source for the
emission which has continued without sensible (if any) abatement for 6000 years.—
(May 4, 1854.) [For a reversal of this conclusion founded on a thermo-dynamic
proof that "the mean specific heat of the Sun's mass is probably more than ten
times, and less than ten thousand times, that of water, see my paper on "the Age
of the Sun's Heat" in Macmillan's *Magazine* for March, 1862, republished as
Appendix (E) of Thomson and Tait's *Natural Philosophy*, 2nd Edition, Part II.,
1883. W. T. Nov. 9, 1882.]

  * [Note of Nov. 9, 1882. Mayer, as I have since learned, had previously
suggested this hypothesis.]

  † To make the argument perfectly conclusive, it would have to be assumed that
meteors not only are, but have been, always falling to the earth for some immense
period of time. The conclusion, however, appears sufficiently probable with the
facts we know.

bodies are at any instant such as to correspond to elliptical or circular orbits round the Sun, the effects of the resisting medium would gradually bring them in to strike his surface. Also, it is easy to prove dynamically that meteors falling into the Sun, whatever may have been their previous state of motion, must enter his atmosphere, or strike his surface, with, on the whole, immensely greater relative velocities than those with which meteors falling to the earth enter the earth's atmosphere, or strike the earth's surface. Now, Joule has shown what enormous quantities of heat must be generated from this relative motion in the case of meteors coming to the earth; and by his explanation* of "falling stars," has made it all but certain that, in a vast majority of cases, this generation of heat is so intense as to raise the body in temperature gradually up to an intense white heat, and cause it ultimately to burst into sparks in the air (and burn if it be of metallic iron) before it reaches the surface. Such effects must be experienced to an enormously greater degree before reaching his surface, by meteors falling to the Sun, if, as is highly probable, he has a dense atmosphere; or they would take place yet more intensely on striking his solid or liquid surface, were they to reach it still possessing great velocities. Hence, it is certain that *some* heat and light radiating from the Sun is due to meteors. It is excessively probable that there is much more of this from any part of the Sun's surface than from an equal area of the earth's, because of the enormously greater action that an equal amount of meteoric matter would produce in entering the Sun, and because the Sun, by his greater attraction, must draw in meteoric matter much more copiously with reference to equal areas of surface. We should have no right then, as was done till Mr Waterston brought forward his theory, to neglect meteoric action in speculating on solar heat, unless we could prove, which we certainly cannot do, that its influence is insensible. It is in fact not only proved to exist as a cause of solar heat, but it is the only one of all conceivable causes which we know to exist from independent evidence.

* See *Philosophical Magazine*, May 1848, for reference to a lecture in Manchester, on the 28th April, 1847, in which Mr Joule said, that " the velocity of a meteoric stone is checked by the atmosphere and its *vis viva* converted into heat, which at last becomes so intense as to melt the body and dissipate it in fragments too small probably to be noticed in their fall to the ground, in most cases." [See also above, Vol. I. Art. XLIX. Part III.]

To test the possibility of this being the *principal or the sole cause* of the phenomenon, let us estimate at what rate meteoric matter would have to fall on the Sun, to generate as much heat as is emitted. According to Pouillet's data*, ·06 of a thermal unit [pound-water-degree] centigrade is the amount of heat incident per second on a square foot directly exposed to solar radiation at the earth's distance from the Sun, which being 95,000,000 miles, and the Sun's radius being 441,000 miles, we infer that the rate of emission of heat from the Sun is

$$06 \times \left(\frac{95,000,000}{441,000}\right)^2 = 2781 \text{ thermal units per}$$

second per square foot of his surface.

The mechanical value of this (obtained by multiplying it by Joule's equivalent, 1390) is

$$83\cdot4 \times \left(\frac{95,000,000}{441,000}\right)^2 = 386,900 \text{ ft. lbs}$$

Now if, as Mr Waterston supposes, a meteor either strikes the Sun, or enters an atmosphere where the luminous and thermal excitation takes place, *without having previously experienced any sensible resistance*, it may be shown dynamically (the velocity of rotation of the Sun's surface, which at his equator is only a mile and a quarter per second, being neglected) that the least relative velocity which it can have is the velocity it would acquire by solar gravitation in falling from an infinite distance, which is equal to the velocity it would acquire by the action of a constant force equal to its weight at the Sun's surface, operating through a space equal to his radius. The force of gravity at the Sun's surface being about 28 times that at the earth's surface, this velocity is

$$\sqrt{\frac{2 \times 28 \times 32\cdot2 \times 441,000}{5280}} = 390 \text{ miles per second;}$$

and its mechanical value per pound of meteoric matter is

$$28 \times 441,000 \times 5280 = 65,000,000,000 \text{ ft. lbs.}$$

Hence the quantity of meteoric matter that would be required, according to Mr Waterston's form of the Gravitation Theory, to strike the Sun per square foot is 0·000060 pounds per second (or about a pound every five hours). At this rate the surface

---

* *Mémoire sur la Chaleur Solaire*, &c. Paris, 1838; see *Comptes Rendus*, July 1838; or Pouillet, *Traité de Physique*, Vol. ii.

would be covered to a depth of thirty feet in a year, if the density of the deposit is the same as that of water, which is a little less than the mean density of the Sun*. A greater rate of deposit than this could not be required, if the hypothesis of no resistance, except in the locality of resistance with luminous reaction, were true ; but a less rate would suffice if, as is probable enough, the meteors in remote space had velocities relative to the Sun not incomparably smaller than the velocity calculated above as due to solar gravitation.

But it appears to me that the hypothesis of no sensible resistance until the " Sun's atmosphere" is reached, or the Sun's surface struck, is not probable † ; because if meteors were falling into the Sun in straight lines, or in parabolic or hyperbolic paths, in anything like sufficient quantities for generating all the heat he emits, the earth in crossing their paths would be, if not intolerably pelted, at least struck much more copiously by meteors than we can believe it to be from what we observe ; and because the meteors we see appear to come generally in directions corresponding to motions which have been elliptic or circular, and rarely if ever in such directions as could correspond to previous parabolic, hyperbolic, or rectilineal paths towards the Sun. If this opinion and the first mentioned reason for it be correct, the meteors containing the stores of energy for future Sun light must be principally within the earth's orbit : and we actually see them there as the "Zodiacal Light," an illuminated shower or rather tornado of stones (Herschel, § 897). The inner parts of this tornado are always getting caught in the Sun's atmosphere, and drawn to his mass by gravitation. The bodies in all parts of it, in consequence of the same actions, must be approaching the Sun, although but very gradually ; yet, in consequence of their comparative minuteness, much more rapidly than the planets. The outer edge of the zodiacal light appears to reach to near the earth at present (Herschel, § 897) ; and in past times it may be that the earth has been in a dense enough part of it to be kept hot, just as the Sun is now, by drawing in meteors to its surface.

---

* This is rather more than double the estimate Mr Waterston has given. The velocity of impact which he has taken is 545 miles per second, in the calculation of which, unless I am mistaken, there must be some error.

† For a demonstration that it is not possible, see Addition, No. 1.

According to this form of the gravitation theory, a meteor would approach the Sun by a very gradual spiral, moving with a velocity very little more than that corresponding to a circular path at the same distance, until it begins to be much more resisted, and to be consequently rapidly deflected towards the Sun; then the phenomenon of ignition commences; after a few seconds of time all the dynamical energy the body had at the commencement of the sudden change is converted into heat and radiated off; and the mass itself settles incorporated in the Sun. It appears, therefore, that the velocity which a meteor loses in entering the Sun is that of a satellite at his surface, which (being $\frac{1}{\sqrt{2}}$ of that due to gravitation from an infinite distance) is 276 miles per second.   The mechanical value (being half that of a body falling to the Sun from a state of comparatively slow motion in space) is about 32,500,000,000 ft. lb. per pound of meteoric matter; hence the fall of meteors must be just twice that which was determined above according to Mr Waterston's form of the theory, and must consequently amount to 3800 lbs. annually per square foot. If, as was before supposed, the density of the deposit is the same as that of water, the whole surface would be covered annually to a depth of 60 feet, from which the Sun would grow in diameter by a mile in 88 years. It would take 4000 years at this rate to grow a tenth of a second in apparent diameter, which could scarcely be perceived by the most refined of modern observations, or 40,000 years to grow 1″, which would be utterly insensible by any kind of observation (that of eclipses included) unassisted by powerful telescopes. We may be confident, then, that the gradual augmentation of the Sun's bulk required by the meteoric theory to account for this heat, may have been going on in time past during the whole existence of the human race, and yet could not possibly have been discovered by observation, and that at the same rate it may go on for thousands of years yet without being discoverable by the most refined observations of modern astronomy. It would take, always at the same rate, about 2,000,000 years for the Sun to grow in reality as much as he appears to grow from June to December by the variation of the earth's distance, which is quite imperceptible to ordinary observation. This leaves for the speculations of geologists on ancient natural history a wide enough range of time with a Sun not sensibly less than our present luminary : Still more, the meteoric theory affords the simplest possible explanation of past changes of climate on the earth. For a time the earth may have been kept melted by the heat of meteors striking it. A period may have followed when the earth was not too hot for vegetation, but was still kept, by the heat of meteors falling through its atmosphere, at a much higher temperature than at present, and illuminated in all regions, polar as well as equatorial, before the existence of night and day. Lastly; although a very little smaller, the Sun may have been at some remote period much hotter than at present by having a more copious meteoric supply.

A dark body of dimensions such as the Sun, in any part of space,. might, by entering a cloud of meteors, become incandescent as intensely in a few seconds as it could in years of continuance of the same meteoric circumstances ; and on again getting to a position in space comparatively free from meteors, it might almost as suddenly become dark again. It is far from improbable that this is the explanation of the appearance and disappearance of bright stars, and of the strange variations of brilliancy of others which have caused so much astonishment*.

The amount of matter, drawn by the Sun in any time from surrounding space, would be such as in $47\frac{1}{2}$ years to amount to a mass equal to that of the earth. Now there is no reason whatever to suppose that 100 times the earth's mass drawn into the Sun, would be missed from the zodiacal light (or from meteors revolving inside the orbit of Mercury, whether visible as the "zodiacal light" or not) ; and we may conclude that there is no difficulty whatever in accounting for a constancy of solar heat during 5000 years of time past or to come. Even physical astronomy can raise no objection by showing that the Sun's mass has not experienced such an augmentation; for according to the form of the gravitation theory which I have proposed, the added matter is drawn from a space where it acts on the planets with very nearly the same forces as when incorporated in the Sun. This form of the gravitation theory then, which may be proved to require a greater

---

* The star which Mr Hind discovered in April 1848, and which only remained visible for a few weeks, during which period it varied considerably in appearance and brightness, but was always of a "ruddy" colour, may not have experienced meteoric impact enough to make its surface more than red hot.

mass of meteoric matter to produce the solar heat than would be required on any other assumption that could be made regarding the previous positions and motions of the meteors, requires not more than it is perfectly possible does fall into the Sun. Hence I think we may regard the adequacy of the meteoric theory to be fully established.

Let us now consider how much chemical action would be required to produce the same effects, with a view both to test the adequacy of the theory that the Sun is merely a burning mass without a supply of either fuel or dynamical energy from without, and to ascertain the extent to which, in the third theory, the combustion of meteors may contribute, along with their dynamical energies, to the supply of solar heat. Taking the former estimate [p. 6 above], 2781 thermal units centigrade, or 3,869,000 foot-lbs. as the rate per second of emission of energy from a square foot of the Sun's surface, equivalent to 7000 horse power*, we find that more than ·42 of a lb. of coal per second, or 1500 lbs. per hour would be required to produce heat at the same rate. Now if all the fires of the whole Baltic fleet were heaped up and kept in full combustion, over one or two square yards of surface, and if the surface of a globe all round had every square yard so occupied, where could a sufficient supply of air come from to sustain the combustion ? yet such is the condition we must suppose the Sun to be in, according to the hypothesis now under consideration, at least if one of the combining elements be oxygen or any other gas drawn from the surrounding atmosphere. If the products of combustion were gaseous, they would in rising check the necessary supply of fresh air ; or if they be solid or liquid (as they might be wholly or partly if the fuel be metallic) they would interfere with the supply of the elements from below. In either or in both ways the fire would be choked, and I think it may be safely affirmed that no such fire could keep alight for more than a few minutes, by any conceivable adaptation of air and fuel. If then the Sun be a burning mass, it must be more analogous to burning gunpowder than to a fire burning in air ; and it is quite conceivable that a solid mass, containing within itself all the elements required for combustion, *provided the products of combustion are*

[* Note of Nov. 11, 1882. This is sixty-seven times the rate per unit of radiant surface, at which energy is emitted from the incandescent filament of the Swan electric lamp when at the temperature which gives about 240 candles per horse power.]

permanently gaseous*, could burn off at its surface all round, and actually emit heat as copiously as the Sun. Thus an enormous globe of gun-cotton might, if at first cold, and once set on fire round its surface, get to a permanent rate of burning, in which any internal part would become heated by conduction, sufficiently to ignite, only when nearly approached by the diminishing surface. It is highly probable indeed that such a body might for a time be as large as the Sun, and give out luminous heat as copiously, to be freely radiated into space, without suffering more absorption from its atmosphere of transparent gaseous products† than the light of the Sun actually does experience from the dense atmosphere through which it passes. Let us therefore consider at what rate such a body, giving out heat so copiously, would diminish by burning away. The heat of combustion could probably not be so much as 4000 thermal units per pound of matter burned‡, the greatest thermal equivalent of chemical action yet ascertained falling considerably short of this. But 2781 thermal units (as found above) are emitted per second from each square foot of the Sun; hence there would be a loss of about ·7 of a pound of matter per square foot per second. Such a loss of matter from every square foot, if of the mean density of the Sun (a little more than that of water), would take off from the mass a layer of about ·5 of a foot thick in a minute, or of about 55 miles thick in a year. At the same rate continued, a mass as large as the Sun is at present would burn away in 8000 years. If the Sun has been burning at that rate in past time, he must have been of double diameter, of quadruple heating power, and of eight-fold mass, only 8000 years ago. We may quite safely conclude then that the Sun does not get its heat by chemical action among particles of matter primitively belonging to his own mass, and we must therefore look to the meteoric theory for fuel, even if we retain the idea of a fire. Now, according to Andrews, the heat of combustion of a pound of iron in oxygen gas is 1301 thermal units, and of a pound of potassium in chlorine 2655; a pound of potassium in oxygen 1700 according to Joule; and carbon in oxygen, according to various observers, 8000. The

---

* On this account gunpowder would not do.

† These would rise and be regularly diffused into space.

‡ Both the elements that enter into combination are of course included in the weight of the burning matter.

greatest of these numbers, multiplied by 1390 to reduce to foot-pounds, expresses only the 6000th part, according to Mr Waterston's theory, and, according to the form of the Gravitation Theory now proposed, only the 3000th part, of the least amount of dynamical energy a meteor can have on entering the region of ignition in the Sun's atmosphere. Hence a mass of carbon entering the Sun's atmosphere, and there burning with oxygen, could only by combustion give out heat equal to the 3000th part of the heat it cannot but give out from its motion. Probably no kind of known matter (and no meteors reaching the earth have yet brought us decidedly new elements) entering the Sun's atmosphere from space, whatever may be its chemical nature, and whatever its dynamical antecedents, could emit by combustion as much as $\frac{1}{1000}$ of the heat inevitably generated from its motion. It is highly probable that many, if not all, meteors entering the Sun's atmosphere do burn, or enter into some chemical combination with substances which they meet. Probably meteoric iron comes to the Sun in enormous quantities, and burns in his atmosphere just as it does in coming to the earth. But (while probably nearly all the heat and light of the sparks which fly from a steel struck by a flint is due to combustion alone) only $\frac{1}{18000}$ part of the heat and light of a mass of iron entering the Sun's atmosphere or $\frac{1}{6}$th of the heat and light of such a meteor entering our own, can possibly be due to combustion. Hence the combustion of meteors may be quite disregarded as a source of solar heat.

At the commencement of this communication, it was shown that the heat radiated from the Sun is either taken from a stock of primitive solar heat, or generated by chemical action among materials originally belonging to his mass, or due to meteors falling in from surrounding space. We saw that there are sufficient reasons for utterly rejecting* the first hypothesis; we have now proved that the second is untenable; and we may consequently conclude that the third is true, or that meteors falling in from space give rise to the heat which is continually radiated off by the Sun. We have also seen that no appreciable portion of the heat thus produced is due to chemical action, either between the meteors and substances which they meet at the Sun, or among elements of the meteors themselves; and that whatever may have been their original positions or motions relatively to one another

* [See note on page 3.]

or to the Sun, the greater part of them fall in gradually from a state of approximately circular motion, and strike the Sun with the velocity due to half the potential energy of gravitation lost in coming in from an infinite distance to his surface. The other half of this energy goes to generate heat very slowly and diffusely in the resisting medium. Many a meteor, however, we cannot doubt, comes in to the Sun at once in the course of a rectilineal or hyperbolic path, without having spent any appreciable energy in the resisting medium; and, consequently, enters the region of ignition at his surface with a velocity due to the descent from its previous state of motion or rest, and there converts both the dynamical effect of the potential energy of gravitation, and the energy of its previous motion, if it had any, into heat which is instantly radiated off to space. But the reasons stated above make it improbable that more than a very small fraction of the whole solar heat is obtained by meteors coming in thus directly from extra-planetary space.

In conclusion, then, the source of energy from which solar heat is derived is undoubtedly meteoric. It is not any intrinsic energy in the meteors themselves, either potential, as of mutual gravitation or chemical affinities among their elements; or actual, as of relative motions among them. It is altogether dependent on mutual relations between those bodies and the Sun. A portion of it, although very probably not an appreciable portion, is that of motions relative to the Sun, and of independent origin. The principal source, perhaps the sole appreciably efficient source, is in bodies circulating round the Sun at present inside the earth's orbit, and probably seen in the sunlight by us and called "the Zodiacal Light." The store of energy for future sunlight is at present partly dynamical, that of the motions of these bodies round the Sun; and partly potential, that of their gravitation towards the Sun. This latter is gradually being spent, half against the resisting medium, and half in causing a continuous increase of the former. Each meteor thus goes on moving faster and faster, and getting nearer and nearer the centre, until some time, very suddenly, it gets so much entangled in the solar atmosphere, as to begin to lose velocity. In a few seconds more, it is at rest on the Sun's surface, and the energy given up is vibrated in a minute or two across the district where it was gathered during so many ages, ultimately to penetrate as light the remotest regions of space.

Explanation of Tables.

The following Tables exhibit the principal numerical data regarding the Mechanical Energies of the Solar System.

In Table I., the mass of the Earth is estimated on the assumption that its mean

TABLE I. *Forces and Motions in the Solar System.*

	Masses in pounds.	Distances from the Sun's centre, in miles.	Forces of attraction towards the Sun, in terrestrial pounds.	Velocities, in miles per second.
Sun	$4,230,000,000 \times 10^{21}$	(surface) 441,000	$28 \cdot 61$ per lb. of matter	(equator) $1 \cdot 27$
Imaginary solid planet close to the Sun	1×10^{21}	441,000	$286,100 \times 10^{17}$	277
Mercury	870×10^{21}	36,800,000	$35,710 \times 10^{17}$	$30 \cdot 36$
Venus	$10,530 \times 10^{21}$	68,700,000	$124,200 \times 10^{17}$	$22 \cdot 22$
Earth	$11,920 \times 10^{21}$	95,000,000	$73,490 \times 10^{17}$	$18 \cdot 89$
Mars	$1,579 \times 10^{21}$	144,800,000	$4,211 \times 10^{17}$	$15 \cdot 28$
Jupiter	$4,037,000 \times 10^{21}$	494,800,000	$919,400 \times 10^{17}$	$8 \cdot 28$
Saturn	$1,208,000 \times 10^{21}$	906,200,000	$81,855 \times 10^{17}$	$6 \cdot 11$
Uranus	$201,490 \times 10^{21}$	1,822,000,000	$3,377 \times 10^{17}$	$4 \cdot 31$
Neptune	$236,380 \times 10^{21}$	2,854,000,000	$1,615 \times 10^{17}$	$3 \cdot 44$
		Distances from Earth's centre.	Attraction towards Earth in terrestrial pounds.	Velocities relatively to Earth's centre, in miles.
Moon	136×10^{21}	237,000	378×10^{17}	$0 \cdot 615$
Earth's equator		3,956	1 per lb. of matter.	$0 \cdot 291$

density is five times that of water, and the other masses are shown in their true proportions to that of the Earth, according to data which Professor Piazzi Smyth has kindly communicated to the author.

In Table II., the mechanical values of the rotations of the Sun and Earth are computed on the hypothesis, that the moment of inertia of each sphere is equal the

TABLE II. Mechanical Energies of the Solar System.

	Potential Energy of gravitation to Sun's surface.		Actual Energy relatively to Sun's centre.	
	In foot-pounds.	Equivalent to supply of Solar Heat, at the present rate of radiation for a period of	In foot-pounds.	Equivalent to supply of Solar Heat, at the present rate of radiation for a period of
Sun	0	0	$976{,}000 \times 10^{30}$	116 yrs. 6 days
Imaginary planet, of 10^{21} lb. of matter, close to the Sun			333×10^{29}	1·44 ...
Mercury	57×10^{33}	6 yrs. 214 days	347×10^{30}	15·2 ...
Venus	697×10^{33}	83 ... 227 ...	$2{,}252 \times 10^{30}$	98·5 ...
Earth	790×10^{33}	94 ... 303 ...	$1{,}843 \times 10^{30}$	80·7 ...
Mars	105×10^{33}	12 ... 252 ...	160×10^{30}	7·0 ...
Jupiter	$268{,}800 \times 10^{33}$	32,240	$119{,}580 \times 10^{30}$	14 yrs. 144 ...
Saturn	$80{,}440 \times 10^{33}$	9,650	$19{,}580 \times 10^{30}$	2 ... 127 ...
Uranus	$13{,}430 \times 10^{33}$	1,610	$1{,}625 \times 10^{30}$	71·2 ...
Neptune	$15{,}750 \times 10^{33}$	1,890	$1{,}217 \times 10^{30}$	53·3 ...
	To the Earth's surface.		Relatively to Earth's centre.	
Moon	$2{,}846 \times 10^{27}$	3·0 hours	$\left.\begin{array}{c}2{,}347 \times 10^{25}\\14{,}810 \times 10^{25}\end{array}\right\}=6{\cdot}4 \times 10^{26}$ ft. tons.	$\left.\begin{array}{c}1{\cdot}48 \text{ minutes}\\9{\cdot}03 \dots\end{array}\right\}$
Earth (rotation)				
Total	$380{,}000 \times 10^{33}$	45,589 years	$1{,}114{,}004 \times 10^{30}$	134 years.

square of its radius multiplied by only one-third of its mass, instead of two-fifths of its mass as would be the case if its matter were of uniform density. These two estimates are only introduced for the sake of comparison with other mechanical values shown in the Table, not having been used in the reasoning.

The numbers in the last column of Table II., showing the times during which the Sun emits quantities of heat mechanically equivalent to the Earth's motion in its orbit, and to its motion of rotation, were first communicated to the Royal Society on the 9th January, 1852, in a paper "On the Sources Available to Man for the production of Mechanical Effect." [Vol. I. Art. LVIII.] These, and the other numbers in the same column, are the only part of the numerical data either shown in the Tables, or used directly or indirectly in the reasoning on which the present theory is founded, that can possibly require any considerable correction; depending as they do on M. Pouillet's estimate of Solar Heat in thermal units. The extreme difficulties in the way of arriving at this estimate, notwithstanding the remarkably able manner in which they have been met, necessarily leave much uncertainty as to the degree of accuracy of the result. But even if it were two or three times too great or too small (and there appears no possibility that it can be so far from the truth), the general reasoning by which the Theory of Solar Heat at present communicated is supported, would hold with scarcely altered force.

The mechanical equivalent of the thermic unit, by which the Solar radiation has been reduced to mechanical units is Mr Joule's result—1390 foot-pounds for the thermal unit centigrade—which he determined by direct experiment with so much accuracy, that any correction it may be found to require can scarcely amount to $\frac{1}{200}$ or $\frac{1}{300}$ of its own value.

ADDITIONS (May 9, 1854), No. I.   *Conclusion of Physical Astronomy against the Extra-planetary Meteoric Theory.*

Meteors which when at great distances possessed, relatively to the centre of gravity of the solar system, velocities not incomparably smaller than the velocity due to gravitation to the Sun's surface, must strike the surfaces of the earth and of the other planets not incomparably less frequently than equal areas of the Sun's surface, and with not incomparably smaller velocities, and consequently must generate heat at the surfaces of the earth and other planets not incomparably less copiously than at equal areas of the Sun's surface. But the whole heat emitted from any part of the Sun's surface is incomparably greater than all that is generated by meteors on an equal area of the earth's surface, and therefore is incomparably greater than all that can be generated at his own surface by meteors coming in with velocities exceeding considerably the velocity due to his attraction from an infinite distance. Hence upon the extra-planetary Meteoric Theory of Solar Heat the quantity of matter required to fall in cannot be much, if at all, less than that required upon the hypothesis that the work done by the Sun's attraction is equal to the mechanical value of the heat emitted from his surface, and must therefore be, as found above, about ·000060 of a pound per square foot per second, or 1900 lbs. per square foot in a year. The mean density of the Sun being about $1\frac{1}{4}$ times that of water, the matter in a pyramidal portion from his centre to a square foot of his surface is about

$$\tfrac{1}{3} \times 441{,}000 \times 5280 \times 1\tfrac{1}{4} \times 64 = 62{,}100{,}000{,}000 \text{ lb.}$$

and the whole annual addition of meteoric matter to the Sun would therefore be

$$\frac{1900}{62{,}100{,}000{,}000} = \frac{1}{32{,}400{,}000}$$

of his own mass. In about six thousand years the Sun would therefore be augmented by $\frac{1}{5000}$ in mass from extra-planetary space. Since the time occupied by each meteor in falling to the Sun from any distance would be much less than the periodic time of a planet revolving at that distance, and since the periodic times of the most distant of the planets is but a small fraction of 6000 years, it follows that the chief effect on the motions of the planetary system produced during such a period by the attraction of the matter falling in would be that depending simply on the augmentation of the central force. To determine this, let M be the Sun's mass at any time t, measured from an epoch 6000 years ago; ω the Earth's mean angular velocity, and a its mean distance at the same time; and $2h$ the constant area described by its radius vector per second. Then we have—

$$\omega^2 a = \frac{M}{a^2} \text{ (centrifugal force)},$$

$$\omega a^2 = h \text{ (equable description of areas)};$$

from which we deduce,

$$a = \frac{h^2}{M},$$

and

$$\omega = \frac{M^2}{h^3}$$

Now, if M_0 denote the mass of the sun at the epoch from which time is reckoned; since the annual augmentation is about $\frac{1}{32400000}$ of the mass itself, we have

$$M = M_0 \left(1 + \frac{t}{32,400,000}\right),$$

and

$$M^2 = M_0^2 \left(1 + \frac{2t}{32,400,000}\right).$$

Hence, if Ω_0 and Ω_T denote the angular velocities at the epoch and at the present time, T; the angular velocity, which is uniformly accelerated during the interval, will have a mean value, Ω, expressed as follows:—

$$\Omega = \tfrac{1}{2}(\Omega_0 + \Omega_T) = \Omega_T \left\{1 - \tfrac{1}{2}\frac{\Omega_T - \Omega_0}{\Omega_T}\right\} = \Omega_T \left(1 - \frac{T}{32,400,000}\right);$$

and if Θ denote the angle described in the time T, we have

$$\Theta = \Omega_T \left(T - \frac{T^2}{32,400,000}\right).$$

To test this conclusion for the case of the earth, let T' denote the number of revolutions round the Sun in the time T. Then, if the unit in which T is measured be the time of a revolution with the angular velocity Ω_T, we have

$$T' = T - \frac{T^2}{32,400,000}.$$

Thus, if T be 4000 years, we have

$$T' = 4000 - \frac{16,000,000}{32,400,000} = 3999\tfrac{1}{2};$$

or only $3999\tfrac{1}{2}$ actual years in a period of 4000 times the present year. Similarly, we should find a loss of $\tfrac{1}{8}$ of a year on a period of 2000 years ago; that is, of about a month and a-half since the Christian era. Thus, if we reckon back about 2000 times the number of days at present in the year, we should find seasons, new and full moons, and eclipses, a month and a half later than would be if the year had been constantly what it is. Now we have abundant historical evidence that there is no such dislocation as this, either in the seasons, or in the lunar phenomena; and it follows that the central attracting mass of the solar system does not receive the augmentation required by the extra-planetary meteoric theory of solar heat. But the reasoning in the preceding paper establishes, with very great probability, a meteoric theory of

Solar Heat; and we may therefore conclude that the meteors supplying the Sun with heat have been for thousands of years far within the Earth's orbit.

No. II. *Friction between Vortices of Meteoric Vapour and the Sun's Atmosphere the immediate Cause of Solar Heat.*

It has been shown that the meteors which contribute the energy for Solar Heat must be for thousands of years within the Earth's orbit before falling to the Sun. But a meteor could not remain for half a year there, unless it were revolving round the Sun, with at each instant the elements of a circular or elliptic orbit. Hence, meteors, on their way in to the Sun, must revolve, each thousands of times round him, in orbits which, whatever may have been their primitive eccentricities, must tend to become more and more nearly circular as they become smaller by the effects of the resisting medium. The resistance must be excessively small, even very near the Sun; since a body of such tenuity as a comet, darting at the rate of 365 miles per second within one-seventh of his radius from his surface, comes away without sensible loss of energy. If, as is probable, the atmosphere of that part of space is carried in a *vortex* round the Sun by the meteors and other planets, it may be revolving at nearly the same rates as these bodies at different distances in the principal plane of the solar system; but we cannot conceive it to be revolving in any locality more rapidly than a planet at the same distance. At one-seventh of the Sun's radius from his surface, this would be about 258 miles per second; and, therefore, a comet approaching so near the Sun, could not have a less velocity relatively to the resisting medium than 107 miles per second, and, if going against the stream, might have as great a relative velocity as 623 miles. On the other hand, the great body of the meteors circulating round the Sun, and carrying the resisting medium along with them, may be moving through it with but small relative velocities; the smaller for each individual meteor, the smaller its dimensions. The effects of the resistance must, therefore, be very gradual in bringing the meteors in to the Sun, even when they are very near his surface; and we cannot tell how many years, or centuries, or thousands of years, each meteor, according to its dimensions, might revolve within a fraction of the Sun's

radius from his surface, before falling in, if it continued solid ; but
we may be sure that it would so revolve long enough to take,
in its outer parts at least, nearly the temperature of that portion
of the space ; and, therefore, probably, unless it be of some
substance infinitely less volatile than any terrestrial or meteoric
matter known to us, long enough to be wholly·converted into
vapour : (the mere fact of a comet* escaping from so near the
Sun as has been stated, being enough to show that there is, at
such a distance, no sufficient atmospheric pressure to prevent
evaporation with so high a temperature).   Even the planet
Mercury, if the Sun is still bright when it falls in, will, in all
probability, be dissipated in vapour long before it reaches the
region of intense resistance; instead of (as it would inevitably do
if not volatile) falling in solid, and in a very short time (perhaps a
few seconds) generating three years' heat, to be radiated off in a
flash which would certainly scorch one half of the earth's surface,
or perhaps the whole, as we do not know that such an extensive
disturbance of the luminiferous medium would be confined by
the law of rectilineal propagation.   Each meteor, when volatilized,
will contribute the actual energy it had before evaporation to a
vortex of revolving vapours, approaching the sun spirally to supply
the place of the inner parts, which, from moving with enormously
greater velocities than the parts of the Sun's surface near them,
first lose motion by intense resistance, emitting an equivalent
of radiant heat and light, and then, from want of centrifugal force,
fall in to the Sun, and, consequently, become condensed to a
liquid or solid state at his surface, where they settle.   The latent
heat absorbed by the meteors in evaporation, and afterwards
partially emitted in their condensation at a higher temperature, is
probably as insensible, in comparison with the heat of friction,
as it has been shown the heat of any combustion or chemical
action they can experience must be, or as we have tacitly assumed

---

* That a comet may escape with only a slight loss by evaporation, if the resistance is not too great to allow it to escape at all, is easily understood, when we
consider that it cannot be for more than a few hours exposed to very intense heat
(not more than two or three hours within a distance equal to the Sun's radius from
his surface). If it consist of a cloud of solid meteors, the smallest fragments may
be wholly evaporated immediately; but all whose dimensions exceed some very
moderate limit of a few feet would unless kept back by the resisting medium and
made to circulate round the Sun until evaporated, get away with only a little boiled
off from their surfaces.

the heat is which is taken and kept by the meteors themselves in approaching from cold space to lodge permanently in the Sun. We may conclude that the Sun's heat is caused, not by solids striking him, or darting through his atmosphere, but by friction in an atmosphere of evaporated meteors, drawn in and condensed by gravitation while brought to rest by the resistance of the Sun's surface.   The quantity of meteoric matter required, if falling in solid, would, as we have seen, be such that half the work done by Solar Gravitation on it, in coming from an infinite distance, is equal to the energy of heat emitted from the Sun, and would, therefore, amount to a pound every 2·3 hours per square foot of the Sun's surface; and it will be the same as this, notwithstanding the process of evaporation and condensation actually going on, if, as appears probable enough, the velocity of the vortex of vapour immediately external to the region of intense resistance in all latitudes be nearly equal to that of a planet close to the Sun.

No. III.   *On the Distribution of Temperature over the Sun's Surface.*

Not only the larger planets, but the great mass of meteors revolving round the Sun, appear to revolve in planes nearly coinciding with his equator, and therefore such bodies, if solid when drawn in to the Sun, would strike him principally in his equatorial regions, and would cause so much a more copious radiation of heat from those regions than from any other parts of his surface, that the appearance would probably be a line or band of light, instead of the round, bright disc which we see.   The nearly uniform radiation which actually takes place from different parts of the Sun's surface appears to be sufficiently accounted for by the distillation of meteors, which, we have seen, must, in all probability, take place from an external region of evaporation at a considerable distance (perhaps several times his radius) inwards to his surface where they are condensed.   Whatever be the dynamical condition of the luminous atmosphere of intense resistance, it is clear that there must be a very strong tendency to an equality of atmospheric pressure over the probably liquid surface of the Sun, and that the temperature of the surface must be everywhere kept near that of the physical equilibrium between the vapours and the liquid or solid into which they are distilling. A lowering of temperature in any part would therefore imme-diately increase the rate of condensation of vapour into it, and so bring a more copious influx of meteoric matter with dynamical energy to supply the deficiency of heat. The various deviations from uniformity which have been observed in the Sun's disc are probably due to eddies which must be continually produced throughout the atmosphere of intense resistance between his surface (which at the equator revolves only at the rate of 1·3 miles per second) and the great vortex of meteoric vapour, which a few miles outside revolves at the rate of 277 miles per second about the equatorial regions, and (if not at the same) certainly at enormously great rates a few miles from the Sun's surface in other localities. Such eddies may ordinarily be seen as the streaks which have been compared to "the streamers of our northern lights" (Herschel, § 387), and when any one of them sends a root down to the Sun's surface it may cause one of the "minute dark dots or pores" which have been observed, and which, when attentively watched, are found to be always changing in appear-ance (Herschel). A great rotatory storm, like the tropical hurri-canes in the earth's atmosphere, may occasionally result from smaller eddies accidentally combining, or from some disturbing cause originating at once an eddy on a much larger scale than usual, and may traverse the Sun's surface, preventing the dis-tillation of meteoric vapour over a great area, and consequently checking both the supply of dynamical energy for radiant heat in the luminous atmosphere of resistance, and the torrents of con-densed meteoric vapours falling to the surface below it. The consequence would be, that the meteoric rain (Herschel's "cloudy stratum,") would be cleared away for a certain space under the central parts of the storm by falling down to the liquid or solid surface, and the luminous atmosphere would lose intensity over a larger space bounded very irregularly by a region of minor eddies, which would cause varying streaks of light. These are exactly the circumstances assumed by Sir William Herschel to account for the great spots with their dark centres surrounded by sharply terminated penumbræ inside the abrupt ragged boundaries of the bright surface, and the branching luminous streaks or "faculæ" in the bright surface outside in their neighbourhood.—(Sir John Herschel's *Astronomy*, § 388.)

No. IV. (Added August 15, 1854.) *On the Age of the Sun.*

The moment of the Sun's rotatory motion (according to the hypothesis mentionod above in the "Explanation of the Tables" regarding the moment of inertia of his mass) is one-third of his mass multiplied by his radius, multiplied by the linear velocity of his equator; and is therefore equal to that of a planet at his surface having a mass equal to $\dfrac{1\cdot 27}{3 \times 277} = \dfrac{1}{650}$ of his own mass. This is equal to the quantity of meteoric matter which would fall in during 25,000 years, at the present rate; and therefore 25,000 years is the time the Sun would take to acquire his actual motion of rotation, by the incorporation of meteors, if these bodies were each revolving in the plane of his equator immediately before entering the region of intense resistance. But it has been shown to be probable that a great space round the Sun is occupied by a vortex of evaporated meteors, and that the incorporation of meteoric matter takes place in reality by the condensation of vapour in a stratum close to his surface all round. It appears not improbable that the tangential velocity of this vortex immediately external to the radiant region of intense resistance may be found to be, in all solar latitudes, very nearly that of a planet close to the Sun. If it be so, the moment of the motion communicated to the Sun by any mass of meteoric matter will be $\dfrac{3\cdot 14159\ldots\ldots}{4}$ of what would be communicated by the incorporation of an equatorial planet of equal mass: as much as $\frac{1}{510}$ of the Sun's mass would have to fall in to produce his present rotation: and 32,000 years would be the time in which this would take place, at the present rate of meteoric incorporation as estimated above.

It will be a very interesting hydrodynamical problem, to fully investigate the motion of the meteoric vortex; and among results to be derived from it will be strict estimates of the contribution to the Sun's rotatory motion, and of the quantity of heat generated, by any amount of meteoric matter in becoming incorporated. With these, and with an accurate determination of the rate at which the Sun radiates heat, we should be able to fix, with certainty, the augmentation of his velocity of rotation actually taking place at present from year to year, and to estimate the time during which the existing rotation would be acquired by meteoric incorporation going on always at the present rate and in the present manner. Whatever this time (which I shall call T years, to avoid circumlocution below) may be, it probably will not be found to differ very widely from the preceding estimate of 32,000 years.

Now, from the fact that the Sun's equator, the planets' orbits, and the Zodiacal Light, all lie nearly in one plane, it appears highly probable that the Sun's present motion has really been acquired by the incorporation of meteors. It is certain that the present manner and rate of meteoric action cannot have been going on for more than the indicated period (T), without giving the Sun a greater rotatory motion than he has, unless (which is very improbable) he were previously rotating in a contrary direction round the same axis : and, at only the present rate, it cannot have been going on for less than that time, unless the Sun has been created with a rotatory motion round his present axis, or has acquired such a motion from some independent mechanical action. The actual rate of Solar radiation in time past may, for all we now know, have been sometimes much greater and sometimes much less than at present ; and there probably has been a time before, when meteors in abundance fell direct to the Sun from extra-planetary space, some getting stopped on their way by the Earth, and illuminating it by friction in its atmosphere and impact at its surface. But the kind of meteoric action now going on, has in all probability produced neither more nor less than T times the quantity of heat now emitted from the Sun in one year. All things considered, it seems not improbable that the Earth has been efficiently illuminated by the Sun alone for not many times more or less than 32,000 years.

As for the future, it will be a most interesting problem to determine the mass of the Zodiacal Light (that is, matter external to the Sun's mass, and within the Earth's orbit), by the perturbations it may probably enough be discovered to produce in the motions of the visible planets. It could scarcely, I think, amount to $\frac{1}{50}$ of the Sun's mass (probably not to nearly as much), without producing such perturbations as could not have been overlooked in the present state of astronomical science ; and we have seen that meteors amounting to $\frac{1}{5000}$ of the Sun's mass, must, at the present estimated rate, fall in in 3000 years. I conclude that Sunlight cannot last as at present for 300,000 years.

The continual acceleration of the Sun's rotatory motion, which the preceding theory indicates, must, sooner or later, be tested by direct observation. The rate of acceleration (which for many thousands of years past and to come must remain sensibly constant, if the solar radiation continues so), is such that the angular velocity is increased annually by $\frac{1}{T}$ of its present value. If T be 32,000, according to the preceding conjectural estimate, the effect in 53 years would amount to diminishing the period of the Sun's revolution by an hour; and the actual effect cannot, according to the theory, be incomparably greater or less. It is just possible that a careful comparison of early with recent observations on the apparent motions of the dark spots may demonstrate this variation; but as some of the most accurate of recent observations of this kind have led to estimates of the period of revolution* differing from one another by as much as 8 hours, it is more probable that, unless some way be discovered for taking into account the motions of the spots themselves with reference to the mass, centuries will elapse before direct evidence can be had either for or against the anticipated acceleration of the Sun's rotatory motion.

APPENDIX.

(Extract from Presidential Address to British Association at Edinburgh in 1871. Published for that year.)

The old nebular hypothesis supposes the solar system, and other similar systems through the universe which we see at a distance as stars, to have originated in the condensation of fiery nebulous matter. This hypothesis was invented before the discovery of thermo-dynamics, or the nebulæ would not have been supposed to be fiery; and the idea seems never to have occurred

		Days.	Hours.	Minutes.
* According to Böhm ...	...	25	12	30
,, Laugier	...	25	8	10
,, Petersen	...	25	4	30

(See *Encyc. Brit.*, 8th edit., Vol. IV., p. 87.) The discrepancies are probably due to proper motions of the spots, which, from the explanation given above in Addition III., may be expected to be very considerable.

to any of its inventors or early supporters that the matter, the condensation of which they supposed to constitute the Sun and stars, could have been other than fiery in the beginning. Mayer first suggested that the heat of the Sun may be due to gravitation : but he supposed meteors falling in to keep always generating the heat which is radiated year by year from the Sun. Helmholtz, on the other hand, adopting the nebular hypothesis, showed in 1854 that it was not necessary to suppose the nebulous matter to have been originally fiery, but that mutual gravitation between its parts may have generated the heat to which the present high temperature of the Sun is due. Further he made the important observations that the potential energy of gravitation in the Sun is even now far from exhausted ; but that with further and further shrinking more and more heat is to be generated, and that thus we can conceive the Sun even now to possess a sufficient store of energy to produce heat and light, almost as at present, for several million years of time future. It ought, however, to be added that this condensation can only follow from cooling, and therefore that Helmholtz's gravitational explanation of future Sun-heat amounts really to showing that the Sun's thermal capacity is enormously greater, in virtue of the mutual gravitation between the parts of so enormous a mass, than the sum of the thermal capacities of separate and smaller bodies of the same material and same total mass. Reasons for adopting this theory, and the consequences which follow from it, are discussed in an article " On the Age of the Sun's Heat," published in *Macmillan's Magazine* for March 1862*.

For a few years Mayer's theory of solar heat had seemed to me probable ; but I had been led to regard it as no longer tenable, because I had been in the first place driven, by consideration of the very approximate constancy of the Earth's period of revolution round the Sun for the last 2000 years, to conclude that " The " principal source, perhaps the sole appreciably effective source of " Sun-heat, is in bodies circulating round the Sun at present " inside the Earth's orbit†" ; and because Le Verrier's researches on the motion of the planet Mercury, though giving evidence of a sensible influence attributable to matter circulating as a great

* [Reprinted as Appendix (E.) to Thomson and Tait's *Natural Philosophy*, Vol. I. Part II., Second Edition (1882).]

† Quoted from "Mechanical Energies, &c." above.

number of small planets within his orbit round the Sun, showed that the amount of matter that could possibly be assumed to circulate at any considerable distance from the Sun must be very small; and therefore "if the meteoric influx taking place at "present is enough to produce any appreciable portion of the heat "radiated away, it must be supposed to be from matter circulating "round the Sun, within very short distances of his surface. The "density of this meteoric cloud would have to be supposed so "great that comets could scarcely have escaped as comets actually "have escaped, showing no discoverable effects of resistance, after "passing his surface within a distance equal to one-eighth of his "radius. All things considered, there seems little probability in "the hypothesis that solar radiation is compensated to any appre- "ciable degree, by heat generated by meteors falling in, at "present; and, as it can be shown that no chemical theory is "tenable*, it must be concluded as most probable that the Sun is "at present merely an incandescent liquid mass cooling†".

Thus on purely astronomical grounds was I long ago led to abandon as very improbable the hypothesis that the Sun's heat is supplied dynamically from year to year by the influx of meteors. But now spectrum analysis gives proof finally conclusive against it.

Each meteor circulating round the Sun must fall in along a very gradual spiral path, and before reaching the Sun must have been for a long time exposed to an enormous heating effect from his radiation when very near, and must thus have been driven into vapour before actually falling into the Sun. Thus, if Mayer's hypo- thesis were correct, friction between vortices of meteoric vapours and the Sun's atmosphere would be the immediate cause of solar heat; and the velocity with which these vapours circulate round equatorial parts of the Sun must amount to 435 kilometres per second. The spectrum test of velocity applied by Lockyer showed but a twentieth part of this amount as the greatest observed relative velocity between different vapours in the Sun's atmo- sphere.

---

* "Mechanical Energies" &c.
† "Age of the Sun's Heat" referred to above.

ART. LXVII. NOTE ON THE POSSIBLE DENSITY OF THE LUMINI-
FEROUS MEDIUM AND ON THE MECHANICAL VALUE OF A
CUBIC MILE* OF SUNLIGHT.

[From *Edin. Royal Soc. Trans.*, Vol. XXI. Part I., May, 1854; *Phil. Mag.* IX.
1854; *Comptes Rendus*, XXXIX. Sept. 1854.]

THAT there must be a medium forming a continuous material
communication throughout space to the remotest visible body is
a fundamental assumption in the undulatory Theory of Light.
Whether or not this medium is (as appears to me most probable)
a continuation of our own atmosphere, its existence is a fact that
cannot be questioned, when the overwhelming evidence in favour of
the undulatory theory is considered; and the investigation of its
properties in every possible way becomes an object of the greatest
interest. A first question would naturally occur, What is the absolute density of the luminiferous ether in any part of space? I am
not aware of any attempt having hitherto been made to answer this
question, and the present state of science does not in fact afford
sufficient data. It has, however, occurred to me that we may assign
an inferior limit to the density of the luminiferous medium in interplanetary space by considering the mechanical value of sunlight
as deduced in preceding communications to the Royal Society
[Art. LXVI. above] from Pouillet's data on solar radiation, and
Joule's mechanical equivalent of the thermal unit. Thus the
value of solar radiation per second per square foot at the earth's
distance from the sun, estimated at ·06 of a thermal unit centigrade, or 83 foot-pounds, is the same as the mechanical value of
sunlight in the luminiferous medium through a space of as many
cubic feet as the number of linear feet of propagation of light per
second. Hence the mechanical value of the whole energy, actual
and potential, of the disturbance kept up in the space of a cubic

* [Note of Dec. 22, 1882. The brain-wasting perversity of the insular inertia
which still condemns British Engineers to reckonings of miles and yards and feet
and inches and grains and pounds and ounces and acres is curiously illustrated by
the title and numerical results of this Article.]

foot at the earth's distance from the sun*, is $\dfrac{83}{192000 \times 5280}$, or $\dfrac{819}{10^7}$ of a foot-pound. The mechanical value of a cubic mile of sunlight is consequently 12050 foot-pounds, equivalent to the work of one-horse power for a third of a minute. This result may give some idea of the actual amount of mechanical energy of the luminiferous motions and forces within our own atmosphere. Merely to commence the illumination of three cubic miles, requires an amount of work equal to that of a horse-power for a minute; the same amount of energy exists in that space as long as light continues to traverse it; and, if the source of light be suddenly stopped, must be emitted from it before the illumination ceases†. The matter which possesses this energy is the luminiferous medium. If, then, we knew the velocities of the vibratory motions, we might ascertain the density of the luminiferous medium ; or, conversely, if we knew the density of the medium, we might determine the average velocity of the moving particles. Without any such definite knowledge, we may assign a superior limit to the velocities, and deduce an inferior limit to the quantity of matter, by considering the nature of the motions which constitute waves of light. For it appears certain that the amplitudes of the vibrations constituting radiant heat and light must be but small fractions of the wave lengths, and that the greatest velocities of the vibrating particles must be very small in comparison with the velocity of propagation of the waves. Let us consider, for instance, plane polarized light, and let the greatest velocity of vibration be denoted by v; the distance to which a particle vibrates on each side of its position of equilibrium, by A; and the wave length, by λ. Then if V denote the velocity of propagation of light or radiant heat, we have

$$\frac{v}{V} = 2\pi \frac{A}{\lambda} ;$$

---

* The mechanical value of sunlight in any space near the sun's surface must be greater than in an equal space at the earth's distance, in the ratio of the square of the earth's distance to the square of the sun's radius, that is, in the ratio of 46,400 to 1 nearly. The mechanical value of a cubic foot of sunlight near the sun must, therefore, be about 0038 of a foot-pound, and that of a cubic mile 560,000,000 foot-pounds.

† Similarly we find 15000 horse-power for a minute as the amount of work required to generate the energy existing in a cubic mile of light near the sun.

and therefore if A be a small fraction of λ, v must also be a small fraction (2π times as great) of V. The same relation holds for circularly polarized light, since in the time during which a particle revolves once round in a circle of radius A, the wave has been propagated over a space equal to λ. Now the whole mechanical value of homogeneous plane polarized light in an infinitely small space containing only particles sensibly in the same phase of vibration, which consists entirely of potential energy at the instants when the particles are at rest at the extremities of their excursions, partly of potential and partly of actual energy when they are moving to or from their positions of equilibrium, and wholly of actual energy when they are passing through these positions, is of constant amount, and must therefore be at every instant equal to half the mass multiplied by the square of the velocity the particles have in the last-mentioned case. But the velocity of any particle passing through its position of equilibrium is the greatest velocity of vibration, which has been denoted by v; and, therefore, if ρ denote the quantity of vibrating matter contained in a certain space, a space of unit volume for instance, the whole mechanical value of all the energy, both actual and potential, of the disturbance within that space at any time is $\frac{1}{2}\rho v^2$. The mechanical energy of circularly polarized light at every instant is (as has been pointed out to me by Professor Stokes) half actual energy of the revolving particles and half potential energy of the distortion kept up in the luminiferous medium; and, therefore, v being now taken to denote the constant velocity of motion of each particle, double the preceding expression gives the mechanical value of the whole disturbance in a unit of volume in the present case. Hence it is clear, that for any elliptically polarized light the mechanical value of the disturbance in a unit of volume will be between $\frac{1}{2}\rho v^2$ and ρv^2, if v still denote the greatest velocity of the vibrating particles. The mechanical value of the disturbance kept up by a number of coexisting series of waves of different periods, polarized in the same plane, is the sum of the mechanical values due to each homogeneous series separately, and the greatest velocity that can possibly be acquired by any vibrating particle is the sum of the separate velocities due to the different series. Exactly the same remark applies to coexistent series of circularly polarized waves of different periods. Hence the mechanical value is certainly less than *half* the mass multiplied into the square of the greatest velocity acquired by a particle, when the disturbance consists in the superposition of different series of plane polarized waves; and we may conclude, for every kind of radiation of light or heat except a series of homogeneous circularly polarized waves, that *the mechanical value of the disturbance kept up in any space is less than the product of the mass into the square of the greatest velocity acquired by a vibrating particle in the varying phases of its motion.* How much less in such a complex radiation as that of sunlight and heat we cannot tell, because we do not know how much the velocity of a particle may mount up, perhaps even to a considerable value in comparison with the velocity of propagation, at some instant by the superposition of different motions chancing to agree; but we may be sure that the product of the mass into the square of an ordinary maximum velocity, or of the mean of a great many successive maximum velocities of a vibrating particle, cannot exceed in any great ratio the true mechanical value of the disturbance. Recurring, however, to the definite expression for the mechanical value of the disturbance in the case of homogeneous circularly polarized light, the only case in which the velocities of all particles are constant and the same, we may define the mean velocity of vibration in any case as such a velocity that the product of its square into the mass of the vibrating particles is equal to the whole mechanical value, in actual and potential energy, of the disturbance in a certain space traversed by it; and from all we know of the mechanical theory of undulations, it seems certain that this velocity must be a very small fraction of the velocity of propagation in the most intense light or radiant heat which is propagated according to known laws. Denoting this velocity for the case of sunlight at the earth's distance from the sun by v, and calling W the mass in pounds of any volume of the luminiferous ether, we have for the mechanical value of the disturbance in the same space,

$$\frac{W}{g} v^2,$$

where g is the number 32·2, measuring in absolute units of force, the force of gravity on a pound. Now we found above, from observation, $\frac{83}{V}$ for the mechanical value, in foot-pounds, of a cubic foot of sunlight; and therefore the mass, in pounds, of a cubic foot of the ether, must be given by the equation,

$$W = \frac{32 \cdot 2 \times 83}{v^2 V}.$$

If we assume $v = \frac{1}{n} V$, this becomes

$$W = \frac{32 \cdot 2 \times 83}{V^3} \times n^2 = \frac{33 \cdot 2 \times 83}{(192000 \times 5280)^3} \times n^2 = \frac{n^2}{3899 \times 10^{20}};$$

and for the mass, in pounds, of a cubic mile we have

$$\frac{32 \cdot 2 \times 83}{(192000)^3} \times n^2 = \frac{n^2}{2649 \times 10^9}.$$

It is quite impossible to fix a definite limit to the ratio which v may bear to V; but it appears improbable that it could be more, for instance, than $\frac{1}{50}$, for any kind of light following the observed laws. We may conclude that probably a cubic foot of the luminiferous medium in the space traversed by the earth contains not less than $\frac{1}{1560 \times 10^{17}}$ of a pound of matter, and a cubic mile not less than $\frac{1}{1060 \times 10^6}$.

If the mean velocity of the vibrations of light within a spherical surface concentric with the sun and passing through the earth were equal to the earth's velocity—a very tolerable supposition— since this is $\frac{1}{10170}$ of the velocity of light, the whole mass of the luminiferous medium within that space would be $\frac{1}{30000}$ of the earth's mass, since the mechanical value of the light within it, being as much as the sun radiates in about 8 minutes, is about $\frac{1}{15000}$ of the mechanical value of the earth's motion. As the mean velocity of the vibrations might be many times greater than has been supposed in this case, the mass of the medium might be considerably less than this [thirty-thousandth of the earth's mass]; but we may be sure it is not incomparably less, not 100,000 times as small for instance. On the other hand, it is worth remarking that the preceding estimate shows that what we know of the mechanical value of light renders it in no way probable that the masses of luminiferous medium in interplanetary spaces, or all round the sun in volumes of which the linear dimensions are comparable with the dimensions of the planets' orbits, are otherwise than excessively small in comparison with the masses of the planets.

But it is also worth observing that the luminiferous medium is enormously denser than the continuation of the terrestrial atmosphere would be in interplanetary space, if rarefied according to Boyle's law always, and if the earth were at rest in a space of constant temperature with an atmosphere of the actual density at its surface*. Thus the mass of air in a cubic foot of distant space several times the earth's radius off, on this hypothesis, would be $\dfrac{1\,\text{lb.}}{442 \times 10^{345}}$; while there cannot, according to the preceding estimate, be in reality less than $\dfrac{1\,\text{lb.}}{156 \times 10^{18}}$, which is 28×10^{326} times as much, of matter in every cubic foot of space traversed by the earth.

ART. LXVIII. Aperçu sur des recherches relatives aux effets des courants électriques dans des conducteurs inégalement échauffés, et à d'autres points de la thermo-électricité.

[*Comptes Rendus*, XXXIX. July, 1854.]

The substance of this Article is contained in pp. 460—463, of Vol. I. Art. LI.

---

* "Newton has calculated (*Princ.* III., p. 512) that a globe of ordinary density at the earth's surface, of one inch in diameter, if reduced to the density due to the altitude above the surface of one radius of the earth, would occupy a sphere exceeding in radius the orbit of Saturn."—(Herschell's *Astronomy*, Note on § 559.) It would (on the hypothesis stated in the text) we may now say occupy a sphere exceeding in radius millions of millions of times the distances of any stars of which the parallaxes have been determined. A pound of the medium, in the space traversed by the earth, cannot occupy more than the bulk of a cube 1000 miles in side. The earth itself, in moving through it, cannot displace less than 250 pounds of matter.

ART. LXIX. ON MECHANICAL ANTECEDENTS OF MOTION, HEAT, AND LIGHT.

[From *Brit. Ass. Rep.*, Part II. 1854; *Edin. New Phil. Jour.* I. 1855; *Comptes Rendus*, XL. 1855.]

THIS communication was opened with some general explanations regarding mechanical energy, and the terms which have been introduced to designate the various forms under which it is manifested. Any piece of matter, or any group of bodies, however connected, which either is in motion, or can get into motion without external assistance, has what is called mechanical energy. The energy of motion may be called either "dynamical energy" or "actual energy*." The energy of a material system at rest, in virtue of which it can get into motion, is called "potential energy." The author showed the use of these terms, explained the idea of a *store of energy*, and conversions and transformations of energy, by various illustrations. A stone at a height, or an elevated reservoir of water, has potential energy. If the stone be let fall, its potential energy is converted into actual energy during its descent, exists entirely as the actual energy of its own motion at the instant before it strikes, and is transformed into heat at the moment of coming to rest on the ground. If the water flow down by a gradual natural channel, its potential energy is gradually converted into heat by fluid friction, according to an admirable discovery made by Mr Joule of Manchester above twelve years ago, which has led to the greatest reform that physical science has experienced since the days of Newton. From that discovery, it may be concluded with certainty that heat is not matter, but some kind of motion among the particles of matter; a conclusion established, it is true, by Sir Humphrey Davy and Count Rumford at the end of last century, but ignored by even the highest

---

* [A few years later, in advocating a restoration of the original and natural nomenclature,—"mechanics the science of machines,"—"dynamics the science of force," I suggested (instead of statics and dynamics the two divisions of mechanics according to the then usual nomenclature) that statics and kinetics should be adopted to designate the two divisions of dynamics. At the same time I gave, instead of "dynamical energy," or "actual energy," the name "kinetic energy" which is now in general use to designate the energy of motion. W. T., Lensfield Cottage, Cambridge, May 15, 1883.]

scientific men during a period of more than forty years. Mr Joule, by a series of well planned and executed experiments, ascertained that a pound of water would have its temperature increased by $1°$ (Fahrenheit) if it kept all the heat that would be generated by its descent through 772 feet; that is, the "actual" or "dynamical" energy of as much heat as raises the temperature of a pound of water $1°$ is an exact equivalent for the potential energy of a pound of matter 772 feet above the ground. Mr Joule also fully established the relations of equivalence among the energies of chemical affinity, of heat, of combination or combustion, of electrical currents in the galvanic battery, of electrical currents in magneto-electric machines, of engines worked by galvanism, and of all the varied and interchangeable manifestations of calorific action and mechanical force which accompany them. These researches, with the theory of animal heat and motion in relation to the heat of combustion of the food, and the theory of the phenomena presented by shooting stars, due to the same penetrating investigator, have afforded to the author of the present communication the chief groundwork for his speculations.

The heat emitted by animals, and the mechanical effects which they produce, are transformations of the energy of chemical affinity with which the food consumed by them combines with the oxygen they inhale. The heat, sound, and mechanical effects produced by the explosion of gunpowder are, all together, equivalent to the energy of chemical affinity between the different substances of which gunpowder is composed. The potential energy of war is contained in the stores of gunpowder and food brought into the field. The gunpowder carried by artillery and infantry contains all the potential energy ordinarily brought into action by those two arms of the service. The men's food, and the forage for the horses, contain the stores of potential energy drawn upon in a charge of cavalry. Artillerymen, foot soldiers, sailors, steamers with their engines, guns, swords, are only means and appliances by which the potential energy contained in the stores of gunpowder and food is directed to strike the blows by which the desired effects are produced.

The heat and mechanical actions of animals are transformations of the potential energy of their food mechanically equivalent to the heat that would be got by burning it. The food of animals is either vegetable, or animal fed on vegetable, or ultimately vegetable after several removes. Now; except mushrooms and other funguses, all of which can grow in the dark, are nourished by organic food like animals, and absorb oxygen and exhale carbonic acid like animals; all known vegetables get the greater part of their substance, certainly all their combustible matter, from the decomposition of carbonic acid and water, absorbed by them from the air and soil. The separation of carbon and of hydrogen from oxygen in these decompositions is an energetic effect, equivalent to the heat of recombination of those elements by combustion or otherwise. The beautiful discovery of Priestley, and the subsequent researches of Sennebier, De Saussure, Sir Humphrey Davy, and others, have made it quite certain that those decompositions of water and carbonic acid only take place naturally in the day time, and that light falling on the green leaves, either from the sun or from an artificial source, is an essential condition, without which they are never effected. There cannot be a doubt but that it is the dynamical energy of the luminiferous vibrations which is here efficient in forcing the particles of carbon and hydrogen away from those of oxygen, towards which they are attracted with such powerful affinities; and that luminiferous motions are reduced to rest, to an extent exactly equivalent to the potential energy thus called into being. Whether or not the coolness of green fields and fresh foliage is to any sensible extent due to this cause, it is quite certain that sun-heat is put out of existence as heat, in the growth of plants in any locality, and that just as much heat, neither more nor less, is emitted from fires in which the whole growth of any period of time is burned. Coal, composed as it is of the relics of ancient vegetation, derived its potential energy from the light of distant ages. Wood fires give us heat and light which has been got from the sun a few years ago. Our coal fires and gas lamps bring out for our present comfort heat and light of a primeval sun which have lain dormant as potential energy beneath seas and mountains for countless ages.

We must look then to the sun as the source from which the mechanical energy of all the motions and heat of living creatures, and all the motion, heat, and light derived from fires and artificial flames is supplied. The natural motions of air and water derive their energy partly, no doubt, from the sun's heat, but partly also from the earth's rotatory motion and the relative motions and mutual forces between the earth, moon, and sun. If we except the heat derivable from the combustion of native sulphur, and of meteoric iron, every kind of motion (heat and light included) that takes place naturally, or that can be called into existence through man s directing powers on this earth, derives its mechanical energy either from the sun's heat or from motions and forces among bodies of the solar system.

In a speculation recently communicated to the Royal Society of Edinburgh, the author has shown that the sun's heat is probably* due to friction in the atmosphere between his surface and a vortex of vapours, fed externally by the evaporation of small planets, in a region of very high temperature round the sun, which they reach by gradual spiral paths, and falling in torrents of meteoric rain, down from the luminous atmosphere of intense resistance, to the sun's surface.

A continuation of the inquiry raises the question, from what source do the planets, large and small, derive the mechanical energy of their motions? This is a question, to the answering of which mechanical reasoning may legitimately be applied : for we know that from age to age the potential energy of the mutual gravitation of those bodies is gradually expended, half in augmenting their motions, and half in generating heat ; and we may trace this kind cf action either backwards or forwards; backwards for a million of million of years with as little presumption as forwards for a single day. If we trace them forwards, we find that the end of this world as a habitation for man, or for any living creature or plant at present existing in it, is *mechanically inevitable ;* and if we trace them backwards according to the laws of matter and motion, certainly fulfilled in all the actions of nature which we have been allowed to observe, we find that a time must have been when the earth, with no sun to illuminate it, the other bodies known to us as planets, and the countless smaller planetary masses at present seen as the zodiacal light, must have been indefinitely remote from one another and from all other solids in space. All such conclusions are subject to limitations, as we do not know at what moment a creation of matter or energy may have given a beginning, beyond which mechanical speculations cannot lead us. If in purely mechanical science we are ever liable to forget this limitation, we ought to be reminded of it by con-

* [For correction of this conclusion see Appendix to Art. LXVI., pp. 25—27 above.]

sidering that purely mechanical reasoning shows a time when the
earth must have been tenantless; and teaches us that our own
bodies, as well as all living plants and animals, and all fossil
organic remains, are organized forms of matter to which science
can point no antecedent except the Will of a Creator, a truth
amply confirmed by the evidence of geological history. But if
duly impressed with this limitation to the certainty of all speculations regarding the future and pre-historical periods of the past,
we may legitimately push them into endless futurity, and we can
be stopped by no barrier of past time, without ascertaining at
some finite epoch a state of matter derivable from no antecedent
by natural laws. Although we can conceive of such a state of all
matter, or of the matter within any limited space, and have cases
of it in the arbitrary distributions of temperature, prescribed at
" initial " in the theory of the conduction of heat (see* *Cambridge
Mathematical Journal*, Vol. IV. 1843, or Art. XI. of Vol. I. of
Mathematical and Physical Papers), yet we have no indications
whatever of natural instances of it, and in the present state of
science we may look for mechanical antecedents to every natural
state of matter which we either know or can conceive at any past
epoch however remote.

It is by tracing backwards the motions which are at present
observed, according to the known laws of motion and heat, with
no limit as to time, that the author arrives at the conclusion
that the bodies now constituting our solar system have been at
infinitely greater distances from one another in space than they
are now. He remarked that the nebular theory, as ordinarily
stated, assuming as it does a previously gaseous state of matter, is
not only untrue, but the reverse of the truth, according to the
views now brought forward, since these show evaporation as a
necessary consequence of heat generated by collisions and friction,
and the general past and present tendency of matter is seen to
be the conglomeration of solids and liquids accompanied by a
gradual increase of the quantity of gaseous fluid occupying space.

Prof. Helmholtz, in a most interesting popular lecture on transformations of natural forces, delivered on the 7th of February last

* " Note on some points in the Theory of Heat," a short article in which it was
shown how to test the age of a distribution of heat, by applying a certain criterion
of convergence to its expression in the infinite series characteristic of the external
circumstances of the body in which it is given.

at Königsberg, has estimated that, if the particles at present constituting the sun's mass have been drawn together by mutual gravitation, from a state of infinite diffusion, as supposed in the nebular theory, not however a gaseous state, as ordinarily supposed, but a state in which the particles exercise no mutual action except that of gravitation, the whole heat generated must have amounted to about 28,000,000 thermal units Centigrade per pound of the sun's mass. This estimate would not, as the author of the present paper shows, require any change, whether we assume, as the antecedent condition of the solar mass, a state of infinite diffusion, or a state of aggregation in solid masses of any dimensions small compared with his present dimensions, and separated from one another at comparatively great distances; provided always there has been no relative motion among them except what is generated by mutual gravitation. If, then, the whole mass of the sun has grown by the process which, according to the author's theory of solar heat (*certain* as regards a part, whether or not it may be sufficient to account for the whole of the radiation of solar heat), we know to be augmenting it at present, there must have been generated, in the whole process of conglomeration, the quantity of heat stated above; a quantity which amounts to about 20,000,000 times as much as is at present radiated off in one year. The author gave reasons for believing that this heat must have been nearly all radiated off immediately on being generated, and that enough of it has not been retained in the conglomerated mass to be the store from which the heat at present radiated is drawn*

That the present solar radiation is supplied chiefly from a store of heat contained in the mass, whether created there or generated mechanically by the impact of meteors which have fallen in during remote periods of past time, appears very improbable [most probable].

On the contrary, there must in all probability be some [is certainly no] agency continually supplying [any approach to a

---

* [Reasons which soon led me to a reversal of this conclusion were given in an article "On the Age of the Sun's Heat," published in *Macmillan's Magazine* for March 1862, and re-published as Appendix (E) to Thomson and Tait's *Natural Philosophy*, Part II. Second Edition (1883). See also Appendix to "Mechanical Energies of the Solar System (Article LXVI. pp. 25—27 above) where will be found justification of the amendments on the text of the present page, now inserted in square brackets. W. T. May 27, 1883.]

sufficiency of] heat to compensate the loss constantly experienced by radiation from the sun; and that agency, as the author has shown elsewhere, can be no other than the mechanical action of masses coming from a state of very rapid motion round the sun to rest on his surface.

The author showed how a system of solid bodies, large and small, initially at rest and at great distances from one another, may, by their mutual gravitations, and by the resistance their motions must experience in the gaseous atmosphere, evaporated from them by the heat of their collisions after a vast period of time, come into a state of motion, heat, and light, analogous to the present condition of our solar system and the stars. The origin of rotatory motion is explained by showing that different systems starting from rest will influence one another so as to acquire contrary rotatory motions, without any aggregate of rotatory momentum being acquired by the whole. Any system or group beginning to concentrate round one principal mass, after having thus acquired a momentum of rotatory motion, will acquire from it, in a certain stage of advancement, just such approximately circular motions as those of the planets, the particles of the zodiacal light and the satellites of our solar system, and such rotatory motions as the central and other masses are known to have, all chiefly in one direction.

In considering the question whether all the heat and motion existing in matter have their origin in that action by which their amount is at present being increased, it is shown that, unless their entire actual energy [kinetic energy] exceeds a certain definite limit, namely, the value of the whole potential energy of gravitation that would be spent in drawing all the particles of matter from a state of infinite diffusion into their present positions, it is quite possible they may be so produced; or that *the potential energy of gravitation may be in reality the ultimate created antecedent of all the motion, heat, and light at present in the universe.*

ART. LXX.   ELEMENTARY DEMONSTRATIONS OF PROPOSITIONS IN
THE THEORY OF MAGNETIC FORCE.

[*Phil. Mag.*, April, 1855.

ELECTROSTATICS AND MAGNETISM, Article XXXVII.]

ART. LXXI.   ON THE MAGNETIC MEDIUM AND ON THE EFFECTS
OF COMPRESSION.

[*Phil. Mag.*, April and Dec. 1855, and Jan. 1856.

ELECTROSTATICS AND MAGNETISM, Article XXXVIII.]

ART. LXXII.   COMPENDIUM OF THE FOURIER MATHEMATICS FOR
THE CONDUCTION OF HEAT IN SOLIDS, AND THE MATHE-
MATICALLY ALLIED PHYSICAL SUBJECTS OF DIFFUSION OF
FLUIDS AND TRANSMISSION OF ELECTRIC SIGNALS THROUGH
SUBMARINE CABLES.

[From the *Encyclopædia Britannica*, new edition (1880), being Appendix
and § 82 of Article "Heat," and *Quarterly Journal of Mathematics*, Vol. I.,
March, 1856.]

[THE most important part of this article (the system of
formulas I. to XVII.) is taken from a private mathematical
memorandum-book, under dates September and October, 1850.
Though first published in 1880, it is reproduced in this place
because it constituted part of the substance of the paper on the
conduction of electricity and heat referred to at the commencement of Art. LXXIII. below, as intended for communication to the
Royal Society, and because the substance of it may be found
useful for mathematical readers of Article LXXIII. and others
which follow it on the "Theory of the Electric Telegraph."]

Let v be the temperature at any point P specified by ξ, η, ζ
according to any system of three sets of plane or curved orthogonal
surfaces used for coordinates. Let $\lambda d\xi$, $\mu d\eta$, $\nu d\zeta$ be the lengths
of the edges of the infinitesimal rectangular parallelepiped having
P for its centre, and its sides parts of the six surfaces $\xi - \frac{1}{2}d\xi$,
$\xi + \frac{1}{2}d\xi$, $\eta - \frac{1}{2}d\eta$, $\eta + \frac{1}{2}d\eta$, $\zeta - \frac{1}{2}d\zeta$, $\zeta + \frac{1}{2}d\zeta$.

The rates of variation of temperature per unit of length in the directions corresponding to the variations of ξ, η, ζ are respectively

$$-\frac{1}{\lambda}\frac{dv}{d\xi}, \qquad -\frac{1}{\mu}\frac{dv}{d\eta}, \qquad -\frac{1}{\nu}\frac{dv}{d\zeta}.$$

Hence the fluxes across three infinitesimal rectangles having their edges parallel to the three pairs of sides of the parallelepiped, and each having its centre at P, are respectively

$$-\frac{k}{\lambda}\frac{dv}{d\xi}\cdot\mu\nu d\eta d\zeta, \qquad -\frac{k}{\mu}\frac{dv}{d\eta}\nu\lambda d\zeta d\xi, \qquad -\frac{k}{\nu}\frac{dv}{d\zeta}\lambda\mu d\xi d\eta.$$

Hence the excess of the quantities of heat conducted in to the parallelepiped above those conducted out across the three pairs of faces is

$$\left\{\frac{d}{d\xi}\left(k\frac{\mu\nu}{\lambda}\frac{dv}{d\xi}\right)+\frac{d}{d\eta}\left(k\frac{\nu\lambda}{\mu}\frac{dv}{d\eta}\right)+\frac{d}{d\zeta}\left(k\frac{\lambda\mu}{\nu}\frac{dv}{d\nu}\right)\right\}d\xi d\eta d\zeta...(1).$$

The effect of this gain of heat is to warm the matter of the parallelepiped at a rate per unit of time equal to the rate of gain of heat divided by $c\lambda\mu\nu d\xi d\eta d\zeta$, the thermal capacity of the matter.

Hence

$$\frac{dv}{dt}=\frac{1}{c\lambda\mu\nu}\left\{\frac{d}{d\xi}\left(k\frac{\mu\nu}{\lambda}\frac{dv}{d\xi}\right)+\frac{d}{d\eta}\left(k\frac{\nu\lambda}{\mu}\frac{dv}{d\eta}\right)+\frac{d}{d\zeta}\left(k\frac{\lambda\mu}{\nu}\frac{dv}{d\zeta}\right)\right\}...(2).$$

This, for the case of the uniform motion of heat ($dv/dt = 0$), was first given by Lamé, to whom the generalized system of curvilinear coordinates for a point is due ("Mémoire sur les Lois de l'Equilibre des Fluides Ethéres," *Journal de l'École Polytechnique*, vol. iii., cahier xxiii.). He deduced it from Fourier's equation [(3) below] in terms of plane rectangular coordinates by a laborious transformation. Equation (2) was first given, proved as above, as the direct expression of Fourier's fundamental law of conduction, by W. Thomson (*Cambridge Mathematical Journal*, Nov. 1843). [Art. IX. Vol. I.]

For plane rectangular coordinates we have $\lambda = \mu = \nu = 1$, and if we put x, y, z for ξ, η, ζ in this case, (2) becomes

$$\frac{dv}{dt}=\frac{1}{c}\left\{\frac{d}{dx}\left(k\frac{dv}{dx}\right)+\frac{d}{dy}\left(k\frac{dv}{dy}\right)+\frac{d}{dz}\left(k\frac{dv}{dz}\right)\right\}\quad........(3),$$

which is Fourier's celebrated fundamental equation. From it we may deduce by transformation the proper forms of the corresponding equation for polar coordinates; but they are more easily got direct from the equation (2) for generalized coordinates. Thus for ordinary polar coordinates r, θ, ϕ we have, if we take these for ξ, η, ζ respectively,

$$\lambda = 1, \quad \mu = r, \quad \nu = r \sin \theta.$$

Hence (2) becomes

$$\frac{dv}{dt} = \frac{1}{cr^2} \left\{ \frac{d}{dr} \left(kr^2 \frac{dv}{dr} \right) + \frac{1}{\sin \theta} \frac{d}{d\theta} \left(k \sin \theta \frac{dv}{d\theta} \right) + \frac{1}{\sin^2 \theta} \frac{d}{d\varphi} \left(k \frac{dv}{d\phi} \right) \right\}..(4).$$

If k be constant, and we we put $k/c = \kappa$, this becomes

$$\frac{dv}{dt} = \frac{\kappa}{r^2} \left\{ \frac{d}{dr} \left(r^2 \frac{dv}{dr} \right) + \frac{1}{\sin \theta} \frac{d}{d\theta} \left(\sin \theta \frac{dv}{d\theta} \right) + \frac{1}{\sin^2 \theta} \frac{d^2 v}{d\phi^2} \right\} \cdots (5);$$

or

$$\frac{du}{dt} = \kappa \left\{ \frac{d^2 u}{dr^2} + \frac{1}{\sin \theta} \frac{d}{d\theta} \left(\sin \theta \frac{du}{d\theta} \right) + \frac{1}{\sin^2 \theta} \frac{d^2 u}{d\phi^2} \right\} \text{ where } u = vr \ldots (6).$$

If again we take for the coordinates r, ϕ, z (polar coordinates in the plane perpendicular to z being denoted by r, ϕ), we have $\lambda = 1$, $\mu = r$, $\nu = 1$, and so we find

$$\frac{dv}{dt} = \frac{1}{c} \left\{ \frac{d}{rdr} \left(kr \frac{dv}{dr} \right) + \frac{d}{rd\phi} \left(\frac{k}{r} \frac{dv}{d\phi} \right) + \frac{d}{dz} \left(k \frac{dv}{dz} \right) \right\} \ldots \ldots (7).$$

For the case of k constant we may take it outside the brackets in each of these equations, as we have already done in (4); thus (2) becomes

$$\frac{dv}{dt} = \frac{k}{c} \left(\frac{d^2 v}{dx^2} + \frac{d^2 v}{dy^2} + \frac{d^2 v}{dz^2} \right) \ldots \ldots \ldots \ldots \ldots (8);$$

or, with κ for k/c, the diffusivity (§§ 81, 82),

$$\frac{dv}{dt} = \kappa \left(\frac{d^2 v}{dx^2} + \frac{d^2 v}{dy^2} + \frac{d^2 v}{dz^2} \right) \ldots \ldots \ldots \ldots \ldots (9).$$

It is this restricted form which, with the further restriction that c be constant, is most generally recognized as Fourier's equation of conduction, and it is for it, with these restrictions, that his brilliant solutions were given. These solutions are available for practical use by limiting the range of temperature within which any one solution is continuously applied to a range of temperature within which the values of k and c are each nearly enough constant. We may expect $10°$ or $20°$ C. on each side of the mean temperature to be practically not too wide a range for any case, judging from copper and iron (§ 80), the only substances for which hitherto we have any information as to variations of both k and c with temperature.

Each of the following expressions I......XVII. for v satisfies (9) or its equivalent (6), as the reader will readily verify for himself. The special condition corresponding to the peculiar character of the particular solution is specially noted in each case.

I. Instantaneous simple point source; a quantity Q of heat suddenly generated at the point $(0, 0, 0)$ at time $t = 0$, and left to diffuse through an infinite homogeneous solid.

Every other solution is obtainable from this by summation.

$$\left. v = \frac{Q\epsilon^{-r^2/4\kappa t}}{8\pi^{\frac{3}{2}}(\kappa t)^{\frac{3}{2}}} \atop r^2 = x^2 + y^2 + z^2 \right\} \quad\quad ....(10).$$

where

Verify that $\int_{-\infty}^{\infty} \int_{-\infty}^{\infty} \int_{-\infty}^{\infty} v\,dx\,dy\,dz = 4\pi \int_0^{\infty} vr^2\,dr = Q$; and that $v = 0$ when $t = 0$; unless also $x = 0, y = 0, z = 0$.

Remark that

$$\frac{dv}{dt} = 0 \quad \text{when} \quad t = \frac{r^2}{6\kappa} ....(11).$$

II. Constant simple point-source, rate q :

$$v\left[= q\int_0^{\infty} dt\, \frac{\epsilon^{-r^2/4\kappa t}}{8\pi^{\frac{3}{2}}(\kappa t)^{\frac{3}{2}}} \right] = \frac{q}{4\pi\kappa r} \quad .. ..(12).$$

The formula within the brackets shows how this obvious solution is derivable from (10).

III. Continued point-source; rate per unit of time, at time t, an arbitrary function, $f(t)$:

$$v = \int_0^{\infty} d\chi f(t - \chi) \frac{\epsilon^{-r^2/4\kappa\chi}}{8\pi^{\frac{3}{2}}(\kappa\chi)^{\frac{3}{2}}} ....(13).$$

IV. Time-periodic simple point-source, rate per unit of time at time t, $q \sin 2nt$:

$$v\left[= q\int_0^{\infty} d\chi \sin 2n\,(t - \chi) \frac{\epsilon^{-r^2/4\kappa\chi}}{8\pi^{\frac{3}{2}}(\kappa\chi)^{\frac{3}{2}}} \right] = \frac{q}{4\pi\kappa r} \epsilon^{-(n/\kappa)^{\frac{1}{2}}r} \sin\left[2nt - (n/\kappa)^{\frac{1}{2}}r\right].$$

Verify that v satisfies (6); also that $-4\pi\kappa r^2 \dfrac{dv}{dr} = q \sin 2nt$ where $r = 0$.

V. Instantaneous spherical surface source; a quantity Q suddenly generated over a spherical surface of radius a, and left to diffuse outwards and inwards:

$$v = Q\frac{\epsilon^{-(r-a)^2/4\kappa t} - \epsilon^{-(r+a)^2/4\kappa t}}{8\pi^{\frac{3}{2}}ar\,(\kappa t)^{\frac{1}{2}}} \quad\ldots\ldots\ldots\ldots\ldots(15).$$

To prove this most easily, verify that it satisfies (6); and farther verify that

$$4\pi\int_0^\infty vr^2 dr = Q;$$

and that $v = 0$ when $t = 0$, unless also $r = a$.

Remark that (15) becomes identical with (10) when $a = 0$; remark farther that (15) is obtainable from (10) by integration over the spherical surface.

VI. Constant spherical surface source; rate per unit of time from the whole surface, q:

$$\left.\begin{aligned}
v\left[= q\int_0^\infty dt\frac{\epsilon^{-(r-a)^2/4\kappa t} - \epsilon^{-(r+a)^2/4\kappa t}}{8\pi^{\frac{3}{2}}ar\,(\kappa t)^{\frac{1}{2}}}\right]\\
= q/4\pi\kappa r, \text{ where } r > a\\
= q/4\pi\kappa a, \text{ where } r < a
\end{aligned}\right\}\ \ldots\ldots\ldots(16).$$

and

The formula within the brackets shows how this obvious solution is derivable from (15).

VII. Time-periodic spherical surface source; rate per unit of time, at time t, from whole surface, $q \sin 2nt$: taking ν to denote n/κ.

$$\left.\begin{aligned}
v\left[= q\int_0^\infty d\chi \sin 2n\,(t-\chi)\frac{\epsilon^{-(r-a)^2/4\kappa\chi} - \epsilon^{-(r+a)^2/4\kappa\chi}}{8\pi^{\frac{3}{2}}ar\,(\kappa\chi)^{\frac{1}{2}}}\right]\\
= \frac{A}{r}\,\epsilon^{-\nu^{\frac{1}{2}}r}\sin\left[2nt - \nu^{\frac{1}{2}}r + B\right], \text{ where } r > a,\\
= \frac{C}{r}\left\{\epsilon^{-\nu^{\frac{1}{2}}r}\sin\left[2nt - \nu^{\frac{1}{2}}r + D\right] - \epsilon^{\nu^{\frac{1}{2}}r}\sin\left[2nt + \nu^{\frac{1}{2}}r + D\right]\right\}\\
\text{where } r < a
\end{aligned}\right\}(17),$$

A, B, C, D being constants determined by the conditions that

$$v_{r>a} - v_{r<a} = 0,$$

and $$-4\pi a^2 \left\{ \left(\frac{dv}{dr}\right)_{r>a} - \left(\frac{dv}{dr}\right)_{r<a} \right\} = q \sin 2nt,$$

when the two values of r exceed a and fall short of a by infinitely small differences. Verify that v satisfies (6). Also that v is finite when $r = 0$.

VIII. Fourier's "Linear Motion of Heat;" instantaneous plane-source; quantity per unit surface, σ:

$$v = \frac{\sigma \epsilon^{-x^2/4\kappa t}}{2\pi^{\frac{1}{2}}(\kappa t)^{\frac{1}{2}}} \dots\dots\dots\dots\dots\dots\dots(18).$$

Verify that this satisfies (9) for the case of v independent of y and z, and that

$$\int_{-\infty}^{\infty} v\,dx = \sigma.$$

Remark that (18) is obtainable from (15) by putting $Q/4\pi a^2 = \sigma$, and $a = \infty$; or directly from (10) by integration over the plane.

IX. "Linear Motion of Heat;" time-periodic plane-source; rate per unit of area, per unit of time, at time t, $\sigma \sin 2nt$:

$$\left. \begin{aligned} v &\left[= \sigma \int_0^\infty d\chi \sin 2n\,(t-\chi)\, \frac{\epsilon^{-x^2/4\kappa\chi}}{2\pi^{\frac{1}{2}}(\kappa\chi)^{\frac{1}{2}}} \right] \\ &= \frac{\sigma}{(2\nu)^{\frac{1}{2}} 2\kappa}\, \epsilon^{-\nu^{\frac{1}{2}}x} \sin\left(2nt - \nu^{\frac{1}{2}}x - \tfrac{1}{4}\pi\right) \text{ where } x > 0 \\ &= \frac{\sigma}{(2\nu)^{\frac{1}{2}} 2\kappa}\, \epsilon^{+\nu^{\frac{1}{2}}x} \sin\left(2nt + \nu^{\frac{1}{2}}x - \tfrac{1}{4}\pi\right) \text{ where } x < 0 \\ &\quad\quad \nu \text{ denoting } n/\kappa \end{aligned} \right\} \quad (19).$$

Verify that v satisfies $\dfrac{dv}{dt} = \kappa \dfrac{d^2v}{dx^2}$, which is what (9) becomes when v is independent of y and z; also that

$$-2\kappa \frac{dv}{dx} = \sigma \sin 2nt \text{ when } x \text{ is infinitely small positive,}$$

and $$+2\kappa \frac{dv}{dx} = \sigma \sin 2nt \text{ when } x \text{ is infinitely small negative.}$$

X. " Linear Motion of Heat;" space-periodic simple harmonic solid source, with plane isothermal surfaces. Initial distribution, $v = V \sin ax$, when $t = 0$. Solution for any value of t

$$v \left[= V \int_{-\infty}^{\infty} d\xi \sin \alpha\, (x - \xi)\, \frac{\epsilon^{-\xi^2/4\kappa t}}{2\pi^{\frac{1}{2}} (\kappa t)^{\frac{1}{2}}} \right] = V \epsilon^{-\kappa a^2 t} \sin ax \ldots (20).$$

Modifying the integral within the brackets to make it appear as an analytical expression belonging to the general theory of images, for the case of a single infinite row of images, and equating the result to the right-hand member, we see that

$$\frac{1}{4\kappa t} \int_0^{\frac{2\pi}{a}} d\xi \, \frac{\sin}{\cos} \alpha\, (x - \xi) \sum_{i=-\infty}^{i=\infty} \epsilon^{-(\xi + 2i\pi/a)^2/4\kappa t} = \epsilon^{-\kappa a^2 t} \frac{\sin}{\cos} ax \ldots (21),$$

i being any integer.

It is obvious that for a we may substitute $j\alpha$ in the second member, and in the factor $\frac{\sin}{\cos} \alpha\, (x - \xi)$ under the integral sign of the first member without altering it elsewhere, j being any integer: thus we have

$$\left. \begin{array}{l} \dfrac{1}{4\kappa t} \displaystyle\int_0^{\frac{2\pi}{a}} d\xi \, \dfrac{\sin}{\cos} j\alpha\, (x - \xi)\, S = \epsilon^{-\kappa j^2 a^2 t} \dfrac{\sin}{\cos} j\alpha x \\[2mm] \text{where} \qquad S = \displaystyle\sum_{i=-\infty}^{i=\infty} \epsilon^{-(\xi + 2i\pi/a)^2/4\kappa t} \end{array} \right\} \ldots \ldots (22).$$

XI. " Linear Motion of Heat ;" space-periodic arbitrary solid source, isothermals plane. Initial distribution, $v = f(x)$, when $t = 0$, f denoting an arbitrary periodic function, period l; so that $f(x + il) = f(x)$, i being any integer. Two solutions (A), (B).

(A) derived synthetically from (18):

$$v = \frac{1}{2 (\pi \kappa t)^{\frac{1}{2}}} \int_0^l d\xi f(x - \xi)\, S \ldots \ldots (23)^*.$$

where

$$S = \sum_{i=-\infty}^{i=\infty} \epsilon^{-(\xi + il)^2/4\kappa t} \ldots \ldots (24).$$

(B) derived analytically and synthetically from (20).

* This is Fourier's (i) of Art. 374 for the case of $f(x)$ a periodic function.

Find A_0, A_1, A_2, ..., B_1, B_2, &c., by the harmonic analysis, to satisfy the condition

$$\left.\begin{aligned} fx = A_0 + A_1 \cos\theta + A_2 \cos 2\theta + \text{etc.} \\ + B_1 \sin\theta + B_2 \sin 2\theta \;\&\text{c.} \end{aligned}\right\} \quad\dots\dots(25).$$

where $\qquad\qquad \theta = \dfrac{2\pi x}{l}$

Then

$$v = A_0 + \sum_{j=1}^{j=\infty} \epsilon^{-\kappa\left(\frac{2j\pi}{l}\right)^2 t}\left(A_j \cos\frac{2j\pi x}{l} + B_j \sin\frac{2j\pi x}{l}\right)\dots(26).$$

XII. Uniform row of simple instantaneous plane sources. Two solutions (A) and (B).

(A), from XI. (A) (23):

$$v = \frac{\sigma}{2\,(\pi\kappa t)^{\frac12}} \sum_{i=-\infty}^{i=\infty} \epsilon^{-(x+il)^2/4\kappa t}\dots\dots\dots\dots\dots (27).$$

The No. 2 diffusion curve of § 82 is the representation of the first term $(i = 0)$ of this formula.

(B), from XI. (B), (25) and (26):

$$v = \frac{\sigma}{l}\left\{ 1 + 2\sum_{j=1}^{j=\infty} \epsilon^{-\kappa\left(\frac{2j\pi}{l}\right)^2 t} \cos\frac{2j\pi x}{l}\right\}\dots\dots\dots(28).$$

The comparison of these two solutions is very interesting physically, and useful arithmetically. To facilitate the comparison, put

$$l\,/\,2\,(\pi\kappa t)^{\frac12} = q,\ \text{ and }\ x\,/\,2\,(\pi\kappa t)^{\frac12} = p\dots\dots\dots(29).$$

the two solutions become

$$\frac{lv}{\sigma} = q \sum_{i=-\infty}^{i=\infty} \epsilon^{-\pi(p+iq)^2} = 1 + 2\sum_{j=1}^{j=\infty} \epsilon^{-\frac{j^2\pi}{q^2}} \cos\frac{2j\pi p}{q}\dots\dots(30).$$

The equation between the second and third member, virtually due to Fourier, is also an interesting formula of Jacobi's, *Fundamenta Nova Theoriæ Functionum Ellipticarum*, as was long ago pointed out by Cayley[*]. Each formula is a series which converges

---

[*] *Quarterly Journal of Mathematics* for 1857; note by Cayley on a Paper by W. Thomson entitled "On the Calculation of Transcendents of the form

$$\int_0^x \epsilon^{-x^2} f(x)\,dx.\text{''}$$

[This Paper and Cayley's note to it, are reprinted as the concluding portion of the present Article.]

for every value of t however small or however great; the first, (27) the more rapidly the less is t; the second, (28) the more rapidly the greater is t. For the case of $t = l^2/4\kappa\pi$, and $x = 0$ (that is, $q = 1$ and $p = 0$), the two series become identical. For the more comprehensive case of $p = 0$, but q unrestricted, the comparison gives the following very curious arithmetical theorem:—

$$q\,\frac{1 + 2\left(\epsilon^{-\pi q^2} + \epsilon^{-4\pi q^2} + \epsilon^{-9\pi q^2} + \&c.\right)}{1 + 2\left(\epsilon^{-\pi/q^2} + \epsilon^{-4\pi/q^2} + \epsilon^{-9\pi/q^2} + \&c.\right)} = 1 \ldots\ldots(31);$$

or, $f(q) = f(q^{-1})$, if $f(q) = q^{\frac{1}{2}}\{1 + 2\left(\epsilon^{-\pi q^2} + \epsilon^{-4\pi q^2} + \epsilon^{-9\pi q^2} + \&c.\right)\}\ldots(31').$

When $t \lessgtr l^2/4\kappa\pi$ (or $q \gtrless 1$) the first solution (27) converges with so great suddenness that three terms suffice for most practical purposes; when $t \gtrless l^2/4\kappa\pi$ (or $q \lessgtr 1$) the second solution (28) converges with so great suddenness that *one* term (after the constant first term) suffices for most practical purposes. Thus by using the solution (27) for all values of t from zero to something less than $l^2/4\kappa\pi$, and (28) for all values greater than the greatest for which (27) is used, we have an exceeding rapid convergence and easy calculation to find v for any values of x and t. These formulas, thus used, have been of great practical value in calculating what is now known as the arrival curve of signals through a submarine cable [p. 72, below], and in designing instruments to record it automatically and allow its telegraphic meaning to be read [Art. LXXXV. below], or without recording it to allow its meaning to be read by watching the motions of a spot of light [p. 105, below]. It is clear that (27) and (28) express the potential at a point at distance x from one end of a cable of length $\frac{1}{2}l$, at time t from an instant when a quantity $\frac{1}{2}\sigma c$ of electricity has been suddenly communicated to that end of the cable, both ends being always kept insulated after that instant (as is done practically by an exceedingly short contact with one pole of a voltaic battery, the other being kept to earth). The value of v for $x = \frac{1}{2}l$, and for all values of t from 0 to ∞, represents the rise of the potential at the remote end towards the limiting value σ/e, towards which the potential rises through the conductor.

XIII. (X. in three dimensions.) Space triple periodic solid source; in other words $v = V\sin\alpha x \sin\beta y \sin\gamma z$, when $t = 0$.

Solution for any value of t,

$$v\left[=V\int_{-\infty}^{\infty}\int_{-\infty}^{\infty}\int_{-\infty}^{\infty}d\xi d\eta d\zeta\sin\alpha(x-\xi)\sin\beta(y-\eta)\sin\gamma(z-\zeta)\frac{-\frac{\xi^2+\eta^2+\zeta^2}{4\kappa t}}{8\pi^{\frac{3}{2}}(\kappa t)^{\frac{3}{2}}}\right]$$

$$=V\sin\alpha x\sin\beta y\sin\gamma z\epsilon^{-\kappa(\alpha^2+\beta^2+\gamma^2)t}\dots\dots\dots\dots\dots\dots\dots(32).$$

Remark that, as an analytical expression for the present case of the general theory of triply-multiple images, the triple integral within the brackets may be written

$$\left.V\int_0^{\frac{2\pi}{\alpha}}\int_0^{\frac{2\pi}{\beta}}\int_0^{\frac{2\pi}{\gamma}}d\xi d\eta d\zeta\sin\alpha(x-\xi)\sin\beta(y-\eta)\sin\gamma(z-\zeta)\frac{S}{8\pi^{\frac{3}{2}}(\kappa t)^{\frac{3}{2}}},\right\}$$

where $\quad S=\sum_{i=-\infty}^{=\infty}\sum_{j=-\infty}^{j=\infty}\sum_{k=-\infty}^{k=\infty}\epsilon^{-\frac{\left(\xi+\frac{2i\pi}{\alpha}\right)^2+\left(\eta+\frac{2j\pi}{\beta}\right)^2+\left(\xi+\frac{2k\pi}{\gamma}\right)^2}{4\kappa t}}$ $\left.\vphantom{\int}\right\}$ (33),

i, j, k, being any positive or negative integers.

XIV. and XV. (X. and XI. in three dimensions.) The formulas may be written down by inspection; from I., with X. and XI. for guides. The analytical theorem thus obtained, corresponding to (30), in three dimensions, is interesting to pure mathematicians.

XVI. Harmonic solutions. Any distribution of heat, whether in an infinite or a bounded solid, which keeps its type unchanged in subsiding towards uniformity, when left without positive or negative sources, except such as are required to fulfil a proper boundary condition, is called a harmonic distribution, provided the temperature does not increase to infinity in any direction. The boundary condition, if the solid is bounded, is essentially that the rate of emission from the surface at every point of it varies in simple proportion to the temperature, and at such a rate per 1° of temperature at each part of the surface as the solution requires. X. and XIII. are examples. The general condition for a harmonic solution is

$$v=f(t)\,F(x,\,y,\,z)\dots\dots\dots\dots\dots(34)\,;$$

and this tried in (9) gives

$$\frac{1}{f(t)}\,\frac{df(t)}{dt}=\kappa\frac{\dfrac{d^2F}{dx^2}+\dfrac{d^2F}{dy^2}\dfrac{d^2F}{dz^2}}{F}\dots\dots\dots\dots(35).$$

The first member being independent of x, y, z, and the second being independent of t, the common value of the two must be independent of x, y, z, t, that is to say, must be an absolute constant. Let it be denoted by $-\rho$; we have

$$f(t) = C\epsilon^{-\rho t} \dots\dots\dots\dots\dots\dots(36),$$

and

$$\frac{d^2F}{dx^2} + \frac{d^2F}{dy^2} + \frac{d^2F}{dz^2} + \frac{\rho}{\kappa} F = 0 \dots\dots\dots\dots(37),$$

or in terms of polar coordinates, by (6),

$$\left. \begin{array}{c} \dfrac{d^2U}{dr^2} + \dfrac{1}{r^2\sin\theta} \dfrac{d}{d\theta}\left(\sin\theta \dfrac{dU}{d\theta}\right) + \dfrac{1}{r^2\sin^2\theta} \dfrac{d^2U}{d\phi^2} + \dfrac{\rho}{\kappa} U = 0 \\[2mm] \text{where} \qquad U = \dfrac{F}{r} \end{array} \right\} (38).$$

Of this we have a spherical harmonic solution,

$$U = [A\phi_i(r) + B\psi_i(r)]S_i \dots\dots\dots\dots\dots(39),$$

where S_i denotes a spherical surface harmonic of order i, and $\phi_i(r)$, $\psi_i(r)$, two particular solutions of the equation

$$\frac{d^2u}{dr^2} + \left[\frac{\rho}{\kappa} - \frac{i(i+1)}{r^2}\right] u = 0 \dots\dots\dots\dots(40).$$

Then (36) and (35) give finally

$$v = \epsilon^{-\rho t}\frac{A\phi_i(r) + B\psi_i(r)}{r} S_i \dots\dots\dots\dots(41),$$

This solution is, in its generality, applicable to an infinite solid occupying all space except a hollow round the origin. The solid may of course be bounded externally also by a finite closed surface. If there be no hollow, A/B must fulfil the condition that $[A\phi_i(r) + B\psi_i(r)]/r$ is finite when $r = 0$. If there are two boundaries, concentric spherical surfaces, with their common centre at the origin of coordinates, the boundary condition obviously requires uniform emissivity over each, but not necessarily equal for the two. If the two emissivities be denoted by h and h', and the radii of the surfaces by a (outer) and a' (inner), the boundary conditions are

$$\left. \begin{array}{c} \dfrac{dv}{dr} = -hv, \text{ when } r = a, \\[2mm] \dfrac{dv}{dr} = h'v, \text{ when } r = a' \end{array} \right\} \dots\dots\dots\dots(42).$$

and

From these we may find h and h', so as to let the harmonic character of the solution be fulfilled in the subsidence. Or if h and h' be given, we have in (41) two equations which determine the two unknown quantities A/B and ρ. Eliminating A/B, we thus find a single transcendental equation for ρ, which is proved to have no imaginary or negative roots, and an infinite number of real positive roots, each $> \kappa i\,(i + 1)/r^2$. In the case of $i = 0$, or temperature independent of ϕ and θ, (40) gives

$$u = A \cos r \sqrt{\frac{\rho}{\kappa}} + B \sin r \sqrt{\frac{\rho}{\kappa}}.$$

For this case the transcendental equation for determining values for ρ is very simple, and its roots are calculated numerically with great ease. With the further restriction of no central hollow, we must have $A = 0$, so that u/r may be finite when $r = 0$. This case was fully investigated by Fourier, and very beautifully worked out in his fifth chapter. The more general problem of a solid sphere, with any given initial distribution of temperature, without the restriction, of temperature independent of θ and ϕ, was solved first we believe by Poisson in the 11th chapter of his *Théorie Mathématique de la Chaleur*, in terms of the formulas (36), (38), (40) above.

XVII. The equation of the transference of heat in terms of columnar coordinates, (7) above, affords naturally another beautiful case of harmonic solution. Assume

$$v = \epsilon^{-\rho t} u_i \, {\sin \atop \cos} \, i\phi \, {\sin \atop \cos} \, mz \dots\dots\dots\dots(43);$$

we find by (7)

$$\frac{d^2 u_i}{dr^2} + \frac{1}{r} \frac{du_i}{dr} + \left(\frac{\rho}{\kappa} - m^2 - \frac{i^2}{r^2}\right) u_i = 0 \dots\dots \dots\dots(44).$$

The treatment of this equation and its integral (obviously derivable by i differentiations from u_0, which is a Bessel's function) for the full solution of the thermal problem is most interesting, and very instructive and suggestive in respect to pure analysis. It was splendidly worked out for the case of $m = 0$ and $i = 0$ by Fourier in his 6th chapter, "The Motion of Heat in a Solid Cylinder," truly a masterpiece of art. When it was printed in 1821, and published after having with the rest of Fourier's work been buried alive for fourteen years in the Archives of the French

Academy, and when Bessel found in it so thorough an investigation and so strikingly beautiful an application of the "Besselsche Function," we can imagine the ordinary feeling towards those "qui ante nos nostra dixerunt" reversed into the pleasure of genuine admiration.

Thermal diffusivity [being § 82 of the Encyclopædia Article]. When the effect of heat conducted across any part of a body, in heating the substance on one side, or leaving the substance on the other side cooler, is to be reckoned, it is convenient to measure the thermal conductivity in terms, not of the ordinary general gramme-water-unit of heat, but of a special unit,—the quantity required to raise unit bulk of the substance by 1°. In other words if k be the conductivity in terms of any thermal unit, and c the thermal capacity of unit bulk of the substance, it is k/c, not merely k, that expresses the quality of the substance on which the phenomenon chiefly depends. We therefore propose to give to k/c the name of "thermal diffusivity" (or simply "diffusivity") when heat is understood to be the subject, while still using the term "thermal conductivity" to denote the conducting power defined without restriction as to the thermal unit employed. It is interesting and important to remark that "diffusivity" is essentially to be reckoned in units of area per unit of time, and that its "dimensions" are L^2/T. Its regular C. G. S. reckoning is therefore in square centimetres per second. In the Article DIFFUSION the relation between diffusion of heat and diffusion of matter is explained. We have added diffusion of electricity through a submarine cable, which has been shewn[1] to follow the same law as the "linear" diffusion of heat, as Fourier calls the diffusion of heat when the isothermal surfaces are parallel planes. The curves of the following diagram and Tables A, B, and C show, in a practically useful way, the result in the course of the times noted, of from fractions of a second to thousands of millions of years, of linear diffusion of two different qualities in an infinite line, from an initial condition in which there is sudden transition from one quality to the other, in the thoroughly practical cases specified in the accompanying explanations.

[1] *Proc. Roy. Soc.*, May, 1855, Wm. Thomson, "On the Theory of the Electric Telegraph." [Article LXXIII. below.]

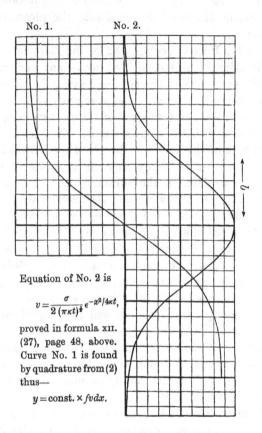

No. 1.   No. 2.

Equation of No. 2 is

$$v = \frac{\sigma}{2\,(\pi\kappa t)^{\frac{1}{2}}}\,e^{-x^2/4\kappa t},$$

proved in formula XII.
(27), page 48, above.
Curve No. 1 is found
by quadrature from (2)
thus—

$$y = \text{const.} \times \int v\,dx.$$

DIAGRAM OF DIFFUSION.

Curve No. 1 shews temperature; or quantity of substance in solution; or
potential in the conductor of a submarine cable through which electricity is
diffusing. Curve No. 2 shews rate per unit of distance of variation of the
temperature, or of the quantity of substance in solution. Vertical ordinates are
actual distances through the medium. Horizontal ordinates represent temperature
or quantity of diffusing substances in No. 1 curve, and rate of variation of temperature or of diffusing substance or of electric potential in No. 2 curve.

DIFFUSIONS.—TABLE A.

Substance.	Time in seconds from the commencement of the Diffusion until the Condition represented by the Curves on the Actual Scale ($b=2$ centimetres) is reached.
Carbonic acid through air	6·97 seconds.
Heat through hydrogen	·89 of a second.
,, ,, copper	·93 ,,
,, ,, iron................................	5·5 seconds.
,, ,, air.................................	6·25 ,,
,, ,, underground strata...............	100·00 ,,
,, ,, wood	770 ,,
Common salt through water....................	87150 ,,
Electricity through Suez-Aden cable	$1·087 \times 10^{-16}$ of a second.
,, ,, Aden-Bombay cable	$0·739 \times 10^{-16}$,,
,, ,, Persian Gulf cable	$0·635 \times 10^{-16}$,,
,, ,, Atlantic cable	$0·440 \times 10^{-16}$,,
,, ,, French Atlantic cable....	$0·396 \times 10^{-16}$,,
,, ,, Direct U. S. cable.........	$0·340 \times 10^{-16}$,,

DIFFUSIONS (SECULAR).—TABLE B.

Substance.	Time in years from the commencement of the Diffusion until the Condition represented by the Curves on the Scale of $b=20$ kilometres, or 1,000,000 times the Actual Scale, is reached.
Carbonic acid through air [1]..............	222,000 years.
Heat through hydrogen..................	28,000 ,,
,, ,, copper	29,000 ,,
,, ,, iron	174,000 ,,
,, ,, air	198,000 ,,
,, ,, underground strata [2]	317,000 ,,
,, ,, wood	24,700,000 ,,
Common salt through water [3]..........	2,760,000,000 ,,

[1] Instructive as to the proportion of carbonic acid in air at different heights, proving its approximate uniformity to be due to convection, not to diffusion.

[2] Instructive as to geological theories respecting terrestrial temperature.

[3] Instructive as to theories respecting the saltness of the sea.

TABLE C.

Name of Cable.	Time in Seconds from the commencement of the Diffusion until the Condition represented by the Curves on the Scale of $b = 1000$ Nautical Miles, or 92,615,000 times the Actual Scale, is reached.
Suez-Aden.....................................	0·932 of a second.
Aden-Bombay	0·634 ,,
Persian Gulf..................................	0·545 ,,
Atlantic..	0·377 ,,
French Atlantic	0·339 ,,
Direct United States.....................	0·292 ,,

ON THE CALCULATION OF TRANSCENDENTS OF THE
FORM $\int_0^x \epsilon^{-x^2} fx\, dx$.

THE practical solution of many problems regarding the conduction of heat through solids, and the propagation of electricity along telegraph wires, is expressed by a transcendent of the form $\int_0^x \epsilon^{-mx^2} f(x)\, dx$, where fx may be a constant, or some simple algebraic or trigonometrical function. Corresponding to every such problem, there is one regarding a solid or a telegraph wire of essentially limited dimensions, of which the solution is expressible in a converging trigonometrical series, with coefficients of the form $\int_0^a \cos \frac{i\pi x}{a} f(x)\, dx$. By increasing indefinitely the dimension a, we may, as Fourier has shewn, pass from the solution of the latter to that of the former problem. Now if we increase a, *not* infinitely, the trigonometrical series becomes less convergent; but while it is still so rapidly convergent, that three or four terms of it may be amply sufficient for a practical solution, its value becomes sensibly the same as that which it rigorously attains only when a is infinite. We have thus, in the actual physical investigations, a process by which a transcendent of the form stated above, is numerically evaluated by means of a very rapidly converging series, of which, in the most important practical cases, each term is calculated with the greatest ease. This process is exhibited analytically in the following investigation.

Let a be so large a number that ϵ^{-a^2} may be neglected according to the degree of accuracy to which the calculations are to be pushed. Then if we take

$$R = \epsilon^{-(x-a)^2} + \epsilon^{-(x+a)^2} - \{\epsilon^{-(x-2a)^2} + \epsilon^{-(x+2a)^2}\} + \dots \&c \dots \dots (1),$$

the value of R will be insensible, provided that of x does not exceed $\frac{1}{2}a$; and we have

$$\epsilon^{-x^2} = \phi(x) + R \dots \dots \dots \dots (2);$$

if

$$\phi(x) = \dots + \epsilon^{-(x-2a)^2} - \epsilon^{-(x-a)^2} + \epsilon^{-x^2} - \epsilon^{-(x+a)^2} + \epsilon^{-(x+2a)^2} \dots \&c. \ (3);$$

where the second member is a series converging with excessive rapidity on each side of its greatest term, whatever the value of ax. Now the function denoted by $\phi(x)$ is clearly a periodical function of x, fulfilling the conditions

$$\phi(x) = (-1)^i \phi(x + ia) \dots \dots \dots (4),$$

and

$$\phi(x) = \phi(-x) \dots \dots \dots \dots (5).$$

Hence it is expressible in a trigonometrical series of cosines of odd multiples of $\dfrac{\pi x}{a}$; so that we may assume

$$\phi(x) = A \cos\frac{\pi x}{a} + B \cos\frac{3\pi x}{a} + C \cos\frac{5\pi x}{a} + \&c.$$

By multiplying both members by $\cos\dfrac{i\pi x}{a}\,dx$, and integrating between the limits 0 and a, according to the well-known method, we find, for the coefficient of $\cos\dfrac{i\pi x}{a}$ in the series, the value

$$\frac{2}{a}\int_0^a \phi(x)\cos\frac{i\pi x}{a}\,dx.$$

Now, if $F(x)$ denote any function of x whatever, we have

$$\int_{-\infty}^{\infty} F(x)\cos\frac{i\pi x}{a}\,dx =$$

$$\int_0^a \{\dots + F(x-2a) - F(x-a) + F(x) - F(x+a)$$

$$+ F(x+2a)\dots\}\cos\frac{i\pi x}{a}\,dx,$$

provided i be odd. Hence

$$\int_0^a \phi x \cos \frac{i\pi x}{a}\, dx = \int_{-\infty}^{\infty} \epsilon^{-x^2} \cos \frac{i\pi x}{a}\, dx.$$

But

$$\int_{-\infty}^{\infty} \epsilon^{-x^2} \cos qx\, dx = \tfrac{1}{2} \int_{-\infty}^{\infty} \epsilon^{-x^2} \left\{ \epsilon^{qx\sqrt{(-1)}} + \epsilon^{-qx\sqrt{(-1)}} \right\} dx$$

$$= \tfrac{1}{2} \epsilon^{-\frac{q^2}{4}} \int_{-\infty}^{\infty} \left\{ \epsilon^{-(x - \frac{1}{2}q\sqrt{-1})^2} + \epsilon^{-(x + \frac{1}{2}q\sqrt{-1})^2} \right\} dx.$$

Now

$$\int_{-\infty}^{\infty} \epsilon^{-(x \pm \frac{1}{2}q\sqrt{-1})^2}\, dx = \int_{-\infty}^{\infty} \epsilon^{-x^2}\, dx = \pi^{\frac{1}{2}},$$

and, therefore,

$$\int_{-\infty}^{\infty} \epsilon^{-x^2} \cos qx\, dx = \pi^{\frac{1}{2}} \epsilon^{-\frac{q^2}{4}} \dots\dots\dots\dots\dots\dots(6).$$

Hence we obtain

$$\int_0^a \phi(x) \cos \frac{i\pi x}{a}\, dx = \pi^{\frac{1}{2}} \epsilon^{\frac{i^2\pi^2}{a^2}}, \text{ when } i \text{ is odd.}$$

The value of the first member is, of course, zero when i is even, because of the relation (4); and we conclude

$$\phi(x) = \frac{2\pi}{a} \left\{ \epsilon^{-\frac{\pi^2}{4a^2}} \cos \frac{\pi x}{a} + \epsilon^{-\frac{3^2\pi^2}{4a^2}} \cos \frac{3\pi x}{a} + \epsilon^{-\frac{5^2\pi^2}{4a^2}} \cos \frac{5\pi x}{a} + \&c. \right\} \dots(7).$$

This used in equation (2), gives

$$\epsilon^{-x^2} = \frac{2\pi^{\frac{1}{2}}}{a} \left(e \cos \frac{\pi x}{a} + e^9 \cos \frac{3\pi x}{a} + e^{25} \cos \frac{5\pi x}{a} + \&c. \right) + R \left. \right\} \dots(8).$$

where $\qquad\qquad e = \epsilon^{\frac{\pi^2}{4a^2}}$

If, for instance, we take $e = (\cdot 1)^{\frac{1}{40}}$, which gives $\frac{\pi^2}{a^2} = \cdot230258$, $a^2 = 42\cdot863$, and $a = 6\cdot547$, the values of R, for all values of x falling short of $\frac{1}{2}a$, must be less than $\epsilon^{-\frac{1}{4}a^2}$, that is in this case $\frac{1}{\epsilon^{10\cdot716}}$, a very small fraction; and to the degree of accuracy to which this may be neglected, we have *for all values of x short of* $\frac{1}{2}a$,

$$\epsilon^{-x^2} = \frac{2\pi^{\frac{1}{2}}}{6\cdot547}\left\{ (\cdot1)^{\frac{1}{40}}\cos\frac{\pi x}{6\cdot547} + \tfrac{1}{3}(\cdot1)^{\frac{9}{40}}\cos\frac{3\pi x}{6\cdot547} + \tfrac{1}{5}(\cdot1)^{\frac{25}{40}}\cos\frac{5\pi x}{6\cdot547} + \&c.\right\},$$

a series of which seven terms are enough to give the value of the sum to four places of decimals. For small values of x a similar degree of accuracy is obtained by taking $e = (\cdot1)^{\frac{1}{10}}$, which gives a half its former value, so that we have

$$\epsilon^{-x^2} = \frac{2\pi^{\frac{1}{2}}}{3\cdot2735}\left\{ (\cdot1)^{\frac{1}{10}}\cos\frac{\pi x}{3\cdot2735} + \tfrac{1}{3}(\cdot1)^{\frac{9}{10}}\cos\frac{3\pi x}{3\cdot2735} + \right.$$
$$\left. \tfrac{1}{5}(\cdot1)^{\frac{25}{10}}\cos\frac{5\pi x}{3\cdot2735} + \&c.\right\},$$

of which three terms are enough to give the sum to four places of decimals—a formula of great use in practical questions regarding terrestrial temperature, and the electrification of submarine telegraph wires.

The form of expression which has been investigated for ϵ^{-x^2} enables us with great ease to evaluate numerically any integral $\int_0^x \epsilon^{-x^2}\, fx\, dx$, in which fx is a function such that $\int_0^x fx \cos qx\, dx$ can be be readily calculated. For instance, we have

$$\left.\begin{array}{c} \int_0^x \epsilon^{-x^2} dx = \dfrac{2}{\pi^{\frac{1}{2}}}\left(e\sin\dfrac{\pi x}{a} + \tfrac{1}{3}e^9\sin\dfrac{3\pi x}{a} + \tfrac{1}{5}e^{25}\sin\dfrac{5\pi x}{a} + \&c.\right) + S \\[2mm] \text{where}\qquad e = \epsilon^{-\frac{\pi^2}{4a^2}} \end{array}\right\}$$

For all values of x less than $\frac{1}{2}a$, the value of S is insensible, if a be so large that $\epsilon^{-\frac{1}{4}a^2}$ can be neglected.

KREUZNACH, *July* 31, 1855.

NOTE BY A. CAYLEY.—The formula of transformation contained in the equations (3), (7) is easily deduced from the theory of elliptic functions. In fact if, in the notation of the *Fund.*

Nova, K, K' are the complete functions corresponding to the moduli k, k' respectively, and $q = e^{-\frac{\pi K'}{K}}$, then we have (Jacobi, p. 183),

$$\Theta\left(\frac{2Kx}{\pi}\right) = 1 - 2q\cos 2x + 2q^4\cos 4x - 2q^9\cos 6x + \ldots$$

$$H\left(\frac{2Kx}{\pi}\right) = 2\sqrt[4]{(q)}\sin x - 2\sqrt[4]{(q^9)}\sin 3x + 2\sqrt[4]{(q^{25})}\sin 5x - \ldots$$

the formula (16), (Jacobi, p. 175), gives

$$e^{-\frac{\pi u^2}{4KK'}}.\Theta(\iota u, \ k) + \sqrt{\frac{k}{k'}\frac{\Theta(0, k)}{\Theta(0, k')}}\ H(u + K', k'),$$

where, $\iota = \sqrt{-1}$.

And by means of the above expression for $\Theta\left(\frac{2Kx}{\pi}\right)$, and the formulas (7), (8), (10), (Jacobi, p. 184), it is easy to shew that

$$\sqrt{\frac{k}{k'}\frac{\Theta(0, k)}{\Theta(0, k')}} = \sqrt{\frac{K}{K'}}.$$

We have, therefore,

$$\Theta(\iota u, k) = \sqrt{\frac{K}{K'}}\ \epsilon^{-\frac{\pi u^2}{4KK'}}\ H(u + K', k).$$

Whence writing $\frac{2K'x}{a}$, instead of u, and assuming $\frac{\pi K'}{K} = a^2$, i.e.

$q = e^{-\frac{\pi K'}{K}} = e^{-a^2}$, $q' = e^{-\frac{\pi K}{K'}} = e^{-\frac{\pi^2}{a^2}}$, and substituting for the functions Θ, H, their expansions, we find

$$e^{-x^2} - e^{-(x-a)^2} - e^{-(x+a)^2} + e^{-(x-2a)^2} + e^{-(x+2a)^2} - \&c.$$

$$= \frac{2\sqrt{\pi}}{a}\left\{e^{-\frac{\pi^2}{4a}}\cos\frac{\pi x}{a} + e^{-\frac{9\pi^2}{4a^2}}\cos\frac{3\pi x}{a} + e^{-\frac{25\pi^2}{4a^2}}\cos\frac{5\pi x}{a}\ \&c.\right\}$$

which is Professor Thomson's formula of transformation.

ART. LXXIII. ON THE THEORY OF THE ELECTRIC TELEGRAPH.

[From the *Proc. Royal Soc.*, May, 1855.]

THE following investigation was commenced in consequence of a letter received by the author from Prof. Stokes, dated Oct. 16, 1854. It is now communicated to the Royal Society, although only in an incomplete form, as it may serve to indicate some important practical applications of the theory, especially in estimating the dimensions of telegraph wires and cables required for long distances; and the author reserves a more complete development and illustration of the mathematical parts of the investigation for a paper on the conduction of Electricity and Heat through solids, which he intends to lay before the Royal Society on another occasion. [See preceding Art. LXXII.]

Extract from a letter to Prof. Stokes, *dated Largs, Oct.* 28, 1854.

"Let c be the electro-statical capacity per unit of length of the wire; that is, let c be such that clv is the quantity of electricity required to charge a length l of the wire up to potential v. In a note communicated as an addition to a paper in the last June Number of the *Philosophical Magazine*, [Electrostatics and Magnetism, Art. III.] and I believe at present in the Editor's hands for publication, I proved that the value of c is $\dfrac{I}{2 \log \dfrac{R'}{R}}$, if I denote the specific inductive capacity of the gutta-percha, and R, R' the radii of its inner and outer cylindrical surfaces.

"Let k denote the galvanic resistance of the wire in absolute electro-statical measure [per unit of the wire's length] (see a paper 'On the application of the Principle of Mechanical Effect to the Measurement of Electro-motive Forces and Galvanic Resistances,' *Phil. Mag.* Dec. 1851 [Art. LIV. Vol. I. above]).

"Let γ denote the strength at the time t, of the current (also in electro-statical measure) at a point P of the wire at a distance x from one end which may be called O.   Let v denote the potential at the same point P, at the time t.

"The potential at the outside of the gutta-percha may be taken as at each instant rigorously zero (the resistance of the water, if the wire be extended as in a submarine telegraph, being certainly incapable of preventing the inductive action from being completed instantaneously round each point of the wire. If the wire be closely coiled, the resistance of the water may possibly produce sensible effects).

"Hence, at the time t, the quantity of electricity on a length dx of the wire at P will be $vcdx$.

"The quantity that leaves it in the time dt will be

$$dt \frac{d\gamma}{dx} dx.$$

"Hence we must have

$$- cdx \frac{dv}{dt} dt = dt \frac{d\gamma}{dx} dx \dots\dots\dots\dots\dots\dots (1).$$

"But the electromotive force, in electro-static units, at the point P, is

$$- \frac{dv}{dx},$$

and therefore at each instant

$$k\gamma = - \frac{dv}{dx} \dots\dots\dots\dots\dots\dots\dots\dots (2).$$

"Eliminating γ from (1) by means of this, we have

$$ck \frac{dv}{dt} = \frac{d^2v}{dx^2} \dots\dots\dots\dots\dots\dots\dots\dots (3),$$

which is the equation of electrical excitation in a submarine telegraph-wire, perfectly insulated by its gutta-percha covering.

"This equation agrees with the well-known equation of the linear motion of heat in a solid conductor; and various forms of solution which Fourier has given are perfectly adapted for answering practical questions regarding the use of the telegraph-wire. Thus first, suppose the wire infinitely long and communicating with the earth at its infinitely distant end : let the end O be suddenly raised to the potential V (by being put in communication with the positive pole of a galvanic battery, of which the negative pole is in communication with the ground, the resistance of the battery being small, say not more than a few yards of the wire); let it be kept at that potential for a time T; and lastly, let it be put in communication with the ground (i.e. suddenly reduced to, and ever afterwards kept at, the zero of potential). An elementary expression for the solution of the equation in this case is

$$v = \frac{V}{\pi} \int_0^\infty dn \epsilon^{-zn^{\frac{1}{2}}} \frac{\sin\left[2nt - zn^{\frac{1}{2}}\right] - \sin\left[(t - T)\,2n - zn^{\frac{1}{2}}\right]}{n} \ ...(4),$$

where for brevity,

$$z = x\sqrt{kc} \ ....(5)."$$

That this expresses truly the solution with the stated conditions is proved by observing,—1st, that the second member of the equation, (4), is convergent for all positive values of z and vanishes when z is infinitely great; 2ndly, that it fulfils the differential equation (3); and 3rdly, that when $z = 0$ it vanishes except for values of t between 0 and T, and for these it is equal to V. It is curious to remark, that we may conclude, by considering the physical circumstances of the problem, that the value of the definite integral in the second member of (4) is zero for all negative values of t, and positive values of z.

"This solution may be put under the following form,

$$v = \frac{2V}{\pi} \int_{t-T}^{t} d\theta \int_0^\infty dn \epsilon^{-zn^{\frac{1}{2}}} \cos\left(2n\theta - zn^{\frac{1}{2}}\right) \ ....(6)."$$

which is in fact the primary solution as derived from the elementary type $\cos\left(2\pi \dfrac{it}{T} - z\sqrt{\dfrac{\pi i}{T}}\right)\epsilon^{-z\sqrt{\frac{\pi i}{T}}}$ given by Fourier in his investigation of periodic variations of terrestrial temperature.

"This, if T be infinitely small, becomes

$$v = \frac{2V}{\pi} T \int_0^\infty dn \epsilon^{-zn^{\frac{1}{2}}} \cos\left(2nt - zn^{\frac{1}{2}}\right) \ ....(7),$$

which expresses the effect of putting the end O of the wire for an infinitely short time in communication with the battery and immediately after with the ground. It may be reduced at once to finite terms by the evaluation of the integral, which stands as follows :—

when t is positive, $\displaystyle\int_0^\infty dn\epsilon^{-zn^{\frac{1}{2}}} \cos(2nt - zn^{\frac{1}{2}}) = \frac{\pi^{\frac{1}{2}}z}{4t^{\frac{3}{2}}}\epsilon^{-\frac{z^2}{4t}},$

and when t is negative, $\qquad\qquad\qquad = 0.$

And so we have

$$v = T\,\frac{Vz}{2\pi^{\frac{1}{2}}t^{\frac{3}{2}}}\epsilon^{-\frac{z^2}{4t}} \dots\dots\dots\dots\dots(8),$$

or by (6), when T is not infinitely small,

$$v = \frac{Vz}{2\pi^{\frac{1}{2}}}\int_{t-T}^{t} \frac{d\theta}{t^{\frac{3}{2}}}\epsilon^{-\frac{z^2}{4\theta}} \dots\dots\dots\dots\dots(9),$$

or which is the same,

$$v = \frac{Vz}{2\pi^{\frac{1}{2}}}\int_0^T \frac{d\theta}{(t-\theta)^{\frac{3}{2}}}\epsilon^{-\frac{z^2}{4(t-\theta)}} \dots\dots\dots\dots(10).$$

It is to be remarked that in (9) and (10) the limits of the integral must be taken 0 to t (instead of $t - T$ to t, or 0 to T), if it be desired to express the potential at any time t between 0 and T, since the quantity multiplied by $d\theta$ in the second number of (6) vanishes for all negative values of θ.

"These last forms may be obtained synthetically from the following solution, also one of Fourier's elementary solutions :—

$$v = \frac{\epsilon^{-\frac{z^2}{4t}}}{t^{\frac{1}{2}}}\cdot\frac{Q}{\pi^{\frac{1}{2}}}\cdot\sqrt{\frac{k}{c}}\dots\dots\dots\dots(11),$$

which expresses the potential in the wire consequent upon instantaneously communicating a quantity Q of electricity to it at O, and leaving this end insulated. For if we suppose the wire to be continued to an infinite distance on each side of O, and its infinitely distant ends to be in communication with the earth, the same equation will express the consequence of instantly communicating $2Q$ to the wire at O. Now suppose at the same instant a quantity $-2Q$ to be communicated at the point O' at a distance

$\dfrac{\alpha}{\sqrt{kc}}$ on the negative side of O: the consequent potential at any

time t, at a distance $\dfrac{z}{\sqrt{kc}}$ along the wire from O, will be

$$v = \frac{Q}{\pi^{\frac{1}{2}}} \left\{ \frac{\epsilon^{-\frac{z^2}{4t}}}{t^{\frac{1}{2}}} - \frac{\epsilon^{-\frac{(z+\alpha)^2}{4t}}}{t^{\frac{1}{2}}} \right\} \dots\dots\dots\dots(12);$$

and if α be infinitely small, this becomes

$$v = \frac{Q\alpha}{2\pi^{\frac{1}{2}}} \cdot \frac{z \epsilon^{-\frac{z^2}{4t}}}{t^{\frac{3}{2}}} \dots\dots\dots\dots\dots(13),$$

which with positive values of z, expresses obviously the effect of communicating the point O with the positive pole for an infinitely short time, and then instantly with the ground.

"The strength of the current at any point of the wire, being equal to $-\dfrac{1}{k} \cdot \dfrac{dv}{dx}$, as shown above, in equation (2), will vary proportionally to $\dfrac{dv}{dx}$ or to $\dfrac{dv}{dz}$. The time of the maximum electrodynamic effect of impulses such as those expressed by (11) or (13) will be found by determining t, in each case, to make $\dfrac{dv}{dz}$ a maximum. Thus we find

$$t = \frac{z^2}{6} = \frac{kcx^2}{6} *,$$

as the time at which the maximum electrodynamic effect of connecting the battery for an instant at O, and then leaving this point insulated, is experienced at a distance x.

"In these cases there is no regular 'velocity of transmission.' But, on the other hand, if the potential at O be made to vary regularly according to the simple harmonic law ($\sin 2nt$), the phases are propagated regularly at the rate $2\sqrt{\dfrac{n}{kc}}$, as is shown by the well-known solution

$$v = \epsilon^{-zn^{\frac{1}{2}}} \sin\left(2nt - zn^{\frac{1}{2}}\right) \dots\dots\dots\dots(14).$$

---

* We may infer that the retardations of signals are proportional to the squares of the distances, and not to the distances simply; and hence different observers, believing they have found a "velocity of electric propagation," may well have obtained widely discrepant results; and the apparent velocity would, *cæteris paribus*, be the less, the greater the length of wire used in the observation. [Comp. Art. LXXVI. below.]

The effects of pulses at one end, when the other is in connexion with the ground, and the length finite, will be most conveniently investigated by considering a wire of double length, with equal positive and negative agencies applied at its two extremities. The synthetical method founded on the use of the solution (11) appears perfectly adapted for answering all the practical questions that can be proposed.

"To take into account the effect of imperfect insulation (which appears to have been very sensible in Faraday's experiments), we may assume the gutta-percha to be uniform, and the flow of electricity across it to be proportional to the difference of potential at its outer and inner surfaces. The equation of electrical excitation will then become

$$kc\frac{dv}{dt} = \frac{d^2v}{dx^2} - hv \dots\dots\dots\dots\dots(15),$$

and if we assume

$$v = \epsilon^{-\frac{h}{kc}t}\,\phi \dots\dots\dots\dots\dots\dots(16),$$

we have

$$kc\frac{d\phi}{dt} = \frac{d^2\phi}{dx^2} \dots\dots\dots\dots\dots\dots(17),$$

an equation, to the treatment of which the preceding investigations are applicable."

Extract from Letter to Prof. Stokes, *dated Largs, Oct. 30, 1854.*

" An application of the theory of the transmission of electricity along a submarine telegraph-wire, shows how the question recently raised as to the practicability of sending distinct signals along such a length as the 2000 or 3000 miles of wire that would be required for America, may be answered. The general investigation will show exactly how much the sharpness of the signals will be worn down* and will show what maximum strength of current through the apparatus, in America, would be produced by a specified battery action on the end in England, with wire of given dimensions, &c.

" The following form of solution of the general equation

$$kc\frac{dv}{dt} = \frac{d^2v}{dx^2} - hv,$$

---

* See the diagram of curves given below.

which is the first given by Fourier, enables us to compare the
times until a given strength of current shall be obtained, with
different dimensions, &c. of wire:—

$$v = \epsilon^{-\frac{ht}{kc}} . \Sigma A_i \sin\left(\pi \frac{ix}{l}\right) . \epsilon^{-\frac{i^2\pi^2 t}{kcl^2}}.$$

If l denote the length of the wire, and V the potential at the end
communicating with the battery, the final distribution of potential
in the wire will be expressed by the equation

$$v = V \frac{\epsilon^{(l-x)\sqrt{h}} - \epsilon^{-(l-x)\sqrt{h}}}{\epsilon^{l\sqrt{h}} - \epsilon^{-l\sqrt{h}}},$$

which, when $h = 0$, becomes reduced to

$$v = V\left(1 - \frac{x}{l}\right),$$

corresponding to the case of perfect insulation. The final maximum strength of current at the remote end is expressed by

$$\gamma = \frac{V}{kl} . \frac{2l\sqrt{h}}{\epsilon^{l\sqrt{h}} - \epsilon^{-l\sqrt{h}}},$$

or, when $h = 0$, $\qquad \gamma = \frac{v}{kl}.$

Hence if we determine A_i so that

$$\Sigma A_i \sin\left(\pi \frac{ix}{l}\right) = - V \frac{\epsilon^{(l-x)\sqrt{h}} - \epsilon^{-(l-x)\sqrt{h}}}{\epsilon^{l\sqrt{h}} - \epsilon^{-l\sqrt{h}}} \text{ when } x > 0 \text{ and } x < l,$$

the equation

$$v = V \frac{\epsilon^{(l-x)\sqrt{h}} - \epsilon^{-(l-x)\sqrt{h}}}{\epsilon^{l\sqrt{h}} - \epsilon^{-l\sqrt{h}}} + \epsilon^{-\frac{ht}{kc}} \Sigma A_i \sin\left(\pi \frac{ix}{l}\right) \epsilon^{-\frac{i^2\pi^2 t}{kcl^2}}$$

will express the actual condition of the wire at any time t after
one end is put in connexion with the battery, the other being kept
in connexion with the ground.

"We may infer that the time required to reach a stated fraction of the maximum strength of current at the remote end will
be proportional to kcl^2. We may be *sure* beforehand that the
American telegraph will succeed, with a battery sufficient to give
a sensible current at the remote end, when kept long enough in
action; but the time required for each deflection will be sixteen
times as long as would be with a wire a quarter of the length,

such, for instance, as in the French submarine telegraph to Sardinia and Africa. One very important result is, that by increasing the diameter of the wire and of the gutta-percha covering in proportion to the whole length, the distinctness of utterance will be kept constant; for n varies inversely as the square of the diameter, and c (the electro-statical capacity of the unit of length) is unchanged when the diameters of the wire and the covering are altered in the same proportion.

"Hence when the French submarine telegraph is fairly tested, we may make sure of the same degree of success in an American telegraph by increasing all the dimensions of the wire in the ratio of the greatest distance to which it is to extend, to that for which the French one has been tried." It will be an economical problem, easily solved by the ordinary analytical method of maxima and minima, to determine the dimensions of wire and covering which, with stated prices of copper, gutta-percha, and iron, will give a stated rapidity of action with the smallest initial expense.

"The solution derived from the type $\dfrac{\epsilon^{-\frac{x^2}{4t}}}{t^{\frac{1}{2}}}$ may be applied to give the condition of the wire, when one end, E, is kept connected with the ground, and the other, O, is operated on so that its potential may be kept varying according to a given arbitrary function of the time: only this, which I omitted to mention in my last letter, must be attended to: instead of merely considering sources (so to speak) at O and O' (the latter in an imaginary continuation of the wire), we must suppose sources at O, O_1, O_2, &c., and at O', O_1', O_2', &c. arranged according to the general principle of successive images, so that the potential at E may be zero, and that at O may be uninfluenced by all other sources except the source at O itself. Taking $\ldots O_2$, O_1, $O, \cdot O'$, O_1', $O_2'\ldots$ equidistant, we have only to suppose equal sources, each represented by the type

$$\frac{z\epsilon^{-\frac{x^2}{4t}}}{t^{\frac{3}{2}}},$$

to be placed at these points. For the effects of O_1 and O' will balance one another as far as regards the potential at O.

"So will those of O_2 and O_1'.

    ,,   ,,   ,, O_3 and O_2'.

           &c., &c.

And again, O and O' would alone keep the potential at E, zero. So would O_1 and O_1'.

     ,, O_2 and O_2'.

             &c., &c.

Hence if we denote $2lkc$ by a, for brevity, the general solution is

$$v = \frac{1}{2\pi^{\frac{1}{2}}} \int_0^t \frac{d\theta F(\theta)}{(t-\theta)^{\frac{3}{2}}} \left\{ \ldots (z+2a)\, \epsilon^{-\frac{(z+2a)^2}{4(t-\theta)}} + (z+a)\, \epsilon^{-\frac{(z+a)^2}{4(t-\theta)}} \right.$$

$$\left. + z\epsilon^{-\frac{z^2}{4(t-\theta)}} + (z-a)\, \epsilon^{-\frac{(z-a)^2}{4(t-\theta)}} + (z-2a)\, \epsilon^{-\frac{(z-2a)^2}{4(t-\theta)}} + \ldots \right\},$$

where $F(\theta)$ is an arbitrary function such that $F(t)$ expresses the potential sustained at O by the battery.

"The corresponding solution of the equation

$$kc\frac{dv}{dt} = \frac{d^2v}{dx^2} - hv$$

is

$$v = \frac{1}{2\pi^{\frac{1}{2}}} \epsilon^{-\frac{ht}{kc}} \int_0^t \frac{d\theta \epsilon^{\frac{h\theta}{kc}} F\theta}{(t-\theta)^{\frac{3}{2}}} \Sigma_{-\infty}^{\infty} \left\{ (z-ia)\, \epsilon^{-\frac{(z-ia)^2}{4(t-\theta)}} \right\},$$

by which the effect of imperfect insulation may be taken into account."

Extract of Letter from Prof. Stokes to Prof. W. Thomson (dated Nov. 1854).

"In working out for myself various forms of the solution of the equation $\dfrac{dv}{dt} = \dfrac{d^2v}{dx^2}$ under the conditions $v = 0$ when $t = 0$ from $x = 0$ to $x = \infty$; $v = f(t)$, when $x = 0$ from $t = 0$ to $t = \infty$, I found that the solution with a single integral only (and there must necessarily be this one) was got out most easily thus:—

"Let v be expanded in a definite integral of the form

$$v = \int_0^\infty \varpi\,(t, \alpha)\,\sin \alpha x dx,$$

which we know is possible.

"Since v does not vanish when $x = 0$, $\dfrac{d^2v}{dx^2}$ is not obtained by differentiating under the integral sign, but the term $\dfrac{2}{\pi}\,\alpha v_{x=0}$ must be supplied*, so that (observing that $v_{x=0} = f(t)$ by one of the equations of condition) we have

$$\frac{d^2v}{dx^2} = \int_0^\infty \left\{\frac{2}{\pi}\,\alpha f(t) - \alpha^2 \varpi\,.\right\}\sin \alpha x dx.$$

Hence

$$\frac{dv}{dt} - \frac{d^2v}{dx^2} = \int_0^\infty \left\{\frac{d\varpi}{dt} + \alpha^2 \varpi - \frac{2}{\pi}\,\alpha f(t)\right\}\sin \alpha x dx,$$

and the second member of the equation being the direct development of the first, which is equal to zero, we must have

$$\frac{d\varpi}{dt} + \alpha^2 \varpi - \frac{2}{\pi}\,\alpha f(t) = 0,$$

whence

$$\varpi = \epsilon^{-\alpha^2 t}\int_0^t \frac{2}{\pi}\,\alpha f(t)\,\epsilon^{\alpha^2 t} dt,$$

the inferior limit being an arbitrary function of α. But the other equation of condition gives

$$\varpi = \epsilon^{-\alpha^2 t}\int_0^t \frac{2}{\pi}\,\alpha f(t)\,\epsilon^{\alpha^2 t}\,dt = \left(\frac{\pi}{2}\right)^{-1}\alpha\int_0^t \epsilon^{-\alpha^2\overline{t-t'}}f(t')\,dt',$$

therefore

$$v = \left(\frac{\pi}{2}\right)^{-1}\int_0^\infty\int_0^t f(t')\,\alpha\epsilon^{-\alpha^2\overline{t-t'}}\sin \alpha x\,d\alpha\,dt'.$$

But

$$\int_0^\infty \epsilon^{-a\alpha^2}\cos b\alpha\,d\alpha = \frac{1}{2}\left(\frac{\pi}{a}\right)^{\frac{1}{2}}\epsilon^{-\frac{b^2}{4a}},$$

therefore

$$\int_0^\infty \epsilon^{-a\alpha^2}\sin b\alpha\,.\,\alpha d\alpha = -\frac{d}{db}\left\{\frac{1}{2}\left(\frac{\pi}{a}\right)^{\frac{1}{2}}\epsilon^{-\frac{b^2}{4a}}\right\}$$

$$= \frac{\pi^{\frac{1}{2}}b}{4a^{\frac{3}{2}}}\epsilon^{-\frac{b^2}{4a}},$$

* According to the method explained in a paper "On the Critical Values of the Sums of Periodic Series," *Camb. Phil. Trans.* Vol. VIII. p. 533 ["Mathematical and Physical Papers." G. G. Stokes, Vol. I. p. 236.]

whence writing $t - t'$, x, for a, b, and substituting, we have

$$v = \frac{x}{2\pi^{\frac{1}{2}}} \int_0^t (t - t')^{-\frac{3}{2}} \, \epsilon^{\frac{x^2}{4(t-t')}} f(t') \, dt'.$$

"Your conclusion as to the American wire follows from the differential equation itself which you have obtained. For the equation $kc \dfrac{dv}{dt} = \dfrac{d^2v}{dx^2}$ shows that two submarine wires will be similar, provided the squares of the lengths x, measured to similarly situated points, and therefore of course those of the whole lengths l, vary as the times divided by ck; or the time of any electrical operation is proportional to kcl^2.

"The equation $kc \dfrac{dv}{dt} = \dfrac{d^2v}{dx^2} - hv$ gives $h \propto l^{-2}$ for the additional condition of similarity of leakage."

The accompanying set of curves represents the strength of the current through the instrument at the remote end of a wire as it gradually rises, or gradually rises and falls, after the end operated on is put in connexion with one pole of a battery, and either kept so permanently, or detached and put in connexion with the ground after various short intervals of time.

The abscissas, measured on OX, represent the time reckoned from the first application of the battery, and the ordinates, measured parallel to OY, the strength of the current.

The time corresponding to a is equal to $\dfrac{kcl^2}{\pi^2} \log_\epsilon \left(\dfrac{4}{3}\right)$, if l be the length of the wire in feet, k its "resistance" per foot, in electro-statical units, and c its electro-statical capacity per foot (which is equal to $\dfrac{I}{2 \log \dfrac{R}{R'}}$, if I be the electro-statical inductive power of the gutta-percha, probably about 2, and R, R' the radii of its outer and inner surfaces). The principal curve (I) represents the rise of the current in the remote instrument, when the end operated on is kept permanently in connexion with the battery. It so nearly coincides with the line of abscissas at first as to indicate no sensible current until the interval of time corresponding to a has elapsed; although, strictly speaking, the effect at the remote end is instantaneous (i.e. according to data limited as regards knowledge of electricity, to such as those assumed in hydrodynamics when water is treated as if incompressible, or the velocity of sound in it considered infinitely great, which re-

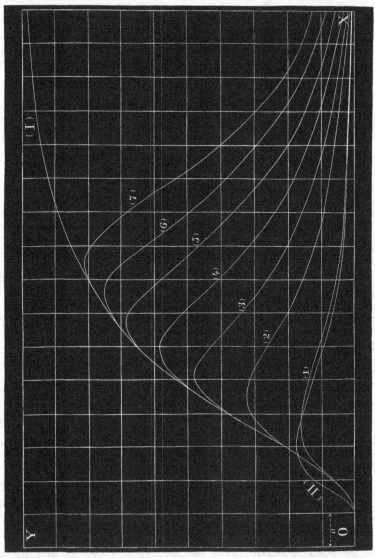

quires instantaneous effects to be propagated through the whole mass of the water, on a disturbance being made in any part of it). After the interval a, the current very rapidly rises, and after about $4a$ more, attains to half its full strength. After $10a$ from the commencement, it has attained so nearly its full strength, that the farther increase would be probably insensible. The full strength is theoretically reached only after an infinite time has passed. The first (1) of the smaller curves represents the rise and fall of the current in the remote instrument when the end operated on is put in connexion with the ground after having been for a time a in connexion with the battery; the second (2) represents similarly the effect of the battery for a time $2a$; the third (3) for a time $3a$ and so on. The curve (II) derived from the primary curve (I) by differentiation (exhibiting in fact the steepness of the primary curve at its different points, as regards the line of abscissas), represents the strength of current at different times through the remote end of the wire, consequent upon putting a very intense battery in communication with the end from which the signal is sent, for a very short time, and then instantly putting this end in communication with the ground. Thus, relatively to one another, the curves (I) and (II) may be considered as representing the relative effects of putting a certain battery in communication for the time a, and a battery of ten or twenty times as many cells for a time $\frac{1}{10}a$ or $\frac{1}{20}a$. If I were to guess what might be called "the retardation," which in the observations between Greenwich and Brussels was found to be about $\frac{1}{10}$th of a second, I should say it corresponded to four or five times a, but this must depend on the kind of instrument used, and the mode of making and breaking contacts with the battery which was followed.

Equation of principal curve (I).
$$y = 10a - 20a\,(e - e^4 + e^9 - e^{16} + \&c.), \text{ where } e = (\tfrac{3}{4})^{x/a};$$
a being half the side of one of the squares.

If $y = f(x)$ denote the equation of the principal wave, and if $f(x)$ be supposed to vanish for all negative values of x, the series of derived curves are represented by the equations

(1) .... $y = f(x) - f(x-a)$,

(2) .... $y = f(x) - f(x-2a)$,

(3) .... $y = f(x) - f(x-3a)$,

....

$$(7) \ldots\ldots\ldots\ldots\ldots\ y = f(x) - f(x - 7a),$$

$$(\text{II})\ldots\ldots\ldots\ldots\ldots\ y = a\,\frac{df(x)}{da}.$$

I think clearly the right way of making observations on telegraph retardations would be to use either Weber's electro-dynamo-
meter, or any instrument of suitable sensibility constructed on
the same principle, that is adapted to show deflections experienced
by a moveable part of a circuit, in virtue of the mutual electro-
dynamic force between it and the fixed part of the same circuit,
due to a current flowing for a very short time through the circuit.
Such an instrument, and an ordinary galvanometer, (showing impulsive deflections of a steel needle,) both kept in the circuit at
the remote end of the telegraph-wire from that at which the signal
is made, would give the values of $\int_0^\infty y^2 dx$, and $\int_0^\infty y\,dx$ (or the area),
for any of the curves; and the ratio of the time a of the diagrams
to the time during which the battery was held in communication
with the wire, might be deduced. The method will lose sensibility
if the battery be held too long in communication, but will be quite
sufficiently precise if this be not more than ten or twenty times a.
I believe there will be no difficulty in applying the method to
telegraph-wires of only twenty or thirty miles long, where no
retardation would be noticed by ordinary observation.   Before,
however, planning any observations of this kind with a view to
having them executed, I wished to form some estimate of the
probable value of a certain element,—the number of electro-
statical units in the electro-magnetic unit of electrical quantity,—
which I hoped to be able to do, from the observation of $\frac{1}{10}$th of a
second as the apparent retardation of signals between Greenwich
and Brussels.   I therefore applied to the Astronomer Royal for
some data regarding the mode of observation on the indications
of the needle, and the dimensions and circumstances of insulation
of the wire; and he was so good as to send me immediately all
the information that was available for my purpose.   This has
enabled me to make the estimate, and so has convinced me that
a kind of experiment which I proposed in a paper on Transient
Electric Currents in the *Philosophical Magazine* for June 1853,
[Art. LXII. Vol. I. above], and which I hope to be able before long
to put in practice, will be successful in giving a tolerably accu-

rate comparison of the electro-statical and electro-dynamic units; and, with a further investigation of the specific inductive capacity of gutta-percha, which will present no difficulty, will enable me to give all the data required for estimating telegraph retardations, without any data from telegraphic operations. This experiment [which I made about five years after the date of the present Article. See " Electrostatics and Magnetism," Art. XVIII.] is simply to put two plane-conducting discs in communication with the two poles of a Daniell's battery (or any other battery of which the electro-motive force is known in electro-magnetic units), and to *weigh* the attraction between them.*

    *   *   *   *   *

It would be easy at any time to make a plan for observing telegraph indications by means of either Weber's electro-dynamo-meter, or an instrument constructed on the same principle, or by measuring thermal effects of intermittent currents, which could be put in practice by any one somewhat accustomed to make observations, and which would give a tolerably accurate determination of the element of time, even in cases where the observable retardation is considerably less than $\frac{1}{10}$th of a second. A single wire in a submarine cable would, as far as regards the physical deductions to be made from this determination, be to be preferred to one of a number of different wires, insulated from one another under the same sheathing. I have little doubt but the Varna and Balaklava wire will be the best yet made for the purpose.

Without knowing exactly what the "retardation" may be, in terms of the element of time "a" of the diagrams, we may judge what the retardation, if similarly estimated, would be found to be in other cables of stated dimensions. Thus, if the retardation in 200 miles of submarine wire between Greenwich and Brussels be $\frac{1}{10}$th of a second, the retardation in a cable of equal and similar transverse section, extending half round the world (14,000 miles), would be

$$\left(\frac{14000}{200}\right)^2 \times \frac{1}{10} = 490 \text{ seconds, or } 8\tfrac{1}{6} \text{ minutes}:$$

* [Here follow, in the original Article, some elaborate estimates and calculations founded on too imperfect data to be worth re-publication now. They are referred to in §§ 316 and 317 of " Electrostatics and Magnetism." W. T. Apr. 3, 1883.]

and in the telegraphic cable (400 miles) between Varna and Balaklava, of which the electro-statical capacity per unit of length
may be about one-half greater than in the other, while the conducting power of the wire is probably the same, the retardation
may be expected to be

$$\left(\frac{400}{200}\right)^2 \times \frac{3}{2} \times \frac{1}{10} = \frac{3}{5} \text{ of a second.}$$

The rate at which distinct signals could be propagated to the
remote end would perhaps be one signal in about a quarter of an
hour in the former case, and nearly two signals in a second in the
latter.

ART. LXXIV. On the Electro-Statical Capacity of a
Leyden Phial and of a Telegraph Wire insulated in
the Axis of a Cylindrical Conducting Sheath.

[*Phil. Mag.*, June supplement, 1855.]

[Electrostatics and Magnetism, Article III.]

[From *Brit. Assoc. Rep.* 1855. Part 2.]

ART. LXXV. ON PERISTALTIC INDUCTION OF ELECTRIC CUR-
RENTS IN SUBMARINE TELEGRAPH WIRES.

RECENT examinations of the propagation of electricity through
wires in subaqueous and subterranean telegraphic cables, have led
to the observation of phenomena of induced electric currents,
which are essentially different from the phenomena (discovered
by Faraday many years ago) of what has hitherto been called
electro-dynamic, or electro-magnetic induction, but which, for
the future, it will be convenient to designate exclusively by the
term electro-magnetic. The new phenomena present a very perfect analogy with the mutual influences of a number of elastic
tubes bound together laterally throughout their lengths, and surrounded and filled with a liquid which is forced through one or
more of them, while the others are left with their ends open or
closed. The hydrostatic pressure applied to force the liquid
through any of the tubes will cause them to swell, and to press
against the others, which will thus, by peristaltic action, compel
the liquid contained in them to move in different parts of them
in one direction or the other. A long solid cylinder of india-
rubber, bored symmetrically in four, six, or more circular passages
parallel to its length, will correspond to an ordinary telegraphic
cable containing the same number of copper wires, separated
from one another only by gutta-percha; and the hydraulic motion
will follow rigorously the same laws as the electrical conduction,
and will be expressed by identical language in mathematics, provided the lateral dimensions of the bores are so small, in comparison with their lengths, or the viscosity of the fluid so great,

that the motions are not sensibly affected by inertia, and are consequently dependent altogether on hydrostatic pressure and fluid friction.   Hence the author considers himself justified in calling the kind of electric action now alluded to, *peristaltic induction*, to distinguish it from the electro-magnetic kind of electro-dynamic induction.   The mathematical treatment of the problem of mutual peristaltic induction is contained in the paper brought before the Section; but the author confined himself in the meeting to mentioning some of the results.   Among others, he mentioned, as being of practical importance, that the experiments which have been made on the transmission of currents backwards and for-wards by the different wires of a multiple cable, do not indicate correctly the degree of retardation that is to be expected when signals are to be transmitted through the same amount of wire laid out in a cable of the full length.   It follows, that expectations as to the working of a submarine telegraph between Britain and America, founded on such experiments, may prove fallacious; and to avoid the chance of prodigious losses in such an under-taking, the author suggested that the working of the Varna and Balaklava wire should be examined.   He remarked that a part of the theory communicated by himself to the Royal Society last May [Art. LXXIII. above] and published in the *Proceedings*, shows that a wire of six times the length of the Varna and Balaklava wire, if of the same lateral dimensions, would give thirty-six times the retardation, and thirty-six times the slowness of action.   If the distinctness of utterance and rapidity of action practicable with the Varna and Balaklava wire are only such as to be not inconvenient, it would be necessary to have a wire of six times the diameter; or better, thirty-six wires of the same dimensions; or a larger number of still smaller wires twisted together, under a gutta-percha cover-ing, to give tolerably convenient action by a submarine cable of six times the length.   The theory shows how, from careful observa-tions on such a wire as that between Varna and Balaklava, an exact estimate of the lateral dimensions required for greater dis-tances, or sufficient for smaller distances, may be made.   Immense economy may be practised in attending to these indications of theory in all submarine cables constructed in future for short distances; and the non-failure of great undertakings can alone be *ensured* by using them in a preliminary estimate.

[From the *Proceedings of the Royal Society*, Vol. VIII., May, 1856.]

ON PERISTALTIC INDUCTION OF ELECTRIC CURRENTS.

RECENT observations on the propagation of electricity through wires in subaqueous and subterranean telegraphic cables have brought to light phenomena of induced electric currents, which, while they are essentially different from the phenomena of what has hitherto been called electro-dynamic induction, are exactly such as might have been anticipated from the well-established theory of electrical equilibrium, had experiment afforded the data of relation between electrostatical and electro-dynamic units wanted for determining what dimensions of wire would be required to render these phenomena sensible to ordinary observation. They present a very perfect analogy with the mutual influences of a number of elastic tubes bound together laterally throughout their lengths, and surrounded and filled with a liquid which is forced through one or more of them, while the others are left with their ends open (*uninsulated*), or stopped (*insulated*), or subjected to any other particular conditions. The hydrostatic pressure applied to force the liquid through any of the tubes will cause them to swell and to press against the others, which will thus, by peristaltic action, compel the liquid contained in them to move, in different parts of them, in one direction or the other. A long solid cylinder of an incompressible elastic solid*, bored out symmetrically in four, six, or more circular passages parallel to its length, will correspond to an ordinary telegraph cable containing the same number of copper wires separated from one another only by gutta-percha: and the hydraulic motion

---

* Such as india-rubber very approximately is in reality.

will follow rigorously the same laws as the electrical conduction, and will be expressed by identical language in mathematics, provided the lateral dimensions of the bores are so small in comparison with their lengths, or the viscosity of the liquid so great, that the motions are not sensibly affected by *inertia*, and are consequently dependent altogether on hydrostatic pressure and fluid friction. The electrical induction now alluded to depends on the electrostatic forces determined by Coulomb; but it would be in one respect a real, and in all respects an apparent, contradiction of terms, to speak of electrostatic induction of electric currents, and I therefore venture to introduce the term *peristaltic* to characterize that kind of induction by which currents are excited in elongated conductors through the variation of electrostatic potential in the surrounding matter. On the other hand, as any inductive excitation of electric motion might be called electro-dynamic induction, it will be convenient to distinguish the kind of electro-dynamic induction first discovered by Faraday, by a distinctive name; and as the term *electro-magnetic*, which has been so applied, appears correctly characteristic, I shall call *electro-magnetic induction* that kind of action by which electric currents are excited, or inequalities of electric potential sustained, in a conductor of electricity, by variations of magnetic or electromagnetic potential, or by absolute or relative motion of the conductor itself across lines of magnetic or electro-magnetic force.

The most general problem of peristaltic induction is to determine the motion of electricity in any number of long conducting wires, insulated from one another within an uninsulated tube of conducting material, when subjected each to any prescribed electrical action at its extremities; without supposing any other condition regarding the sections and relative dispositions of the conductors than—(1), that their lateral dimensions and mutual distances are so small in proportion to their lengths, that the effects of peristaltic induction are paramount over those of electromagnetic induction; and (2), that the section of the entire system of conductors, if not uniform in all parts, varies so gradually as to be sensibly uniform through every part of the length not a very large multiple of the largest lateral dimension. In the present communication I shall only give the general equations of motion by which the physical conditions to be satisfied are expressed for every case; and I shall confine the investigation of solutions to certain cases of uniform and symmetrical arrangement, such as are commonly used in the submarine telegraph cable.

At any time t, let q_1, q_2, q_3, &c. be the quantities of electricity with which the different wires are charged, per unit of length of each, at a distance x from one extremity, O, of the conducting system; and let v_1, v_2, v_3, &c. be the electrostatical potentials in the same parts of those conductors. Let $\varpi_1^{(1)}$, $\varpi_1^{(2)}$, $\varpi_1^{(3)}$, &c., $\varpi_2^{(1)}$, $\varpi_2^{(2)}$, $\varpi_2^{(3)}$, &c., $\varpi_3^{(1)}$, $\varpi_3^{(2)}$, $\varpi_3^{(3)}$, &c. be coefficients, such that the electrostatical potentials (v_1, v_2, &c.), due to stated charges (q_1, q_2, &c.) of the different wires, are expressed by the equations

$$\left.\begin{array}{l} v_1 = \varpi_1^{(1)} q_1 + \varpi_1^{(2)} q_2 + \varpi_1^{(3)} q_3 + \&c. \\ v_2 = \varpi_2^{(1)} q_1 + \varpi_2^{(2)} q_2 + \varpi_2^{(3)} q_3 + \&c. \\ v_3 = \varpi_3^{(1)} q_1 + \varpi_3^{(2)} q_2 + \varpi_3^{(3)} q_3 + \&c. \\ \qquad \&c. \qquad\qquad \&c. \end{array}\right\} \dots\dots\dots(1).$$

If the sections of all the conductors are circular, these coefficients ($\varpi_1^{(1)}$, $\varpi_1^{(2)}$, &c.) may be easily determined numerically to any required degree of accuracy, in each particular case, by the *method of electrostatical images*. The electromotive force per unit of length at the position x will be, in the different wires,

$$\frac{dv_1}{dx}, \ \frac{dv_2}{dx}, \ \frac{dv_3}{dx},$$

respectively, and therefore if γ_1, γ_2, γ_3, &c. denote the strength of current at the same position, and k_1, k_2, k_3, &c. the resistances to conduction per unit of length in the different wires respectively, we have by the law of Ohm, applied to the action of peristaltic electromotive force,

$$k_1\gamma_1 = -\frac{dv_1}{dx}, \ \ k_2\gamma_2 = -\frac{dv_2}{dx}, \ \ k_3\gamma_3 = -\frac{dv_3}{dx} \ \dots\dots(2).$$

Now unless the strength of current be uniform along any one of the wires, the charge of electricity will experience accumulation or diminution in any part of it by either more or less electricity flowing in on one side than out on the other; and the mathematical expression of these circumstances is clearly

$$\frac{dq_1}{dt} = -\frac{d\gamma_1}{dx}, \ \frac{dq_2}{dt} = -\frac{d\gamma_2}{dx}, \ \frac{dq_3}{dt} = -\frac{d\gamma_3}{dx} \ \dots\dots(3).$$

Using in these equations the values of γ_1, γ_2, γ_3, &c. given by (2), and then substituting for v_1, v_2, v_3, &c. their expressions (1), we obtain

$$
\begin{aligned}
\frac{dq_1}{dt} &= \frac{d}{dx}\left\{\frac{1}{k_1}\ \frac{d\left(\varpi_1^{(1)}q_1\right)}{dx} + \frac{1}{k_2}\ \frac{d\left(\varpi_1^{(2)}q_2\right)}{dx} + \frac{1}{k_3}\ \frac{d\left(\varpi_1^{(3)}q_3\right)}{dx} + \&c.\right\} \\
\frac{dq_2}{dt} &= \frac{d}{dx}\left\{\frac{1}{k_1}\cdot\frac{d\left(\varpi_2^{(1)}q_1\right)}{dx} + \frac{1}{k_2}\cdot\frac{d\left(\varpi_2^{(2)}q_2\right)}{dx} + \frac{1}{k_3}\cdot\frac{d\left(\varpi_2^{(3)}q_3\right)}{dx} + \&c.\right\} \\
\frac{dq_3}{dt} &= \frac{d}{dx}\left\{\frac{1}{k_1}\cdot\frac{d\left(\varpi_3^{(1)}q_1\right)}{dx} + \frac{1}{k_2}\cdot\frac{d\left(\varpi_3^{(2)}q_2\right)}{dx} + \frac{1}{k_3}\cdot\frac{d\left(\varpi_3^{(3)}q_3\right)}{dx} + \&c.\right\}
\end{aligned}\quad(4),
$$

which are the general equations of motion required.

It is to be observed that k_1, k_2, &c., $\varpi_1^{(1)}$, $\varpi_1^{(2)}$, $\varpi_2^{(1)}$, &c. will be functions of x if the section of the conducting system is heterogeneous in different positions along it; but in all cases in which each conductor is uniform, and uniformly situated with reference to the others along the whole length, these coefficients will be constant, and the equations become reduced to

$$
\begin{aligned}
\frac{dq_1}{dt} &= \frac{\varpi_1^{(1)}}{k_1}\frac{d^2q_1}{dx^2} + \frac{\varpi_1^{(2)}}{k_2}\frac{d^2q_2}{dx^2} + \frac{\varpi_1^{(3)}}{k_3}\frac{d^2q_3}{dx^2} + \&c. \\
\frac{dq_2}{dt} &= \frac{\varpi_2^{(1)}}{k_1}\frac{d^2q_1}{dx^2} + \frac{\varpi_2^{(2)}}{k_2}\frac{d^2q_2}{dx^2} + \frac{\varpi_2^{(3)}}{k_3}\frac{d^2q_3}{dx^2} + \&c. \\
\frac{dq_3}{dt} &= \frac{\varpi_3^{(1)}}{k_1}\frac{d^2q_1}{dx^2} + \frac{\varpi_3^{(2)}}{k_2}\frac{d^2q_2}{dx^2} + \frac{\varpi_3^{(3)}}{k_3}\frac{d^2q_3}{dx^2} + \&c.
\end{aligned}\quad\ldots\ldots(5).
$$

The most obvious general method of treatment for integrating these equations, is to find elementary solutions by assuming

$$q_1 = A_1 u, \quad q_2 = A_2 u, \quad q_3 = A_3 u, \ldots q_i = A_i u \ldots\ldots(6),$$

where u satisfies the equation

$$\frac{du}{dt} = \kappa\frac{d^2u}{dx^2}\ \ldots\ldots\ldots\ldots\ldots\ldots\ldots\ldots(7).$$

This will reduce the differential equations (5) to a set of linear equations among the coefficients A_1, A_2, $\ldots A_i$, giving by elimination an algebraic equation of the ith degree having i real roots, to determine κ. The particular form of elementary solution of the equation (7) to be used may be chosen from among those given by Fourier, according to convenience, for satisfying the terminal conditions for the different wires.

In thinking on some applications of the preceding theory, I have been led to consider the following general question regarding the mutual influence of electrified conductors :—If, of a system of detached insulated conductors, one only be electrified with a given absolute charge of electricity, *will the potential excited in any one of the others be equal to that which the communication of an equal absolute charge to this other would excite in the first?* I now find that a general theorem communicated by myself to the Cambridge Mathematical Journal and published in the Numbers for November 1842 and February 1843, [Arts. IV. and V. Vol. I. above], but, as I afterwards (Jan. 1845) learned, first given by Green in his Essay on the Mathematical Theory of Electricity and Magnetism (Nottingham, 1828), leads to an affirmative answer to this question.

The general theorem to which I refer is, that if, considering the forces due respectively to two different distributions of matter (whether real, or such as is imagined in theories of electricity and magnetism), we denote by N_1, N_2 their normal components at any point of a closed surface, or group of closed surfaces, S, containing all parts of each distribution of matter, and by V_1, V_2 the potentials at the same point due respectively to the two distributions, and if ds be an element of the surface S, the value of $\iint N_1 V_2 \, ds$ is the same as that of $\iint N_2 V_1 \, ds$ (each being equal to the integral $\iiint R_1 R_2 \cos\theta \, dx \, dy \, dz$ extended over the whole of space external to the surface S, at any point (x, y, z) of which external space the two resultants are denoted by R_1, R_2 respectively, and the angle between their directions by θ). To apply this with reference to the proposed question, let the first distribution of matter consist of a certain charge, q, communicated to one of a group of insulated conductors, and the inductive electrifications of the others, not one of which has any absolute charge ; let the second distribution of matter consist of the electrifications of the same group of conductors when an equal quantity q is given to a second of them, and all the others are destitute of absolute charges ; and let surface S be the group of the surfaces of the different conductors.   Since the potential is constant through each separate conductor, the integral $\iint N_1 V_2 \, ds$ will be equal to the

6—2

sum of a set of terms of the form $[V_2][\iint N_1\, ds]$, where $[V_2]$ denotes the value in any of these conductors of the potential of the second distribution, and $[\iint N_1\, ds]$ an integral including the whole surface of the same conductor, but no part of that of any of the others. Now by a well-known theorem, first given by Green, $[\iint N_1\, ds]$ is equal to $4\pi q$ if q denote the absolute quantity of matter within the surface of the integral (as is the case for the first group of conductors), and vanishes if there be no distribution of matter, or (as is the case with each of the other conductors) if there be equal quantities of positive and negative matter within the surface over which the integral is extended. Hence if $[V_2]_1$ denote the potential in the first conductor due to the second distribution of matter, we have

$$\iint N_1 V_2\, ds = 4\pi\, [V_2]_1\, q.$$

Similarly, we have

$$\iint N_2 V_1\, ds = 4\pi\, [V_1]_2\, q.$$

Hence, by the general theorem, we conclude $[V_2]_1 = [V_1]_2$, and so demonstrate the affirmative answer to the question stated above.

I think it unnecessary to enter on details suited to the particular case of lateral electrostatic influence between neighbouring parts of a number of wires insulated from one another under a common conducting sheath, when uniform or varying electric currents are sent through by them; for which a particular demonstration in geometry of two dimensions, analogous to the demonstration of Green's theorem to which I have referred as involving the consideration of a triple integral for space of three dimensions, may be readily given; but, as a particular case of the general theorem I have now demonstrated, it is obviously true that the potential in one wire due to a certain quantity of electricity per unit of length in the neighbouring parts of another under the same sheath, is equal to the potential in this other, due to an equal electrification of the first.

Hence the following relations must necessarily subsist among the coefficients of mutual peristaltic induction in the general equations given above,

$$\varpi_1^{(2)} = \varpi_2^{(1)}; \quad \varpi_1^{(3)} = \varpi_3^{(1)}; \quad \varpi_2^{(3)} = \varpi_3^{(2)}; \quad \&c.$$

On the Solution of the Equations of Peristaltic Induction in symmetrical systems of Submarine Telegraph Wires.

The general method which has just been indicated for resolving the equations of electrical motion in any number of linear conductors subject to mutual peristaltic influence, fails when these conductors are symmetrically arranged within a symmetrical conducting sheath (and therefore actually in the case of any ordinary multiple wire telegraph cable), from the determinantal equation having sets of equal roots. Regular analytical methods are well known by which the solutions for such particular cases may be derived from the failing general solutions; but it is nevertheless interesting to investigate each particular case specially, so as to obtain its proper solution by a synthetical process, the simplest possible for the one case considered alone. In the present communication, the problem of peristaltic induction is thus treated for some of the most common cases of actual submarine telegraph cables, in which two or more wires of equal dimensions are insulated in symmetrical positions within a cylindrical conducting sheath of circular section.

CASE I.—*Two-wire Cable.*

In the general equations (according to the notation of the first part of this communication) we have $k_1 = k_2$; $\varpi_1^{(1)} = \varpi_2^{(2)}$; and $\varpi_2^{(1)} = \varpi_1^{(2)}$: and it will be convenient now to denote the values of the members of these three equations by k, $1/c$, and f/c respectively; that is, to express by k the galvanic resistance in each wire per unit of length, by c the electrostatical capacity of each per unit of length when the other is prevented from acquiring an absolute charge, and by f the proportion in which this exceeds the electrostatical capacity of each when the other has a charge equal to its own; or in other words, to assume c and f so that

$$\left.\begin{aligned} v_1 &= \frac{1}{c} q_1 + \frac{f}{c} q_2 \\ v_2 &= \frac{f}{c} q_1 + \frac{1}{c} q_2 \end{aligned}\right\} \quad \dots\dots\dots\dots\dots\dots\dots(1),$$

if v_1 and v_2 be the potentials in the two wires in any part of the cable, where they are charged with quantities of electricity respectively q_1 and q_2 per unit of length. The equations of electrical conduction along the two wires then become

$$\left.\begin{aligned} \frac{dv_1}{dt} &= \frac{1}{kc}\left(\frac{d^2v_1}{dx^2} + f\frac{d^2v_2}{dx^2}\right) \\ \frac{dv_2}{dt} &= \frac{1}{kc}\left(f\frac{d^2v_1}{dx^2} + \frac{d^2v_2}{dx^2}\right) \end{aligned}\right\} \quad \dots\dots\dots\dots\dots(2).$$

From these we have, by addition and subtraction,

$$\frac{d\vartheta}{dt} = \frac{1+f}{kc}\frac{d^2\vartheta}{dx^2}, \quad \text{and} \quad \frac{d\omega}{dt} = \frac{1-f}{kc}\frac{d^2\omega}{dx^2} \dots\dots\dots\dots(3),$$

where ϑ and ω are such that

$$v_1 = \vartheta + \omega, \quad v_2 = \vartheta - \omega \dots\dots\dots\dots\dots\dots(4).$$

If both wires reached to an infinite distance in each direction, the conditions to be satisfied in integrating the equations of motion would be simply that the initial distribution of electricity along each must be whatever is prescribed; that is, that

$$\left.\begin{aligned} v_1 &= \phi_1(x), \quad \text{and} \quad v_2 = \phi_2(x) \end{aligned}\right\} \dots\dots\dots\dots(5),$$
when $\qquad t = 0$

ϕ_1 and ϕ_2 denoting two arbitrary functions. Hence, according to Fourier, we have, for the integrals of the equations (3),

$$\left.\begin{aligned} \vartheta &= \sqrt{\frac{kc}{4(1+f)\pi}} \cdot t^{-\frac{1}{2}} \int_{-\infty}^{\infty} \frac{1}{2}\{\phi_1(\xi) + \phi_2(\xi)\} \, \epsilon^{-\frac{kc(\xi-x)^2}{4(1+f)t}} d\xi \\ \omega &= \sqrt{\frac{kc}{4(1-f)\pi}} \; t^{-\frac{1}{2}} \int_{-\infty}^{\infty} \frac{1}{2}\{\phi_1(\xi) - \phi_2(\xi)\} \, \epsilon^{-\frac{kc(\xi-x)^2}{4(1-f)t}} d\xi \end{aligned}\right\} \quad (6),$$

and the solution of the problem is expressed in terms of these integrals by (4).

If now we suppose the cable to have one end at a finite distance from the part considered, for instance at the point O from which x is reckoned, and if at this end each wire is subjected to electric action, so as to make its potential vary arbitrarily with the time, there will be the additional condition

$$v_1 = \psi_1(t), \text{ and } v_2 = \psi_2(t),$$

when $\qquad x = 0$ ....(7),

to be fulfilled. In the other conditions, (5), only positive values of x have now to be considered, but they must be fulfilled in such a way as not to interfere with the prescribed values of the potentials at the ends of the wires; which may be done according to the principle of images, by still supposing the wires to extend indefinitely in both directions, and in the beginning to be symmetrically electrified with contrary electricities on the two sides of O. To express the new condition (7), a form of integral, investigated in a communication to the Royal Society ('Proceedings,' May 10, 1855, p. 385) [Art. LXXIII. above], may be used; and we thus have for the integrals of equations (3),

$$\vartheta = \sqrt{\frac{kc}{4(1+f)\pi}} \left[t^{-\frac{1}{2}} \int_0^\infty \frac{1}{2}\{\phi_1(\xi) + \phi(\xi)\} \{\epsilon^{-\frac{kc(\xi-x)^2}{4(1+f)t}} - \epsilon^{-\frac{kc(\xi+x)^2}{4(1+f)t}}\} d\xi \right.$$
$$\left. + x \int_0^t \frac{1}{2}\{\psi_1(\theta) + \psi_2(\theta)\} \epsilon^{-\frac{kcx^2}{4(1+f)(t-\theta)}} \frac{d\theta}{(t-\theta)^{\frac{3}{2}}} \right]$$

$$\omega = \sqrt{\frac{kc}{4(1-f)\pi}} \left[t^{-\frac{1}{2}} \int_0^\infty \frac{1}{2}\{\phi_1(\xi) - \phi_2(\xi)\} \{\epsilon^{-\frac{kc(\xi-x)^2}{4(1-f)t}} - \epsilon^{-\frac{kc(\xi+x)^2}{4(1-f)t}}\} d\xi \right.$$
$$\left. + x \int_0^t \frac{1}{2}\{\psi_1(\theta) - \psi_2(\theta)\} \epsilon^{-\frac{kcx^2}{4(1-f)(t-\theta)}} \frac{d\theta}{(t-\theta)^{\frac{3}{2}}} \right] \quad (8).$$

Lastly, instead of the cable extending indefinitely on one side of the end O, let it be actually limited at a point E. If the ends of the two wires at E be subjected to electric action, so as to make each vary arbitrarily with the time, the new conditions to be satisfied, in addition to the others, (5) and (7), will be

$$v_1 = \chi_1(t) \text{ and } v_2 = \chi_2(t)$$

when $\qquad x = a$ ....(9),

if χ_1 and χ_2 denote two arbitrary functions, and a the length OE. Or, on the other hand, if they be connected together, so that a current may go from O to E along one and return along the other, the new conditions will be

$$v_1 - v_2 = 0, \quad \frac{d(v_1 + v_2)}{dx} = 0,$$

when $\qquad x = a$ ....(9)'.

Either of these requirements may be fulfilled in an obvious way by the *method of successive images*, and we so obtain the following respective solutions:—

$$\vartheta = \sqrt{\frac{kc}{4(1+f)\pi}} \left[t^{-\frac{1}{2}} \int_0^a \frac{1}{2}\{\phi_1(\xi)+\phi_2(\xi)\} F_{(f)}(\xi,\, t)\, d\xi \right.$$
$$\left. + \int_0^t \frac{d\theta}{(t-\theta)^{\frac{3}{2}}} \left[\frac{1}{2}\{\psi_1(\theta)+\psi_2(\theta)\}\mathcal{F}_{(f)}(x,t-\theta)+\frac{1}{2}\{\chi_1(\theta)+\chi_2(\theta)\}\mathcal{F}_{(f)}(a-x,t-\theta) \right] \right]$$

$$\omega = \sqrt{\frac{kc}{4(1-f)\pi}} \left[t^{-\frac{1}{2}} \int_0^a \frac{1}{2}\{\phi_1(\xi)-\phi_2(\xi)\} F_{(-f)}(\xi,\, t)\, d\xi \right.$$
$$\left. + \int_0^t \frac{d\theta}{(t-\theta)^{\frac{3}{2}}} \left[\frac{1}{2}\{\psi_1(\theta)-\psi_2(\theta)\}\mathcal{F}_{(-f)}(x,t-\theta)+\frac{1}{2}\{\chi_1(\theta)-\chi_2(\theta)\}\mathcal{F}_{(-f)}(a-x,t-\theta) \right] \right]$$
$$\ldots\ldots\ldots\ldots(10),$$

$$\vartheta = \sqrt{\frac{kc}{4(1+f)\pi}} \left[t^{-\frac{1}{2}} \int_0^a \frac{1}{2}\{\phi_1(\xi)+\phi_2(\xi)\} E_{(f)}(\xi,\, t)\, d\xi \right.$$
$$\left. + \int_0^t \frac{1}{2}\{\psi_1(\theta)+\psi_2(\theta)\}\mathfrak{E}_{(f)}(x,\, t-\theta)\frac{d\theta}{(t-\theta)^{\frac{3}{2}}} \right]$$

$$\omega = \sqrt{\frac{kc}{4(1-f)\pi}} \left[t^{-\frac{1}{2}} \int_0^a \frac{1}{2}\{\phi_1(\xi)-\phi_2(\xi)\} F_{(-f)}(\xi,\, t)\, d\xi \right.$$
$$\left. + \int_0^t \frac{1}{2}\{\psi_1(\theta)-\psi_2(\theta)\}\mathcal{F}_{(-f)}(x,\, t-\theta)\frac{d\theta}{(t-\theta)^{\frac{3}{2}}} \right]$$
$$\ldots\ldots\ldots\ldots(10)',$$

where $F,\ \mathcal{F},\ E,\ \mathfrak{E}$ denote for brevity the following functions:—

$$F_{(f)}(\xi,\, t) = \sum_{i=-\infty}^{i=\infty} \left\{ \epsilon^{-\frac{kc(x+2ia-\xi)^2}{4(1+f)t}} - \epsilon^{-\frac{kc(x+2ia+\xi)^2}{4(1+f)t}} \right\}$$

$$\mathcal{F}_{(f)}(x,t-\theta) = \sum_{i=-\infty}^{i=\infty} (x+2ia)\epsilon^{-\frac{kc(x+2ia)^2}{4(1+f)(t-\theta)}} = \frac{(1+f)(t-\theta)}{kc}\{F_{(f)}(\xi,t-\theta)\div\xi\}_{\xi=0}$$

$$E_{(f)}(\xi,\, t) = \sum_{-\infty}^{\infty} (-1)^i \left\{ \epsilon^{-\frac{kc(x+2ia-\xi)^2}{4(1+f)t}} - \epsilon^{-\frac{kc(x+2ia+\xi)^2}{4(1+f)t}} \right\}$$

$$\mathfrak{E}_{(f)}(x,t-\theta) = \sum_{-\infty}^{\infty} (-1)^i(x+2ia)\epsilon^{-\frac{kc(x+2ia)^2}{4(1+f)(t-\theta)}} = \frac{(1+f)(t-\theta)}{kc}\{E_{(f)}(\xi,t-\theta)\div\xi\}_{\xi=0}$$
$$\ldots\ldots\ldots(11).$$

Each of the functions F and E is clearly the difference between two periodical functions of $(\xi - x)$ and $(\xi + x)$; and each of the functions $\mathcal{F}$ and $\mathfrak{E}$ is a periodical function of x simply. The expressions for these four functions, obtained by the ordinary formulas for the expression of periodical functions in trigonometrical series, are as follows :—

$$F_{(f)}(\xi, t) = \frac{2}{a}\sqrt{\frac{4(1+f)\pi t}{kc}} \sum_{i=1}^{i=\infty} \epsilon^{-\frac{i^2\pi^2(1+f)t}{a^2kc}} \sin\frac{i\pi x}{a} \sin\frac{i\pi\xi}{a}$$

$$\mathfrak{F}_{(f)}(x, t-\theta) = \frac{1}{2a^2}\left[\frac{4(1+f)\pi(t-\theta)}{kc}\right]^{\frac{3}{2}} \sum_{1}^{\infty} i\epsilon^{-\frac{i^2\pi^2(1+f)(t-\theta)}{a^2kc}} \sin\frac{i\pi x}{a}$$

$$E_{(f)}(\xi, t) = \frac{2}{a}\sqrt{\frac{4(1+f)\pi t}{kc}} \sum_{1}^{\infty} \epsilon^{-\frac{(2i-1)^2\pi^2(1+f)t}{4a^2kc}} \sin\frac{(2i-1)\pi x}{2a} \sin\frac{(2i-1)\pi\xi}{2a}$$

$$\mathfrak{E}_{(f)}(x, t-\theta) = \frac{1}{4a^2}\left[\frac{4(1+f)\pi(t-\theta)}{kc}\right]^{\frac{3}{2}} \sum_{1}^{\infty}(2i-1)\epsilon^{-\frac{(2i-1)^2\pi^2(1+f)(t-\theta)}{4a^2kc}} \sin\frac{(2i-1)\pi x}{2a}$$

$$\dots\dots\dots\dots(12).$$

Either (11) or (12) may be used to obtain explicit expressions for the solutions (10) and (10)', in convergent series; but of the series so obtained, (11) converge very rapidly and (12) very slowly when t is small; and, on the contrary, (11) very slowly and (12) very rapidly when t is large. It is satisfactory, that, as t increases, the first set of series (11) do not cease to be, before the second set (12) become, convergent enough to be extremely convenient for practical computation.

The solutions obtained by using (12), in (10) and (10)', are the same as would have been found by applying Fourier's ordinary process to derive from the elementary integral $\epsilon^{-mt}\sin nx$ the effects of the initial arbitrary electrification of the wires, and employing a method given by Professor Stokes* to express the effects of the variations arbitrarily applied at the free ends of the wires.

CASE II.—*Three-wire Cable.*

The equations of mutual influence between the wires may be clearly put under the forms

$$cv_1 = q_1 + f(q_2 + q_3), \quad cv_2 = q_2 + f(q_3 + q_1), \quad cv_3 = q_3 + f(q_1 + q_2);$$

---

* See *Cambridge Phil. Trans.* vol. VIII. p. 533, "On the Critical Values of the sums of Periodic Series." [*Mathematical and Physical Papers.* G. G. Stokes, Vol. I. p. 237.]

and the equations of electrical motion along them are then as
follows :—

$$kc\frac{dq_1}{dt} = \frac{d^2q_1}{dx^2} + f\left(\frac{d^2q_2}{dx^2} + \frac{d^2q_3}{dx^2}\right), \quad kc\frac{dq_2}{dt} = \frac{d^2q_2}{dx^2} + f\left(\frac{d^2q_3}{dx^2} + \frac{d^2q_1}{dx^2}\right),$$

$$kc\frac{dq_3}{dt} = \frac{d^2q_3}{dx^2} + f\left(\frac{d^2q_1}{dx^2} + \frac{d^2q_2}{dx^2}\right)$$

If we assume

$$\sigma = q_1 + q_2 + q_3, \quad \omega_1 = 2q_1 - q_2 - q_3, \quad \omega_2 = 2q_2 - q_3 - q_1, \quad \omega_3 = 2q_3 - q_1 - q_2,$$

which give

$$q_1 = \frac{1}{3}\sigma + \omega_1, \quad q_2 = \frac{1}{3}\sigma + \omega_2, \quad q_3 = \frac{1}{3}\sigma + \omega_3,$$

and require that $\omega_1 + \omega_2 + \omega_3 = 0$, we find by addition and subtraction, among the equations of conduction,

$$kc\frac{d\sigma}{dt} = (1 + 2f)\frac{d^2\sigma}{dx^2}$$

and

$$kc\frac{d\omega}{dt} = (1 - f)\frac{d^2\omega}{dx^2},$$

where for ω may be substituted either ω_1, ω_2, or ω_3.

CASE III.—*Four-wire Cable.*

The equations of mutual influence being

$$cv_1 = q_1 + f(q_2 + q_4) + gq_3,$$

and other four symmetrical with this; and the equations of
motion,

$$kc\frac{dq_1}{dt} = \frac{d^2q_1}{dx^2} + f\left(\frac{d^2q_2}{dx^2} + \frac{d^2q_4}{dx^2}\right) + g\frac{d^2q_3}{dx^2},$$

&c.   &c.   &c.,

we may assume

$$q_1 + q_2 + q_3 + q_4 = \sigma, \quad q_1 - q_3 = \omega_1,$$

$$q_1 - q_2 + q_3 - q_4 = \vartheta, \quad q_2 - q_4 = \omega_2;$$

which give

$$q_1 = \frac{1}{4}(\sigma + \vartheta + 2\omega_1); \quad q_2 = \frac{1}{4}(\sigma - \vartheta + 2\omega_2);$$

$$q_3 = \frac{1}{4}(\sigma + \vartheta - 2\omega_1); \quad q_4 = \frac{1}{4}(\sigma - \vartheta - 2\omega_2);$$

and we find from the equations of conduction,

$$kc\frac{d\sigma}{dt}=(1+2f+g)\frac{d^2\sigma}{dx^2}\;;\;\;kc\frac{d\vartheta}{dt}=(1-2f+g)\frac{d^2\vartheta}{dx^2}\;;\;\;kc\frac{d\omega}{dt}=(1-g)\frac{d^2\omega}{dx^2}\;.$$

CASE IV.—*Cable of six wires symmetrically arranged.*

Equations of mutual influence,

$$cv_1=q_1+f(q_2+q_6)+g(q_3+q_5)+hq_4$$
$$\text{\&c.}\qquad\text{\&c.}\qquad\text{\&c.}$$

Equations of conduction,

$$kc\frac{dq_1}{dt}=\frac{d^2q_1}{dx^2}+f\left(\frac{d^2q_2}{dx^2}+\frac{d^2q_6}{dx^2}\right)+g\left(\frac{d^2q_3}{dx^2}+\frac{d^2q_5}{dx^2}\right)+h\frac{d^2q_4}{dx^2}\;.$$

Then assuming

$$q_1+q_2+q_3+q_4+q_5+q_6=\sigma$$
$$q_1-q_4+q_3-q_6+q_5-q_2=\vartheta$$
$$3(q_1+q_4)-\sigma=\omega_1;\;\;\;3(q_3+q_6)-\sigma=\omega_2;\;\;\;3(q_5+q_2)-\sigma=\omega_3;$$
$$3(q_1-q_4)-\vartheta=\rho_1;\;\;\;3(q_3-q_6)-\vartheta=\rho_2;\;\;\;3(q_5-q_2)-\vartheta=\rho_3;$$

which require that

$$\omega_1+\omega_2+\omega_3=0,\;\;\text{and}\;\;\rho_1+\rho_2+\rho_3=0\;;$$

we have

$$kc\frac{d\sigma}{dt}=[1+2(f+g)+h]\frac{d^2\sigma}{dx^2}\;;\;\;kc\frac{d\vartheta}{dt}=[1-2(f-g)-h]\frac{d^2\vartheta}{dx^2}\;;$$

$$kc\frac{d\omega}{dt}=[1-(f+g)+h]\frac{d^2\omega}{dx^2}\;;\;\;kc\frac{d\rho}{dt}=[1+(f-g)-h]\frac{d^2\rho}{dx^2}\;.$$

These equations, integrated by the usual process to fulfil the prescribed conditions, determine σ, ϑ, ω_1, ω_2, ω_3, ρ_1, ρ_2, ρ_3; and we then have, for the solution of the problem,

$$q_1=\frac{1}{6}(\sigma+\vartheta+\omega_1+\rho_1);\;\;q_3=\frac{1}{6}(\sigma+\vartheta+\omega_2+\rho_2);\;\;q_5=\frac{1}{6}(\sigma+\vartheta+\omega_3+\rho_3)\;;$$

$$q_4=\frac{1}{6}(\sigma-\vartheta+\omega_1-\rho_1);\;\;q_6=\frac{1}{6}(\sigma-\vartheta+\omega_2-\rho_2);\;\;q_2=\frac{1}{6}(\sigma-\vartheta+\omega_3-\rho_3).$$

[From the *Athenæum*, Oct. 4, 1856.]

ART. LXXVI. LETTERS ON "TELEGRAPHS TO AMERICA."

Invercloy, Isle of Arran, *Sept.* 24, 1856.

AN account of Mr Wildman Whitehouse's communication to the British Association at Cheltenham on the Submarine Electric Telegraph, which I have only seen a few days ago in your Number for the 30th of August, contains expressions regarding views previously put forward by myself on theoretical grounds, in reply to which I beg to offer a few remarks. Mr Whitehouse's communication not only professes to overturn my theoretical conclusions, but it gives what might at first sight appear to be sufficient experimental evidence of the validity of an ordinary submarine cable for telegraphic communication between this country and America, in opposition to my warning that more than ordinary lateral dimensions of wire or insulating coat might be necessary to allow sufficient rapidity in the communication of intelligence through a conductor so much longer than any hitherto used in practical operations. I therefore think it right to say, that all Mr Whitehouse's experimental results are perfectly consistent with my theory; but at the same time I wish it to be understood that my ground for saying so, is not confidence that a way I now see as possibly leading to an explanation of the apparent discrepancy is the true way, but a knowledge of the theory itself, which, like every *theory*, is merely a combination of established truths. Those who have not made themselves acquainted with the theory, will of course attribute no weight to the expression I am now giving, with no other support than my own consciousness of its truth; and,

were it not that silence on my part might appear to indicate
acquiescence in what has been published as an experimental
demonstration of the falseness of my conclusions, I should prefer
not troubling you with any observations on the subject, until I am
able to offer you a short account of some practical developments of
the theoretical investigation, which I intend before long to communicate to the Royal Society. In the meantime I shall allude no
further to details than to say, that Mr Whitehouse's observations,
as reported, do not show at what speed such a succession of signals
as is required for the letters of a word, can be sent through the
greatest length of wire which he used (something upwards of 1,000,
miles), in a cable of ordinary lateral dimensions.   It is easily seen,
without special experiment, that a continued and uniform succession, of alternate applications of the positive and negative poles of
a battery to one end of an insulated conductor, however long, must
give rise to uniform alternations of currents gradually rising and
falling in the two directions, at the other end connected with the
ground ; and that the more rapid the succession of these alternations the feebler will be the maximum intensities of the alternate
currents at the remote end. Mr Whitehouse's experiments show how
many such alternations may be made per minute, with the battery
he actually employs, at one end, without making the alternate
currents at the other end insensible to his tests ; but more practical experiments than any mentioned in the account of his communication published in the *Athenæum*, are required to show, at
what rate the irregular non-periodic alternations of currents, required to spell out a word or give a message according to any
possible telegraphic code, may be produced at one end by means of
operations performed at the other end of a cable of ordinary lateral
dimensions, and 1,000 miles long.   Capitalists ought to require a
very "matter-of-fact" proof of the attainability of a sufficient
rapidity in the communication of actual messages, by whatever
cable may be proposed, before sinking so large an amount of property in the Atlantic, as would be involved in any cable of ordinary
or of extraordinarily great lateral dimensions, to form an electric
communication between Britain and America.

I remain, &c.,

WILLIAM THOMSON.

[From the *Athenæum*, Nov. 1, 1856.]

Invercloy, Isle of Arran, *Oct.* 24, 1856.

There are some conclusions from my theory of electrical conduction in submarine cables which, if correct, must be found valuable in planning the best dimensions and arrangement of wires
and insulating tubes for telegraphic communication between Britain
and America :—an undertaking at present actually in progress, as
appears from statements in the *Athenæum*. It is, therefore, of
importance that such of your readers as may be interested in the
subject should, without delay, have some evidence to allow them
to judge as to the validity of the objections to those conclusions,
which Mr Wildman Whitehouse made in his communication to
the British Association at Cheltenham [reported *Athen.* August 30],
and which he still maintains in his letter recently addressed to you
[*Athen.* Oct. 11], notwithstanding my confident assertion of their
correctness, and my reference to the theoretical demonstration
[*Proceedings of the Royal Society*, May 10, 1855, and *Philosophical
Magazine*, vol. Jan. to June, 1856 (Art. LXXIII. above)], as sup-
porting this assertion by unassailable evidence [*Athen.* Oct. 4].   On
this account, rather than from any desire to continue in your
columns a controversy which I believe will be easily adjusted by
private correspondence, I offer you the following remarks.

The points of greatest practical importance in my conclusions
are, 1, " the law of squares," as it has been called by Mr W. Whitehouse; and, 2, the influence of lateral dimensions of wire and
insulating coating on the rapidity of signals by a submarine cable
of stated length.   It is on these points chiefly that Mr W. Whitehouse brings his attack, and it is to these that I shall confine my
present defence.

1.   *Law of Squares.*—In his Cheltenham communication, Mr
W. Whitehouse brings forward the following table of results, as
conclusive against the law of squares :—

Amount of Retardation observed at various Distances.
Voltaic current.   Time stated in Parts of a Second.

Mean of 550 obsrvns.	Mean of 110 obsrvns.	Mean of 1840 obsrvns.	Mean of 1960 obsrvns.	Mean of 120 simultaneous observations.	
83 miles.	166 miles.	249 miles.	498 miles.	535 miles.	1020 miles.
·08	·14	·36	·79	·74	1·42

The retardations here shown as the results of undoubtedly trustworthy observations, are not proportional to the squares of the distances; but it depends on the nature of the electric operation performed at one extremity of the wire, and on the nature of the test afforded by the indicating instrument at the other extremity, whether or not any *approach to the law of squares is to be expected* in the observed results. Thus, the following table, which I have derived solely from theory*, shows for different lengths of cable, of certain lateral dimensions, the delays, from the instants of beginning various definitely-specified electrical operations at one extremity, experienced in waiting for the indications by each of two definite tests of electrical effect at the other extremity :—

Durations of battery action.	Times from beginning of battery action till strength of current rises to	150 miles.		300 miles.		600 miles.		1200 miles.		2400 miles.	
		¾ max.	max.	¾ max.	max.	¾ max.	max.	¾ max.	max.	¾ max.	max.
seconds.		seconds.	seconds.	seconds.	seconds.	seconds.	seconds.	seconds.	seconds.	seconds.	seconds.
0		·011	·0164	·044	·0656	·175	·2622	·7	1·049	2·8	4·196
·333		·0378	·333	·145	·354	·353	·49	·84	1·17	2·93	4·27
·667		·0378	·667	·151	·667	·495	·78	1·08	1·45	3·07	4·4
1·000		·0378	1·000	·151	1·000	·564	1·104	1·25	1·70	3·20	4·53
8		·0378	8·000	·151	8·000	·604	8·000	2·416	8·33	6·93	10
16		·0378	16·000	·151	16·000	·604	16·000	2·416	16·00	9·00	17·47
24		·0378	24·000	·151	24·000	·604	24·000	2·416	24·00	9·60	25·33
∞		·0378	∞	·151	∞	·604	∞	2·416	∞	9·664	∞

The electrical operation to which the first horizontal series of numbers in the table refers, is in strictness, an infinitely short application of an infinitely intense electro-motive force; but, practically this condition will be realized to a sufficient degree of approximation for verifying the result, by applying a powerful battery or an electro-magnetic impulse during one-twentieth, or any smaller fraction of the time, shown as the delay of the maximum in each case,—for instance, during $\frac{1}{1200}$ of a second, or any less time, for the 150-mile cable, or during $\frac{1}{5}$ of a second or any less

---

* I have obtained the tabulated results merely by inspection from a diagram of curves [p. 72 above] published with my first communication to the Royal Society [Art. LXXIII. above] on the subject. The decimals will, no doubt, require some correction when exact calculations are made, as will very soon be the case. The lateral dimensions for which the table has been calculated are those of a single wire cable,—copper wire ·065 in. diameter, gutta-percha tube ¼ in. outer diameter.

time for the cable of 2,400 miles. The electrical operations referred to in the other horizontal lines consist of applications of one
pole of a constant battery during the stated times. In every case
the extremity operated on is supposed to be put in connexion with
the earth, at the instant when it is disconnected from the battery.
The other extremity is supposed to be always kept in connexion
with the earth. The conductor of the indicating instrument
through which this connexion is established, is supposed to exercise
but a small resistance compared with that of any number of miles
of the wire, and is assumed to give rise, by electro-magnetic induction, to no delay comparable with that due to the induction of the
other kind (peristaltic induction), in the cable in each case. This
last condition will be far from being fulfilled, except for the
longest cable, when large coils with soft iron cores are used in the
indicating instrument, as any one will be ready to admit who has
seen Faraday's beautiful magneto-optical experiment, and has remarked the gradually rising illumination, consequent on the sudden
completion of the voltaic circuit.

Now, it will be observed that the law of squares is fulfilled
throughout no horizontal line of the table, except the first and the
last. If, for instance, we take the series of retardations of 3/4 of
maximum strength, and of maximum strength, consequent on an
application of the battery for one-third of a second, we have the
following comparative view :—

Lengths	150	300	600	1,200	2,400
Retardations of ¾ maximum...	·0378	·145	·353	·84	2·93
Retardations of maximum ...	·333	·354	·49	1·17	4·27

Here the retardation of the 3/4 maximum is a little short of four
times as much in the 300-mile cable as in the 150; it is only 2½
times as much in the 600 as in the 300; only 2¼ times as much
in the 1,200 as in the 600; and only 3½ times as much in the 2,400
as in the 1,200; instead of being 4 times greater in each of the
longer cables than in the other of half-length, as it would be if the
law of squares were applicable.

By comparing, on the same principle, Mr W. Whitehouse's
results quoted above [*Athen.* Aug. 30, p. 1093], we find very analogous *deviations from the law of squares;* but for a strict verification of their *agreement with theory,* it would be necessary to have a
precise specification of the electrical operations which were per-

formed at one extremity, and of the exact nature of the test of electrical effect afforded by the receiving instrument at the other extremity. The manuscript of the Cheltenham paper, to which, through the courtesy of Mr Whitehouse, I have had access, shows that the operations were *certainly not such as could give effects, on the whole manifesting the law of squares even approximately*, and makes me think it probable that electro-magnetic induction in the receiving instrument, has most sensibly influenced the retardations observed with the shorter lengths of cable, increasing, relatively, those observed with the shortest. Each application of the battery appears to have been maintained for a second, and, therefore, whether this was instantly followed by a reverse application (as appears to have been the case), or by the establishment of a connexion with the ground, the law of squares ought to be verified very closely for all lengths of cable less than 300 miles, if the battery power is arranged so that the maximum strength of the current is in all cases the same. Accordingly, we have a tolerable verification of the law of squares by Mr Whitehouse's observed retardations in the 166-mile and the 249-mile cable, which are respectively ·14 and ·36. These show an even greater increase of retardation in the greater length than according to that law, which would have been exactly verified by the numbers ·14 and ·315. If the indicating instrument does not show its effect until a somewhat larger proportion than 3/4 of the maximum strength is reached, the retardation in 498 miles may be either more or less than twice the retardation in 249 miles ; and if it be only a little more than twice (being ·79 as Mr Whitehouse finds it), it will follow from the theory illustrated by the table, that the retardation in 996 miles ought to be found considerably less than twice the retardation in 498. Thus Mr Whitehouse finds even the retardation in 1,020 miles less than the double of the retardation which he finds in 498 miles. Another cause which may contribute to deviations from the law of squares, and which must most seriously derange the results from agreement, with any easily-calculated indications of theory, is the inconstancy of the galvanic battery employed. It is well known to all who have used the sand-battery, or any other form of battery which evolves gaseous hydrogen during its continuous action, that, in commencing to act, after having been quiescent for any time, it acts with immensely greater electro-motive force than it can keep up, and that during the initial second of its operation the electro-motive force *falls in general with great rapidity*. This circumstance, it is probable, largely influences the phenomena which Mr Whitehouse observed; and if it does, must complicate extremely the character, both relative and absolute, of the electrical operations used in his experiments on different lengths of cable.

It appears then, that in the one case of comparison in which a manifestation of the law of squares could be expected, that law is manifested by Mr Whitehouse's results, and that in all other cases, so far as the data supplied suffice for testing, they agree with the theory of which the law of squares is a part.

To test the theory extensively and strictly by observations with Mr Whitehouse's instruments will, probably, not be easy without using either a constant battery, or a system of measured electro-magnetic impulses. It will, however, I have no doubt, be quite practicable with a sufficient number of cells of Daniell's battery to verify the law of squares with considerable precision for lengths of cable from 498 miles upwards. All that will be necessary will be, to take such battery powers as to give the same permanent current through the cable and instrument in the different cases, that is to say, to make the numbers of cells used, proportional to the lengths of the cable; and in each experiment to hold on the battery connexion, until after the indicating instrument has made its mark. The most favourable relation between the sensitiveness of the indicator and the battery power, to give a sharp test, is such, that the full strength of the permanent current is about four times the strength that just moves the index. But this would probably give too small a retardation in the 498-mile cable, to free the comparison from large influence of electro-magnetic induction in the coils; and it may be that a more satisfactory test would be obtained with the actual instruments, by using battery power capable of maintaining a permanent current more, by only 1/3 or 1/4, than enough to give an indication.

2. *Influence of Lateral Dimensions of Conductor and Insulating Tube on Rapidity of Electrical Action.*—My remedy for inconvenient slowness of action in a submarine cable, is an increased diameter of conductor and of gutta-percha coat, in the same proportions, or in different proportions, indicated by economical considerations calculable from theory, with data as to the price of

copper, gutta percha, and iron, and the relative expense and risk in laying cables of different weights (see my first communication on the subject referred to above). Until the price of the copper comes to be a serious consideration in comparison with the whole expense of making and laying the cable, an increase in the diameter of the conductor alone, up to a certain limit, without any increase in the thickness of the gutta-percha tube, may be the cheapest remedy. Accordingly, in my statement quoted by Mr Whitehouse, either one thick copper-wire, or (as is preferable for mechanical reasons) "a number of small wires twisted together" is recommended to obviate inconvenient slowness of utterance. Mr Whitehouse tests this proposal by three wires connected at their ends, so as to afford a triple conducting channel, *but separated throughout their lengths by their gutta-percha coats.* Now it is perfectly clear, that an electrical impulse through three wires so arranged, cannot differ from the sum of three separate electrical impulses of one-third strength each through any one of the wires, except in virtue of mutual peristaltic induction. With these very wires, Mr Whitehouse found the effects of mutual peristaltic induction to be extremely slight, requiring some of his finest tests to be shown at all. How then he found "the retardation nearly twice as great with the triple wire as in one of the wires alone," is an anomaly which it is not my part to explain. I may be allowed to suggest as a possible explanation, that he used only battery power enough to give the same permanent current through the triple wire as through the single wire, in the comparative experiments; and that the triple conductor exhausted the high initial electro-motive force much more rapidly during the first quarter of a second than the single wire. In reality, when justice is done to the triple wire, it will be found to transmit the action *a very little more rapidly than one wire alone*, in consequence of the diminution of electrostatic capacity in each wire, which is caused by mutual induction among the three, when equal neighbouring parts of them are similarly and equally electrified at each instant. But my law regarding the effect of increasing the diameter of one wire or of using *a number of fine wires twisted together*, is neither proved or disproved by experiments on the combined effects of a number of wires about as widely separated from one another as from their common iron sheath.

Neither on the two points to which I have especially adverted, nor on any other, will it be possible, I believe, to find the theory at fault; and Mr Whitehouse has too effectively proved himself to be a sincere searcher for truth, to allow me to doubt that he will sooner or later fully admit the correctness of my conclusions, although they have hitherto appeared to him to be at variance with the phenomena he has observed.

On the other hand, I have much pleasure in admitting that he has quite answered my objections to experiments on the retardation of effects of a current sent through a certain length of wire backwards and forwards in a "multiple" cable, as trustworthy for indicating the retardations to be expected in the same length of wire laid out singly to an untried distance. I am very glad, too, to perceive that he has succeeded in convincing practical men by his experiments, in conjunction with Mr Bright and Mr Morse, that a cable of ordinary lateral dimensions may be not too slow in its action for commercial success in the projected sub-Atlantic telegraph.

I should certainly not have expected that this would be found to be the case *if the Varna and Balaclava telegraph, when pushed to the utmost, had been found to be inconveniently slow;* but although I have never been able to learn any particulars as to the experience of those who have worked that telegraph, I have the best possible reason now for knowing, that if nine letters per second have not been delivered by it, the fault, or rather I should say the deficiency, has not been in the cable but in the instruments. Measurements published since the beginning of the present year, by that most profound and accurate of all experimenters, Wilhelm Weber, have allowed me to reduce his own previous determination of the electrical conductivity of copper, to the proper kind of unit for the telegraph problem, and have so afforded me the data for calculating the retardations of electric signals sent through a copper wire of stated dimensions, insulated in any specified manner. I thus find that the retardation of the maximum strength at one extremity, consequent on a single instantaneous electrical impulse at the other extremity of 400 miles of Black Sea cable*, amounts to only about 117 of a second. The corresponding retardation in a cable of equal lateral dimensions, and 2,400 miles long, would be only about 4·2.

---

* Or a single wire cable; copper wire 1/15·4 of an inch in diameter, in gutta-percha tube 1/4 inch external diameter.

A mode of operating so as to clear a wire rapidly of residual electricity, which I have worked out from theory, and a plan for telegraphic receiving instruments to take the most full advantage of it, which has recently occurred to me, allow me now to feel confident of the possibility of sending a distinct letter every $3\frac{1}{2}$ seconds by such a cable. This, amounting to 17 letters a minute, would give 200 messages of 20 words in the 24 hours, and at 30s. a message would be not a bad return for 1,000,000l. of capital expended. The rate could be augmented to a letter every $2\frac{1}{4}$ seconds, or 280 messages in 24 hours, by augmenting the diameter of copper and gutta-percha tube each in the proportion 5 to 4,—that is to say, by using a copper conductor 1/12·4 of an inch diameter, in a gutta-percha tube 5/16 of an inch outer diameter.

From the letter in the *Daily News* (Oct. 9), I can form no idea of the rate which the experiments on the 2,000 miles of cable promise for the delivery of messages. That letter gives no information whatever which could not have been told ten years ago without experiment*, and which was not actually asserted by myself a year and a half ago,—except this, that the eminent men who directed and assisted in those experiments, saw something that convinced them of the practicability of telegraphing, at a rate commercially advantageous, between Newfoundland and Ireland, by means of a cable of such dimensions and arrangement as that on which they made their experiments. If they can satisfy themselves that the effects of mutual induction between wires in a multiple cable of such a length will not be injurious, and that the rates of telegraphing which they can accomplish are as good as those I have indicated, then a multiple cable, such as that proposed by Mr Whitehouse [*Athen*. Aug, 30, p. 1093], will be undoubtedly satisfactory. But the effects of mutual induction between three or four wires in a cable 2,400 miles long, will be decidedly more embarrassing in rapid operations than they have been found with shorter lengths; and even without the new plans of working to which I allude, may be found so inconvenient that single wire cables will be preferred. If these new plans of wires are to be used, it may be necessary to have only one wire under one iron sheath, as I fear they would not succeed in a multiple cable, although on this point I cannot yet speak with confidence.

---

* See my last letter to the *Athenæum*, published Oct. 4.

I shall only say, in conclusion, that I intend to prepare and publish a table of different sets of dimensions for cables, with estimated rapidity of transmitting messages by each, which those engaged in projecting the Ocean Telegraph may find useful if their experiments leave anything undecided as to the best plan of cable for their purpose, and if they have any confidence in scientific deductions from established principles.

I remain, &c.

WILLIAM THOMSON.

[From the *Proceedings of the Royal Society*, December, 1856. *Philosophical Magazine*, July, 1857.]

ART. LXXVII. ON PRACTICAL METHODS FOR RAPID SIGNALLING BY THE ELECTRIC TELEGRAPH.

I am at present engaged in working out various practical applications of the formulas communicated some time ago in a short article on the "Theory of the Electric Telegraph" (*Proceedings*, May 17, 1855), [Art. LXXIII. above] and I hope to be able very soon to lay the results in full before the Royal Society. In the mean time, as the project of an Atlantic Telegraph is at this moment exciting much interest, I shall explain shortly a telegraphic system to which, in the course of this investigation, I have been led, as likely to give nearly the same rapidity of utterance by a submarine one-wire cable of ordinary lateral dimensions between Ireland and Newfoundland, as is attained on short air or submarine lines by telegraphic systems in actual use.

Every system of working the electric telegraph must comprehend (1) a plan of operating at one extremity, (2) a plan of observing at the other, and (3) a code of letter-signals. These three parts of the system which I propose will be explained in order,—I. for long submarine lines, and II. for air or short submarine lines.

I. *Proposed telegraphic system for long submarine lines.*

1. *Plan of operating.*—This consists in applying a regulated galvanic battery to give, during a limited time, a definite variation of electric potential determined by theory, so as to fulfil the condition of producing an electric effect at the other extremity, which, after first becoming sensible, rises very rapidly to a

maximum, then sinks as rapidly till it becomes again, and continues, insensible.

The principle followed is that pointed out by Fourier, by which we see, that, when the wire is left with both ends uninsulated after any electrical operations whatever have been performed upon it, the distribution of electric potential through it will very quickly be reduced to a harmonic law, with an amplitude falling in equal proportions during equal intervals of time. Unless the electric operations fulfil a certain condition, this ulterior distribution is according to the simple harmonic law (that is, is proportional to the sine of the distance from either extremity, the whole length being reckoned as 180°). The condition which I propose to fulfil is, that the coefficient of the simple harmonic term in the expression for the electrical potential shall vanish. Then, according to Fourier, the distribution will very much more quickly wear into one following a double harmonic law (that is, the sine of the distance from one extremity, the whole length being reckoned as 360°). In this state of electrification the two halves of the wire on each side of its middle point, being symmetrically and oppositely electrified, will discharge into one another, as well as into the earth at their remote extremities; each will be like a single wire of half-length, with the simple harmonic distribution ; and the wire will, on the whole, be discharged as fast as a wire of half the length, or four times as fast as a wire of the whole length, after an ordinary electrification. There is considerable latitude as to the mode of operating so as to fulfil this condition, but the theoretical investigation is readily available for finding the best way of fulfilling it in practice. The result, as I have tested by actual calculation of the electric pulse at the remote end, is most satisfactory. The calculations, and curves exhibiting the electric pulse in a variety of cases, will, I trust, very soon be laid before the Royal Society.

The time and law of operations being once fixed upon, a mechanical contrivance of the simplest kind will give the means of directing a regulated galvanic battery to perform it with exactness, and to any stated degree (positive or negative) of strength. Complete plans of all details I have ready to describe when wanted, and shall very soon be able to state exactly the battery power required for a cable of stated dimensions.

2. *Plan of observation for receiving a message**.—The instrument which I propose is Helmholtz's galvanometer, with or without modification. The time of vibration of the suspended magnet, and the efficiency of the copper damper, will be so arranged, that during the electric pulse the suspended magnet will turn from its position of equilibrium into a position of maximum deflection, and will fall back to rest in its position of equilibrium. The possibility of fulfilling these conditions is obvious from the form of the curve I have found to represent the electric pulse. The observer will watch through a telescope the image of a scale reflected from the polished side of the magnet, or from a small mirror carried by the magnet, and he will note the letter or number which each maximum deflection brings into the middle of his field of view.

3. *Code of letter-signals.*—The most obvious way of completing a telegraphic system on the plans which have been described, is to have the twenty-six letters of the alphabet written on the scale of which the image in the suspended mirror is observed, and to arrange thirteen positive and thirteen negative strengths of electric operation, which will give deflections, positive or negative, bringing one or other of these letters on the reflected scale into the centre of the field of view. But it would be bad economy to give the simple signals to rare letters, and to require double or triple signals for double and triple combinations of frequent occurrence. Besides, by the plans which I have formed, it will, I believe, be easy to make much more than thirteen different positive and thirteen different negative strengths of electric operation, giving unmistakeably different degrees of deflection ; and if so, then many of the most frequent double and triple combinations, as well as all the twenty-six letters of the alphabet singly, might be made by simple signals. But it is also possible (although I believe highly improbable), that in practice only three

---

* [Note of April 3, 1883. This plan, modified and simplified, became developed a year later into the method of reading telegraphic signals by my form of mirror galvanometer, which was first used on the 1858 Atlantic cable during its few weeks of life, and next in 1866 on the finally successful Atlantic cables of 1865 and 1866. From 1866 to the present time it has done a large proportion of the signalling of the great submarine cables ; all the work, in fact, except what has been done by the "siphon recorder" (Art. LXXXV. below) which I introduced in 1870, and which is now in very general use for the main practical work of submarine cables of more than two or three hundred miles length.]

or four, or some number less than thirteen, of unmistakeably different deflections could be produced in the galvanometer at one end by electric operations performed on the other extremity. If so, the whole twenty-six letters could not each have a simple signal, and double signals would have to be chosen for the less frequent letters. Experience must show what number of perfectly distinct simple signals can be made, and I have scarcely a doubt but that it will be much more than twenty-six. Then it will be easy to invent a letter code, which will use these signals with the best economy for the language in which the message is to be delivered. Towards this object I have commenced collecting statistics showing the relative frequency with which the different simple letters, and various combinations of simple letters, occur in the English language, and I must soon have information enough to guide in choosing the best code for a given number of simple signals.

The investigation leading to a measurement of the electro-magnetic unit of electricity in terms of the electro-static unit, published since the commencement of the present year by Kohlrausch and Weber, has given all that is required to deduce from Weber's own previous experiments the measurement of the electric conductivity of copper wire in terms of the proper kind of unit for the telegraph problem. The data required for estimating the rapidity of action in a submarine wire of stated dimensions would be completed by a determination of the specific inductive capacity of gutta-percha, or better still, a direct experiment on the electro-static capacity of a yard or two cut from the cable itself. I have estimated the retardations of various electric pulses, and the practicable rate of transmitting messages by cables 2,400 miles long, and of certain ordinary lateral dimensions, on the assumption that the specific inductive capacity of gutta-percha, measured as Faraday did that of sulphur, shell-lac, &c., is 2, from which it probably does not differ much. These estimates have been published elsewhere (*Athenœum*, Oct. 1856), and I shall not repeat them until I can along with them give a table of estimates for cables of various dimensions, with the uncertainty as to the physical property of gutta-percha either done away with by experiment, or taken strictly into account.

II.   *Plan for rapid self-recording signals by air wires and short submarine cables.*

The consideration of the preceding plans has led me to think of a system of working air lines, and short submarine lines, by which great rapidity of utterance, considerably greater I believe than any hitherto practised, may be attained. I have no doubt but that on this system five or six distinct letters per second, or sixty words per minute, may be readily delivered through air lines and submarine lines up to 100 miles, or perhaps even considerably more, of length, and recorded by a self-acting apparatus, which I shall describe in a communication I hope to make to the Royal Society before its next meeting.

[From the *Proceedings of the Royal Society*, Dec. 11, 1856.   *Philosophical Magazine*, July, 1857.]

ON PRACTICAL METHODS FOR RAPID SIGNALLING BY THE ELECTRIC TELEGRAPH.   (*Second Communication.*)

I.   *Further remarks on proposed method for great distances.*

Since my former communication on this subject I have worked out the determination of operations performed at one extremity of a submarine wire, so adjusted, that when the other extremity is kept constantly uninsulated, the subsidence of the electricity in the wire shall follow the *triple harmonic law* (that is to say, the electrical potential shall ultimately vary along the wire in proportion to the sine of the distance from either end, one-third of the length of the wire being taken as $180°$). The condensation of the electrical pulse at the receiving extremity, due to such operations, is of course considerably greater than that which is obtained from operations leading only to *the double harmonic* as described in my last communication; but experience will be necessary to test whether or not the precision of adjustment in the operations required to obtain the advantages which the theory indicates, can be attained in practice when so high a degree of condensation is aimed at. The theory shows exactly what amount and duration of residual charge in the wire would result from stated deviations from perfect accuracy in the adjustments of the operations; but it cannot be known for certain, without actual trial, within what limit such deviations can be kept in practice. From Weber's experiments on the electric conductivity of copper, and from measurements which I have made on specimens of the cable now in process of manufacture for the Atlantic telegraph, I think it highly probable that, with an alphabet of twenty letters, one letter could be delivered every two seconds between Newfoundland and Ireland (which would give, without any condensed code, six words per minute) on the general plan which I explained in my last communication ; and that no higher battery power than from 150 to 200 small cells of Daniell's (perhaps even considerably less) would be required. Whether or not this system may ultimately be found preferable to the very simple and undoubtedly practicable method of telegraphing invented by Mr Wildman Whitehouse, can scarcely be decided until one or both methods shall have been tested on a cable of the dimensions of the Atlantic cable, either actually submerged or placed in perfectly similar inductive circumstances.

II. *Method for telegraphing through submarine or subterranean lines of not more than 500 miles length.*

The plan which I have proposed to describe for rapid signalling through shorter wires, has one characteristic in common with the plan I have already suggested for the Atlantic telegraph ; namely, that of using different strengths of current for different signals.

But in lines of less than 500 miles, condensed pulses, such as have been described, may be made to follow one another more rapidly than to admit of being read off by an observer watching the image of a scale in a suspended mirror ; and a new plan of receiving and recording the indications becomes necessary.

Of various plans which I have considered, the following seems most likely to prove convenient in practice.

Several small steel magnets (perhaps each about half an inch long) are suspended horizontally by fine threads or wires at different positions in the neighbourhood of a coil of which one end is connected with the line wire and the other with the earth. Each of these magnets is held in a position deflected from the magnetic meridian by two stops on which its ends press; and two other small stops of platinum wire are arranged to prevent

it from turning through more than a very small angle when
actuated by any deflecting force making it leave the first position.
When a current passing through the coil produces this effect on
any one of these magnets, it immediately strikes the last-mentioned stops, and so completes a circuit through a local battery
and makes a mark on prepared electro-chemical paper.   For each
suspended magnet there is a separate style, but of course one
battery is sufficient for the whole printing process.   One set of
the different suspended magnets are so adjusted, that a current in
one direction of any strength falling short of a certain limit makes
only one of them move ; that a current in the same direction, of
strength exceeding this limit but falling short of another limit,
moves another also of the suspended magnets ; and so on for a
succession of different limits of strength of current in one direction.
The remaining set of suspended magnets are adjusted to move
with different strengths of current in the other direction through
the coil.   Without experience it is impossible to say how many
gradations of strength could be conveniently arranged to be thus
distinguished unmistakeably.   I have no doubt, however, that
very moderate applications of electric resources would give at
least three different strengths of current in each direction, which
could with ease and certainty be distinguished from one another by
the test which the suspended magnets afford.   Thus, a signal
of six varieties—one letter of an alphabet of six—could be recorded by almost instantaneous movements of six suspended
magnets, making one, two or three marks by one set of three
styles, or one, two or three marks by another set of three styles,
placed all six beside one another, pressing on a slip of electro-
chemical paper drawn by clockwork, as in the Morse instrument.

In subterranean or submarine lines of less than 100 miles
length, it would be easy, by means of simple battery applications,
followed by connexions with the earth, or by means of simple
electro-magnetic impulses at one end of the wire, to give ten or
twelve of such signals per second without any confusion of utterance at the other end.   The confusion of utterance which would
be experienced in working thus through longer lines would be
easily done away with, in any length up to 500 miles, by following
up each battery application with a reverse application for a shorter
time, or by following up each electro-magnetic impulse by a
weaker reverse impulse, so as approximately to fulfil the condition

(described in my former communication), of reducing the subsidence of the electrification in the wire to the double harmonic
form. It would, I believe, be readily practicable to send distinctly
five or six such signals per second (each a distinct letter of an
alphabet of six) through a wire of 500 miles length in a submarine
cable of ordinary dimensions. To perform the electrical operations
required for sending a message on this system, mechanism might
be had recourse to, and, by the use of perforated slips, as in Bain's
and other systems, it would be easy to work from twelve to
twenty of the six-fold varied signals per second through lines
of less than 100 miles length. Operating by the hand is, however,
I believe, generally preferred for ordinary telegraphing; and no
such speed as the last-mentioned could be attained even by a
skilful operator working with both hands. Six distinct letters or
signs of an alphabet of thirty, could, however, I believe, be delivered per second by the two hands working on a key-board with
twelve keys (perhaps like those of a pianoforte), provided the
keys are so arranged as to fulfil the following conditions :—

(1)   That by simply striking once any one of a first set of six
of the keys, an electric operation of one or other of the six varieties
shall be made twice, the second time commencing at a definite interval (perhaps $\frac{1}{12}$th of a second) later than the first.

(2)   That by striking one or other of the remaining six keys at
the same time, or very nearly at the same time, as one of the first
set, the second operation of the double electric signal will be that
corresponding to the key of the second set which is struck, instead
of being a mere repetition of the operation corresponding to the
key of the first set.

It would certainly be easy to make a key-board to fulfil these
conditions with the aid of some clockwork power. Then by
arranging the thirty-six permutations and doubles of the six
simple signals to represent an alphabet of thirty-six letters and
signs, an experienced operator would have to direct his mind to
only six different letters per second, while executing them by six
double operations with his fingers. That it would be possible to
work by hand at this rate there can be no doubt, when we consider
the marvels of rapid execution so commonly attained by practice
on the pianoforte; and it appears not improbable that in regular
telegraphic work, practised operators of ordinary skill could per-

form from four to six letters with ease per second, or from forty to sixty words per minute, on lines of not more than 100 miles length. The six signals per second, which, according to the preceding estimate, could be distinctly conveyed by a submerged wire of 500 miles in length, could of course be easily performed by the hand, with the aid of a key-board and clockwork power adapted to make the double operations for giving rapid subsidence of electricity in the wire when any one key is touched, and to let the different strengths of current, in one direction or the other, be produced by the different keys.　Thus without a condensed code, thirty words per minute could be telegraphed through subterranean or submarine lines of 500 miles ; and from thirty to fifty or sixty words per minute through such lines, of lengths of from 500 miles to 100 miles.

The rate of from fifty to sixty words per minute could be attained through almost any length of air line, were it not for the defects of insulation to which such lines are exposed.　If the imperfection of the insulation remained constant, or only varied slowly from day to day with the humidity of the atmosphere, the method I have indicated might probably, with suitable adjustments, be made. successful ; and I think it possible that it may be found to answer for air lines even of hundreds of miles' length.　But in either long or short air lines; the strengths of the currents received, at one extremity, from graduated operations performed at the other, might suddenly, in the middle of a message, become so much changed as to throw all the indications into confusion, in consequence of a shower of rain, or a trickling of water along a spider's web.

[From the *Proceedings of the Royal Society*, June, 1857.]

ART. LXXVIII. ON THE ELECTRIC CONDUCTIVITY OF COM-
MERCIAL COPPER OF VARIOUS KINDS.

IN measuring the resistances of wires manufactured for submarine telegraphs, I was surprised to find differences between different
specimens so great as most materially to affect their value in the
electrical operations for which they are designed. It seemed at first
that the process of twisting into wire rope and covering with gutta-
percha, to which some of the specimens had been subjected, must
be looked to to find the explanation of these differences. After,
however, a careful examination of copper-wire strands, some
covered, some uncovered, some varnished with india-rubber, and
some oxidized by ignition in a hot flame, it was ascertained that
none of these circumstances produced any sensible influence on the
whole resistance; and it was found that the wire-rope prepared for
the Atlantic cable (No. 14 gauge, composed of seven No. 22 wires,
and weighing altogether from 109 to 125 grains per foot) conducted
about as well, on the average, as solid wire of the same mass : but,
in the larger collection of specimens which thus came to be tested,
still greater differences in conducting power were discovered than
any previously observed. It appeared now certain that these
differences were owing to different qualities of the copper wire
itself, and it became important to find how wire of the best quality
could be procured. Accordingly, samples of simple No. 22 wire,
and of strand spun from it, distinguished according to the manufactories from which they were supplied, were next tested, and the
following results were obtained :—

Table of relative conducting qualities of single No. 22 Copper wire,
supplied from manufactories A, B, C, D.

	Resistances of equal lengths.	Weights of seven feet.	Resistances reduced to equal conducting masses and lengths.	Conducting power, (reciprocals of resistances) of equal and similar masses.
A	100	121·2 grs.	100	100
B	100·2	125·8 ,,	104·0	96·05
C	111·6	120·0 ,,	110·5	90·5
D	197·6	111·7 ,,	182·0	54·9

The strands spun from wire of the same manufactories showed
nearly the same relative qualities, with the exception of an inversion as regards the manufactories B and D, which I have been led
to believe must have arisen from an accidental change of labels
before the specimens came into my hands.

Two other samples chosen at random about ten days later, out
of large stocks of wire supplied from each of the same four manufactories, were tested with different instruments, and exhibited, as
nearly as could be estimated, the same relative qualities. It seems,
therefore, that there is some degree of constancy in the quality of
wire supplied from the same manufactory, while there is vast
superiority in the produce of some manufactories over that of
others. It has only to be remarked, that *a submarine telegraph constructed with copper wire of the quality of the manufactory* A *of only*
$\frac{1}{21}$ *of an inch in diameter, covered with gutta-percha to a diameter of
a quarter of an inch, would, with the same electrical power, and the
same instruments, do more telegraphic work than one constructed
with copper wire of the quality* D, *of* $\frac{1}{16}$ *of an inch diameter,
covered with gutta-percha to a diameter of a third of an inch,* to
show how important it is to shareholders in submarine telegraph
companies that only the best copper wire should be admitted for
their use. When the importance of the object is recognized, there
can be little difficulty in finding how the best, or nearly the best,
wire is to be uniformly obtained, seeing that all the specimens of
two of the manufactories which have as yet been examined have
proved to be of the best, or little short of the best quality, while
those of the others have been found inferior in nearly constant
proportion.

What is the cause of these differences in electrical quality is a
question not only of much practical importance, but of high

scientific interest. If chemical composition is to be looked to for the explanation, very slight deviations from perfect purity must be sufficient to produce great effects on the electric conductivity of copper; the following being the results of an assay by Messrs. Matthey and Johnson, made on one of the specimens of copper wire which I had found to be of low conducting power :—

Copper .... 99·75
Lead .... ·21
Iron .... ·03
Tin or antimony .... ·01

100·00

The whole stock of wire from which the samples experimented on were taken, has been supplied by the different manufacturers as remarkably pure; and being found satisfactory in mechanical qualities, had never been suspected to present any want of uniformity as to value for telegraphic purposes, when I first discovered the difference in conductivity referred to above. That even the worst of them are superior in conducting power to some other qualities of commercial copper, although not superior to all ordinary copper wire, appears from the following set of comparisons which I have had made between specimens of the No. 22 A wire, ordinary copper wire purchased in Glasgow, fine sheet-copper used in blocks for calico printers, and common sheet-copper.

Lengths of No. 22 A, weighing 17·3 grs. per foot, used as standards.	Conductors tested.	Their weights per foot.	Lengths resisting as much as standards if of equal conductivity.	Lengths found by experiment to resist as much as standards.	Conductivity referred to that of No. 22 A as 100.
inches.		grs.	inches.	inches.	
23·8	Ordinary No. 18 wire	57·5	79·0	73·6	93·2
7·5	Slip of fine sheet-copper	37·6	16·3	9·1	55·8
15·5	Slip of common sheet-copper ...	51·1	45·77	15·6	34·1

To test whether or not the mechanical quality of the metal as to hardness or temper had any influence on the electrical conducting power, the following comparison was made between a piece of soft No. 18 wire, and another piece of the same pulled out and hardened by weights applied up to breaking.

Soft No. 18 copper wire.	No. 18 copper wire, stretched to breaking.	Length found equivalent by experiment.
Weight per foot, 57·5 grs. Length used, 30·8 inches.	Weight per foot, 44·8 grs. Equivalent length, if of equal conductivity, 24·0 inches.	24·0 inches.

The result shows that the greatest degree of brittleness produced by tension does not alter the conductivity of the metal by as much as one half per cent. A similar experiment showed no more sensible effect on the conductivity of copper wire to be produced by hammering it flat. There are, no doubt, slight effects on the conductivity of metals, produced by every application and by the altered condition left after the withdrawal of excessive stress*; and I have already made a partial examination of these effects in copper, iron, and platinum wires, and found them to be in all cases so minute, that the present results as to copper wire are only what was to be expected.

To find whether or not there is any sensible loss of conducting power on the whole due to the spiral forms given to the individual wires when spun into a strand, it would be well worth while to compare very carefully the resistances of single wires with those of strands spun from exactly the same stock. This I have not yet had an opportunity of doing; but the following results show that any deficiency which the strand may present when accurately com-

No. 16 Solid Wire. Pairs of samples in different states of preparation, each 1000 inches long.

Resistances†.			Weight per foot.	Specific resistances reduced to British† absolute measure.
			grs.	
Not covered	$\begin{cases} E_1 \cdot 2036 \\ E_2 \cdot 1995 \end{cases}$	·2015	74·6	11,850,000
Once covered	$\begin{cases} F_1 \cdot 2054 \\ F_2 \cdot 1999 \end{cases}$	·2026	77·55	12,410,000
Twice covered	$\begin{cases} G_1 \cdot 1963 \\ G_2 \cdot 1963 \end{cases}$	·1963	77·2	11,970,000
Thrice covered	$\begin{cases} H_1 \cdot 1893 \\ H_2 \cdot 1916 \end{cases}$	·1904	77·73	11,680,000
Means		·1977	76·78	11,980,000

* See the Bakerian Lecture "On the Electro-dynamic Qualities of Metals," §§ 104, 105, and 150, *Philosophical Transactions* for 1856 [Art. xci. below].

† These resistances were measured, by means of a Joule's tangent galvanometer

pared with solid wire, is nothing in comparison with the differences presented by different samples chosen at random from various stocks of solid wire and strand in the process of preparation for telegraphic purposes.

No. 14 Gauge Strand (seven No. 22 wires twisted together). Pairs of samples in different states of preparation, each 1000 inches long.

Resistances.			Weight per foot.	Specific resistances reduced to British absolute measure.
			grs.	
Not covered	$\begin{cases} K_1 \\ K_2 \end{cases}$	$\begin{cases} \cdot 1595 \\ .1634 \end{cases}$ $\cdot 1614$	115·82	14,750,000
Once covered	$\begin{cases} L_1 \\ L_2 \end{cases}$	$\begin{cases} \cdot 1037 \\ \cdot 1043 \end{cases}$ $\cdot 1040$	109·37	8,964,000
Twice covered	$\begin{cases} M_1 \\ M_2 \end{cases}$	$\begin{cases} \cdot 1426 \\ \cdot 1424 \end{cases}$ $\cdot 1425$	111·95	12,590,000
Thrice covered.........	$\begin{cases} N_1 \\ N_2 \end{cases}$	$\begin{cases} \cdot 1092 \\ \cdot 1085 \end{cases}$ $\cdot 1088$	121·30	10,430,000
Means		$\cdot 1297$	114·61	11,680,000

The specific resistances of the specimens of copper wire from the manufactories A, B, C, D, of which a comparative statement is given in the first Table above, I have estimated in absolute measure by comparing each with F_2, of which the resistance in

with a coil of 400 turns of fine wire, in terms of the resistance of a standard conductor as unity. The resistance of this standard has been determined for me in absolute measure through the kindness of Professor W. Weber, and has been found to be 20,055,000 German units $\left(\dfrac{\text{metre}}{\text{seconds}} \right)$, or 6,580,000 British units $\left(\dfrac{\text{foot}}{\text{seconds}} \right)$. The numbers in the last column, headed "Specific resistances reduced to British measure," express the resistances of conductors composed of ten different qualities of metal, each one foot long and weighing one grain [and may therefore be reduced to C. G. S. units by dividing by 4102, being the product of the number of centimetres in a foot × the number of grains in a gramme × the specific gravity of copper, or 30·48 × 15·432 × 8·72]. It is impossible to over-estimate the great practical value of this system of absolute measurement carried out by Weber into every department of electrical science, after its first introduction into the observations of terrestrial magnetism by Gauss. See "Messungen galvanischen Leitungswiderstände nach einem absoluten Maasse," Poggendorff's *Annalen*, March, 1851. See also the author's articles entitled "On the Mechanical Theory of Electrolysis" [Art. LIII. Vol. I. above], and "Application of the Principle of Mechanical Effect to the Measurement of Electromotive Force, and of Galvanic Resistances in Absolute Units, *Philosophical Magazine*, December, 1851 [Art. LIV. Vol. I. above].

absolute measure is 6,580,000 × ·1999, or 1,316,000. The various
results reduced to specific resistances per grain of mass per foot of
length are collected in the following Table, and shown in order of
quality in connexion with four determinations of specific conductivity by Weber.

Specific Conductivities of specimens of Copper expressed in
British Absolute Measure.

Description of Metal.	Specific resistances.
Copper wire A No. 22 [=1853 C. G. S.]	7,600,000
Wire of electrolytically precipitated copper: Weber (1) ...	7,924,000
Copper wire B No. 22 [=1936 C. G. S.]	7,940,000
Ordinary No. 18 copper wire	8,100,000
Copper wire C No. 22..	8,400,000
Weber's copper wire: Weber (2)	8,778,000
No. 14 strand specimen, once covered	8,960,000
Kirchhoff's copper wire: Weber (3)..............................	9,225,000
No. 14 strand specimen, thrice covered	10,400,000
Jacobi's copper wire: Weber (4) [=2650 C. G. S.]	10,870,000
No. 16 wire specimen, thrice covered	11,700,000
Ditto, twice covered	11,970,000
Ditto, not covered	11,850,000
Ditto, once covered..	12,410,000
No 14 strand specimen, twice covered	12,590,000
Slip of fine sheet-copper	13,600,000
Copper wire D No. 22	13,800,000
No. 14 strand specimen, not covered	14,750,000
Slip of common sheet-copper [=5436 C. G. S.]	22,300,000

ART. LXXIX. ANALYTICAL AND SYNTHETICAL ATTEMPTS TO ASCERTAIN THE CAUSE OF THE DIFFERENCES OF ELECTRIC CONDUCTIVITY DISCOVERED IN WIRES OF NEARLY PURE COPPER.

[From the *Proceedings of the Royal Society*, February, 1860.]

FIVE specimens of copper wire No. 22 gauge, out of a large number which had been put into my hands by the Gutta Percha Company to be tested for electric conductivity, were chosen as having their conductivities in proportion to the following widely different numbers, 42, 71·3, 84·7, 86·4, and 102 ; and were subjected to a most careful chemical analysis by Professor Hofmann, who at my request kindly undertook and carried out what proved to be a most troublesome investigation. The following report contains a statement of the results at which he arrived :—

> "Royal College of Chemistry,
> March 10th, 1858.

" SIR,—I now beg to communicate to you the results obtained in the analysis of the several varieties of copper wire intended for the use of the Transatlantic Telegraph Company, which you forwarded to me for examination.

"I have limited the inquiry to a *minute qualitative* analysis of
the wires, to a very accurate determination of the amount of copper,
and an approximative determination of the amount of oxygen. The
qualitative analysis has been repeated several times with as considerable quantities as the amount of material at my disposal
permitted. The *quantitative* determinations of the copper have
been made with particular care, and after a lengthened scrupulous
inquiry into the limit of accuracy of which the method employed is
capable, I am convinced that the true per-centages of copper
cannot be more than 0·1 per cent. either above or below the means
of the determinations, the details of which I give you in the
Appendix.

"The following Table contains the results furnished by
analysis :—

Conductivity of the wire, in relative measure *.	42.	71·3.	84·7.	86·4.	102.
Qualitative analysis	Copper. Iron. Nickel. Arsenic. Oxygen.	Copper. Iron. Nickel. Oxygen.	Copper. Iron. Nickel (doubtful). Oxygen.	Copper. Iron. Nickel (doubtful). Oxygen.	Copper. Iron. Oxygen.
Per-centage of copper	98·76	99·20	99·53	99·57	99·90
Amount of impurities.	1·24	0·80	0·47	0·43	0·10
	100·00	100·00	100·00	100·00	100·00

"Since it appeared probable that the extraordinary difference
in the conductivity of the several specimens was due rather to non-
metallic impurities than to metallic admixtures, careful experiments
were made in every case for the detection of sulphur. In none of
the specimens was it possible to discover the slightest trace of
sulphur. Qualitative experiments having established on the other

* I have since found [see preceding Art. LXXVIII. foot note pp. 115, 116.]
$10^{-9} \times 131\frac{1}{2}$ as the factor to reduce from this to absolute measure. Thus the conductivities of the five specimens are respectively 55·2, 95·3, 111·4, 113·6, 134·1, in
terms of one one thousand millionth of the British absolute unit.—W. T.

hand the presence of oxygen, probably in the form of suboxide of copper, in every one of the specimens, an attempt was made to ascertain the quantities by determining the loss which the wire after rolling suffered when heated in an atmosphere of hydrogen, and by simultaneously estimating the quantity of water formed.

"In this experiment, the details of which are given in its Appendix, the following numbers were obtained:—

| Conductivity | 42 | 71·3 | 84·7 | 86·4 | 102 |
| Percentage of Oxygen | 0·087 | 0·119 | 0·172 | 0·159 | 0·193 |

" Unfortunately the same reliance cannot be placed upon these numbers as upon the preceding ones, since the method employed involves many sources of error, and want of material precluded the possibility of repeating the experiments.

"From the preceding analysis, it is obvious that the amount of impurities in the several specimens examined is small, varying as it does between 0·10 and 1·24 per cent. The number of foreign constituents also is comparatively small. I should, however, state that the analytical results which I have given do not exclude the presence of exceedingly minute quantities, even of other metals which might have been detected if larger quantities of copper could have been submitted to analysis. Some years ago, Max Duke of Leuchtenberg* examined the black precipitate formed at the anode in the electrotype process, during the decomposition of sulphate of copper by the galvanic current. In this precipitate, of which considerable quantities accumulate by the gradual solution of large quantities of copper passing through the process, he found the following constituents:—

Antimony	9·22	Iron	0·30	Oxygen	24·82
Arsenic	7·40	Nickel	2·26	Sulphur	2·46
Platinum	0·44	Cobalt	0·86	Selenium	1·27
Gold	0·98	Vanadium	0·64	Sand	1·90
Silver	4·54	Tin	33·50		
Lead	0·15	Copper	9·24		

"Of these constituents, the ten first metals were obviously derived from the copper, in which they could have been scarcely detected unless by this accumulative process. Of the remainder of the constituents, the tin in a great measure is derived from the solderings.

* Petersburgh Acad. Bull. vii. p. 218.

"The results obtained in the analysis of the copper wires which you forwarded to me, appear to establish one fact in a satisfactory manner, viz. that the diminution of conductivity observed in certain specimens of copper is due to the presence in these specimens of a certain amount of foreign matter, and not, as it has been supposed, to a peculiar change in the physical condition of the metal; for in the specimens analysed the conductive power rises in the same order as the total amount of impurities diminishes.

"I have, &c.,

(Signed) "A. W. HOFMANN."

"*Professor William Thomson, F.R.S., &c.*"

It appears therefore that in the case of these four specimens, the electric conductivity is in order of purity of the copper; but yet that only extremely small admixtures of other substances are to be found even in those which have but half the conductivity of the best.

On the other hand, I have found by experimenting on artificial alloys, that comparatively large admixtures of lead, iron, silver, and zinc seem to produce sometimes improvement, sometimes little or no sensible influence, and sometimes (as in the case of zinc) an injurious effect on the conductivity of specimens of pure electrotype copper from which the alloys were made. The largeness of the proportion of other metal required to produce any considerable deterioration in comparison with that of the whole amount of impurities which Professor Hofmann's investigation demonstrates in specimens of low quality as to conductivity, is worthy of remark, and seems to indicate that this low quality must be due to other than metallic impurities.

The great difference between the conducting qualities of two specimens of electrotype copper, from which two series of alloys were separately prepared, seems also to indicate some as yet undiscovered cause, as operative in general. I am assured by Messrs. Matthey and Johnson, by whom all the alloys were prepared, that similar methods were followed and equal care bestowed to ensure purity in the two cases.

The results of my measurements of conductivity are shown in the following Tables :—

TABLE I.—Two Series, Nos. 1-10 and Nos. 1-32, of Specimens prepared by Messrs. Matthey and Johnson from pure electrotype copper, and the same alloyed with other metals, as specified.

No of. Spec.	Specification of compound.	Specific conductivity.
	SERIES I.	
1	Pure copper....................	138·5
2	Pure copper alloyed with ·25 per cent. silver	138·5
3	Pure copper alloyed with ·13 per cent. silver	139·5
4	Pure copper alloyed with ·25 per cent. lead	144
5	Pure copper alloyed with ·13 per cent. lead	146
6	Pure copper alloyed with ·25 per cent. tin	131
7	Pure copper alloyed with ·13 per cent. tin	133
8	Pure copper alloyed with ·80 per cent. zinc	125
9	Pure copper alloyed with ·40 per cent. zinc	120·5
10	Pure copper alloyed with 1·40 per cent. zinc	103
	SERIES II.	
1	997·5 copper + 2·5 silver	69·8
2	998·7 copper + 1·3 silver	117·7
3	997·5 copper + 2·5 lead	94·5
4	998·7 copper + 1·3 lead	105·8
5	997·5 copper + 2·5 tin	91·6
6	998·7 copper + 1·3 tin	116·9
7	999 copper + 1 silver	126·7
8	999 copper + 1 lead.............	134·2
9	999 copper + ·5 lead + ·5 silver	128·0
10	Equal parts of 1 and 3	89·3
11	997·5 copper + 2·5 iron	129·7
12	998·7 copper + 1·3 iron	113·7
13	1000 copper + 2·5 protoxide of copper..............................	122·5
14	1000 parts of 3 & + 2·5 protoxide of copper (too brittle to test)	
15	1000 parts of 4 & + 2·5 protoxide of copper	119·7
16	997·5 copper + 2·5 zinc...	108·9
17	995 copper + 2·5 lead + 2·5 zinc	85·1
18	995 copper + 2·5 lead + 2·5 iron	131·5
19	998·7 parts of 11 + 1·3 lead	135·0
20	997·5 parts of 11 + 2·5 zinc	77·6
21	998·7 parts of 11 + 1·3 zinc	95·2
22	997 parts of 11 + 1·3 lead & + 1·3 zinc	117·6
23	992 copper + 8 zinc ....·....	118·9
24	996 copper + 4 zinc	117·0
25	986 copper + 14 zinc	80·2
26	982 copper + 18 zinc	102·3
27	994 copper + 6 zinc	109·5
28	980 copper + 20 aluminium	44·0
29	990 copper + 10 aluminium	128·7
30	995 copper + 5 aluminium	122·5
31	997 copper + 3 aluminium	130·2
32	Pure copper, from which all the above were made	120·9

TABLE II.—First series (10 specimens) arranged in order of
conductivity.

No. of Spec.	Specification of compound.	Specific conductivity.
5	Pure copper alloyed with ·13 per cent. of lead	146
4	Pure copper alloyed with ·25 per cent. of lead	144·5
3	Pure copper alloyed with ·13 per cent of silver	139·5
2	Pure copper alloyed with ·25 per cent. of silver	138·5
1	Pure copper..	138·5
7	Pure copper alloyed with ·13 per cent. of tin	133
6	Pure copper alloyed with ·25 per cent. of tin	131
8	Pure copper alloyed with ·80 per cent. of zinc	125
9	Pure copper alloyed with ·40 per cent. of zinc	120·5
10	Pure copper alloyed with 1·40 per cent. of zinc	103

TABLE III.—Second Series (32 specimens) arranged in order of
conductivity.

No. of Spec.	Specification of compound with manufacturers' description of mechanical quality of wire.	Specific conductivity.
19	998·7 of No. 11 + 1·3 lead : fair	135·0
8	999 copper + 1 lead : fair	134·2
18	995 copper + 2·5 lead + 2·5 iron : very good	131·5
31	997 copper + 3 aluminium : good	130·2
11	997·5 copper + 2·5 iron : not very good	129·7
29	990 copper + 10 aluminium : good	128·7
9	999 copper + ·5 lead + ·5 silver : rather better than No. 8	128·0
7	999 copper + 1 silver : fair	126·7
13	1000 copper + 2·5 protoxide of copper : very bad.................	122·5
30	995 copper + 5 aluminium : very good	122·5
32	Pure copper : very good	120·9
15	1000 of No. 4 + 2·5 protoxide of copper : better than No. 14, but not good	119·7
23	992 copper + 8 zinc : first-rate alloy	118·9
2	998·7 copper + 1·3 silver : fair, but rather frangible	117·7
22	997·5 of no. 11 + 1·3 lead + 1·3 zinc : very good indeed	117·6
24	996 copper + 4 zinc : moderately good	117·0
6	998·7 copper + 1·3 tin : perhaps not quite as good as No. 5 ...	116·9
12	998·7 copper + 1·3 iron : frangible	113·7
27	994 copper + 6 zinc : good	109·5
16	997·5 copper + 2·5 zinc : first-rate alloy.......................	108·9
4	998·7 copper + 1·3 lead : rather better than No. 4	105·8
26	982 copper, 18 zinc : very good	102·3
21	998·7 of No. 11 + 1·3 zinc : very fair	95·2
3	997·5 copper + 2·5 lead : good, but requires care...............	94·5
5	997·5 copper + 2·5 tin : much the same as Nos. 3 and 4..........	91·6
10	Equal parts of Nos. 1 and 3 : bad, frangible....................	89·3
17	995 copper + 2·5 lead + 2·5 zinc : very good	85·1
25	986 copper + 14 zinc : first-rate alloy.........................	80·2
20	997·5 of No. 11 + 2·5 zinc : very fair	77·6
1	997·5 copper 2·5 silver : fair, but rather frangible	69·8
28	980 copper + 20 aluminium : not very good	44·0
14	1000 parts of No. 3 + 2·5 protox. copper ; almost undrawable (too brittle to test).	

The alloys numbered 14 and 15 were prepared with a view to testing the possible effect of a suboxide of copper mixed or combined with the mass. Although they do not seem worse than others of nearly the same metallic composition, it cannot be considered that they demonstrate that oxygen exercises no influence, as the portion of oxide introduced may have been reduced in the melting; and indeed it is quite possible that some accident in the melting may possibly give rise to *oxidation* to a greater or less degree, and may cause some of the irregularities and uncertainties which have been observed. On this I may remark, that although I have found that no mechanical alteration by hammering, twisting, &c. produces any considerable effect of the conductivity of one piece of solid copper, I have not yet found whether or not specimens either good or bad retain their specific qualities after melting.

I may add, that it will be of great importance to ascertain the laws of variation of conductivity with temperature, of different specimens of nearly pure copper differing largely in conductivity. I have hitherto used standards of copper wire in all the relative determinations of conductivity which I have made for different commercial specimens and artificial alloys of copper; and before I found the very large differences of conductivity shown in this and in my preceding communication to the Royal Society (June 15, 1857), it seemed natural to suppose that the relation between specimen and standard would remain constant, or nearly constant, when the temperatures of the two are varied to the same extent. Now, however, it seems scarcely probable that this can be the case, and a rigorous experimental examination of the influence of temperature becomes necessary.

P.S. April 11, 1860.—I append the following extract from evidence which I gave on examination before the Government Committee on submarine telegraphs on the 17th December, 1859, as it bears directly on the subject of the preceding article, and shows what degree of weight may in my opinion be attached to the synthetical attempts which have been described.

(*Chairman.*) Question 2458. Soon after you became a Director of the Atlantic Telegraph Company, was your attention directed to the conductivity of copper?—Yes.

2459. You instituted a series of experiments, did not you, to determine the variation of this quality in different samples of copper?—A number of samples* of copper were, at my request, put into my hands for the purpose of measuring their conductivity in consequence of my having accidentally noticed differences greater than I expected in the conducting power of one or two samples which I had had previously.

2460. Will you be good enough to state the general results at which you ultimately arrived, and your modes of experimenting?— My modes of experimenting did not differ materially from the methods which had been followed by certain other experimenters, especially in Germany, and were in reality all based on Professor Wheatstone's invention of a beautiful method for comparing resistances, to which I have frequently referred as Professor Wheatstone's electric balance.

2461. What were the results at which you arrived?—That different specimens chosen at random from the stock supplied for manufacture differed immensely in conducting power.

2462. Although nominally the same quality of copper?—Yes, although nominally the same quality of copper. All those specimens of wire were supposed to be of the very best quality, the only copper supposed to be good being that which admitted of being drawn into wire suitably for the purpose. A good mechanical quality was necessary to prevent frequent fractures in the wire-

---

* [Note of date June 27, 1883. These were portions of the wire out of which the 1857 Atlantic Cable had been made. On finding in several of them the deficiency of conductivity referred to in the text, I called the attention of the Board of Directors to it, and after much perseverance succeeded in obtaining the insertion of a clause requiring high conductivity, in the contract for 600 miles of new cable ordered in the autumn of 1857, to supply the place of a part of the cable which was lost in the 1857 attempt to establish telegraphic communication between Ireland and America, and to allow a renewal of the attempt in 1858. When the demand for high conductivity was first submitted to the contractors, their reply was a *non possumus*. It was sent back to them with the question: For what addition to the contract price will you undertake it? the answer was £42 per mile for the gutta-percha-covered copper wire strand, instead of £40 per mile which had been the first tender. This offer was accepted by the Board. It was not until practical testing to secure high conductivity had been commenced in the factory, that practical men came thoroughly to believe in the reality of the differences of conductivity in the different specimens of copper wire, all supposed good and supplied for use in submarine cables, which I had pointed out as the result of my laboratory measurements. From that time to the present there has never been a question, on the part of either Companies or Contractors, as to the necessity for the stipulation of "high conductivity;" and a branch of copper manufacture has grown up in the course of these 26 years for producing what is called in the trade "conductivity copper." W. T.]

drawing; and to understand that, I should say that hanks in unbroken lengths amounting to a large mass were always required, the worse metal being found to break before it could be drawn into a hank of a certain size. The mechanical qualities seem to have been satisfactory, but no suspicion whatever was entertained that there were also large differences in electric conducting power. W. Weber had many years before pointed out considerable differences in different specimens of copper wire which he had tested. I found differences much exceeding those, and I did not, as I expected, find any approximation to a uniform average among the different specimens tested; some specimens I found nearly double in their conducting power, compared with others, reckoned according to the weight and length, allowing for the variations of gauge. Calling the best specimen which I had in the summer of 1857, 100, I found many specimens standing at 60 in specific conductivity, many standing at 50, many standing at 80, a few above 90; and so far as I can recollect, the average of a large number of specimens that I then examined may have stood between 60 and 70, but I consider the statement of such an average to be of no value, it is so much a matter of chance. If I had received a dozen specimens of a low quality below the average, or if I had chanced to receive a dozen specimens of a higher quality, the average would have been so much the lower or the higher. I never had an opportunity of measuring the conductivity of 200 or 300 miles of submarine cable; such alone would have given me exact information as to the average for that portion of cable. I may mention that a month or two later, still in the summer of 1857, I received specimens of wire which were in stock for submarine telegraphs,—for some of the Mediterranean telegraphs, I believe,—which stood as low as 43 on that scale; and, lastly, I may mention that I have since met with specimens standing 2 or 3 per cent. above the 100; and an artificial alloy, which I had prepared, stood, so far as I can estimate, as high as 111.

2463. What was that alloy?—The alloy consisted, so far as I can recollect, of copper and ·13 per cent. of lead. I have made experiments upon a series of alloys, in all about 43 or 44, and have recently repeated the examination so as to arrive at accuracy, within certain limits; and I expect, immediately, to be able to communicate to the Royal Society, for publication, the results. A few months ago I sent a provisional list of the specimens, showing the relative conductivity of those alloys, but, possibly, requiring correction as to the absolute conductivity stated. That list was communicated to Mr Latimer Clarke, and, I believe, a copy of it was laid before the Committee.

2464. (*Professor Wheatstone.*) Were you quite certain that you employed pure copper in your experiments?—I could not be quite certain.

2465. The copper might be alloyed with other things than metals; is it not very probable that it might contain some suboxide, and that the mixing of lead afterwards with it might have reduced the suboxide, and therefore have given it a higher conducting power on that account?—That is possible. I cannot say that I am at all satisfied that the experiments which I have made point out distinctly the relation between the ascertained chemical combination and conductivity. I may mention that one of my alloys was made with a suboxide melted with the copper; but the uncertainty of the process of melting the suboxide and the uncertainty as to how much of the oxidation may have disappeared in the melting, prevented me from attributing much weight to the experiment.

2466. (*Chairman.*) What was the result with that alloy; was it a low result, or a high result?—A moderate result; not a low result.

2467. But not a high one?—A somewhat high result; but I may mention that in one series the highest conductivity was found with a mixture of lead and iron; fractions of a per cent. of lead, and fractions of a per cent. of iron mixed with pure copper gave a higher conductivity than a nominally pure copper, with which the alloys were prepared. I must mention further, that in two series the alloys, both prepared by Messrs. Matthey and Johnson, and as I have been assured with equal care, gave results presenting considerable discrepancies; the conductivity of the pure copper in the first stood high, nearly agreeing with the 100 of my first scale, the pure copper of the second series fell considerably below that limit. On this account it appears that even pure copper, carefully prepared by the electrotype process, does not always give us results which show perfectly in point of conductivity; but to make such experiments in a satisfactory manner, it would be necessary to have a thorough chemical investigation, both synthetical and analytical, of the metals used; such a thorough investigation I have not been able to carry out, in consequence of the large expense which it would entail. I may mention that Mr Mathiessen has gone through a series of experiments on alloys, of which the chemical composition has been ascertained with all possible accuracy, and has, I believe, arrived at highly important results relative to electrical conductivity. I have been in communication with him, and have supplied him with a specimen of one of my standards.   He mentions to me that he has obtained specimens conducting better to a considerable extent than the 100 of my first scale.   In that respect he has confirmed what I have myself ascertained, having myself found specimens as high as 111 on that scale.   A number of alloys of definite chemical composition, prepared with great care by Mr Calvert of Manchester, and already tested by him for thermal conductivity and for mechanical properties, have been put into my hands, in order that I may measure their electric conductivities. I hope soon to be able to obtain and publish results for this series of alloys.

ART. LXXX.　Remarks on the Discharge of a Coiled Electric Cable.

[From the *British Association Report*.　Part II.　1859.]

Mr Jenkin had communicated to the author during last February, March, and April a number of experimental results regarding currents through several different electric cables coiled in the factory of Messrs R. S. Newall and Co., at Birkenhead. Among these results were some in which a key connected with one end of a cable, of which the other end was kept connected with the earth, was removed from a battery by which a current had been kept flowing through the cable and instantly pressed to contact with one end of the coil of a tangent galvanometer, of which the other end was kept connected with the earth.　Mr Jenkin and the author remarked that the deflections recorded in these experiments were in the contrary direction to that which the true discharge of the cable would give.　Mr Jenkin repeated the experiments, watching carefully for indications of reverse currents to those which had been previously noted.　It was thus found that the first effect of pressing down the key was to give the galvanometer a deflection in the direction corresponding to the true discharge current, and that this was quickly followed by a reverse deflection generally greater in degree, which latter deflection corresponded to a current in the same direction as that of the original flow through the cable.　Professor Thomson explained this second current, or false discharge, as it has since been sometimes called, by attributing it to mutual electro-magnetic induction between different portions of the coil, and anticipated that no such reversal could ever be found in a submerged cable　The effect of this induction is to produce in those parts of the coil first in-fluenced by the motion of the key, a tendency for electricity to flow in the same direction as that of the decreasing current flowing on through the remoter parts of the coil. Thus, after the first violence of the back flow through the key and galvano-meter, the remote parts of the cable begin, by their electro-mag-netic induction on the near parts, to draw electricity back from the earth through the galvanometer into the cable again, and the current is once more in one and the same direction throughout the cable. The mathematical theory of this action, which is necessarily very complex, is reserved by the author for a more full communication, which he hopes before long to lay before the Royal Society. [See more on this subject in Article LXXXIII. below]

ART. LXXXI. VELOCITY OF ELECTRICITY.

[From Nichol's *Cyclopædia*. Second Edition. 1860.]

THE velocity of the transmission of electricity has been a subject of careful inquiry, and of extremely interesting experiments, on the part of several distinguished Physicists; foremost among whom rank Professor Wheatstone and MM. Fizeau and Gonnelle. The nature of the apparatus used has been alluded to under Chronoscope; and also, at some length, under Light, Velocity of. The chief results are the following :—Mr Wheatstone found that electricity can travel at a rate *one and a-half times* greater than that of light : in other words, "that the electric spark would go round our globe about twelve times in one *second*." According to Fizeau and Gonnelle (whose results are virtually confirmed by Mr Mitchel of Cincinnati), the beginning of an electric *current* may be transmitted along a copper wire at a velocity which is not greater than *three-fifths* the velocity of light.

Still greater discrepancies are shewn by extensions of experiments, with the same object in view, to varied circumstances of insulation and length in the conductors experimented on. For instance, trials in Queenstown Harbour, in July, 1856, when the two portions of the first Atlantic cable on board H.M.S. Agamemnon and the U.S. steam frigate Niagara, were for the first time joined into one conductor, 2,500 statute miles in length, gave about $1\frac{3}{4}$ seconds as the time of transmission of a signal from induction coils, through that length ; corresponding to a velocity of 1,400 miles per second. The following table contains some of the chief results hitherto published as evaluations of the "velocity" of the transmission of electricity :—

	Miles per second.
* Wheatstone in 1834, with copper wire............	288,000
* Walker in America with telegraph iron wire ...	18,780
* O'Mitchell, ditto, ditto	28,524
* Fizeau and Gonnelle (copper wire)	112,680
* Ditto, (iron wire)	62,600
† A. B. G. (copper) London and Brussels telegraph	2,700
† Ditto (copper) London and Edinburgh telegraph	7,600
Induction coils through 2,500 miles Atlantic cable, tested by heavy needle galvanometer, Queenstown, 1857	1,430
Daniell's battery through 3,000 miles Atlantic cable, tested by mirror galvanometer, Devonport, 1858	3,000

Now it is obvious, from the results which have been quoted, that the supposed "velocity" of transmission of electric signals is not a definite constant like the velocity of light, even when one definite substance, copper, is the transmitting medium, but is largely influenced by the circumstances in which the conductor is placed, being, for instance, much greater when the wire is insulated in air on poles than when it is surrounded by gutta percha and iron sheathing, and either submerged, or in coils as on board ship. Further, it is to be remarked that even in conductors, in precisely similar lateral circumstances, the apparent "velocity" is greater the smaller the length of the conductor used. Lastly, we may allude to the fact that some experimenters and writers have maintained, that the velocity of the transmission of electric signals differs with the source of excitation, being on the whole greater, the more intense and sudden is the electric impulse applied; while, on the highest authority, it has been maintained that the velocity of transmission is quite independent of the intensity of the source. Among all these discrepancies between the statements of careful experimenters and writers of high intelligence, how is the truth to be understood? We cannot doubt the general accuracy of their results; and if their state-

---

* Liebig and Kopp's *Report*, 1850 (translated), p. 168.

† *Athenæum*, 14th January, 1854, p. 54.

ments and conclusions are contradictory, we may be sure that an explanation is to be found in a comprehensive theory of the subject. It would carry us much beyond the limits of the present article to do more than sketch very slightly the chief points of electro-dynamic theory which are involved.

In the first place, it must be considered that three properties of electricity, in the present state of science not understood except as quite distinct from one another, are concerned in the transmission of an electric signal along an insulated conductor:— (1), "Charge" or electrical accumulation in a conductor subjected in any way to the process of electrification. (2), "Electro-magnetic induction," or electromotive force, excited in a conductor by variations of electric currents, either in adjacent conductors or in different parts of its own length. (3), Resistance to conduction through a solid. We may illustrate these three properties of electricity in an elongated conductor, such as a telegraph wire, by considering their hydrodynamical analogies for water in a canal or in a tube:—(1), Accumulation of a greater or less quantity of water in any part of the canal or tube. (2), Inertia of the water. (3), Viscosity or fluid friction. If the first did not exist, as would be the case if the water were incompressible, and if it were inclosed in a perfectly rigid channel or tube, completely filled by it, the velocity of water flowing along the canal would necessarily be the same throughout its length. In these imaginary circumstances, if a piston is pressed into the tube at one end, the effect in moving the water must commence simultaneously along the whole length, and the velocity of transmission of a water pressure signal must be infinite (although, of course, the maximum strength of current producible by the force applied cannot be acquired in an instant, because of the inertia of the water). But, in reality, water is somewhat compressible, and therefore, even in a perfectly rigid tube, the immediate consequence of pushing forward a piston at one end, is to cause a condensation, and therefore accumulation of water in the near parts ; and, according to the dynamical theory of sound, the first effect only reaches a distant part of the tube after a finite time, corresponding to a constant velocity, called the velocity of sound, which depends solely on the compressibility and the inertia of the fluid. If the tube inclosing the fluid, instead of being perfectly rigid, be, as every real substance is, to a greater or less degree, somewhat expansible, the transmission of an impulse will be modified—in general retarded—and to a very great degree, if it consist of such a substance as India-rubber, when the compressibility of the water itself will not come sensibly into play, and the first appreciable impulse, received at the remote end, will come in a time depending on the inertia of the fluid and the lateral yielding of the tube, and corresponding to a velocity much inferior to that of sound in the fluid itself. Nearly the same law of motion will be followed under the influence of gravitation, instead of elasticity, by water in a canal, when a quantity of water, not enough at any time to increase or diminish the depth of the canal in any part by a difference considerable in proportion to the whole depth, is admitted or drawn off from either end. If, lastly, the viscosity of the fluid is taken into account in any of these cases—and if, to make the law of resistance to the motion of the fluid through its channel be the same as that of the resistance to conduction of electricity through a solid wire, we suppose the whole interior of the channel to be filled with porous or spongy matter, or to be closely set with transverse barriers, filled with minute apertures, the hydrodynamical problem presents precisely the same elements for a mathematical calculation of the results, and the law of motion is expressed by the same partial differential equation as we have in the electrical problem to determine the laws of the transmission of electric signals through telegraphic lines, either extended on poles through the air, or insulated under the sea in the usual manner of submarine cables. In a line insulated in the air on poles the electrostatic capacity is extremely small, and the transmission of signals follows laws agreeing closely in character with those of the transmission of pressure impulses through water or air contained in a long rigid tube. Accordingly, a definite velocity of propagation of electric impulses, depending on the inertia and the capacity for charge, is to be looked for, as has been done in a first article, published by Kirchoff, on the subject; and a law of extinction, identical with that of sound in a rigid tube, when sensibly influenced by the viscosity of the fluid, will be required to complete the theory by expressing the effect which resistance to conduction produces on the motion of electricity through the wire, as the same able mathematician has found in a subsequent investigation, recently published. This theory points to a velocity of propagation for electric signals in a telegraph wire considerably greater than that of light, and is so far in accordance with Wheatstone's observation; but it must be admitted that the foundation is incomplete on one important point—the electro-magnetic induction which determines what we have called the "electric inertia;" and until this lacuna is filled up, it cannot be considered that we have a precise determination of the velocity of an electrical impulse, or of the time of an electrical oscillation through a telegraph wire in any circumstances.   That it must be so great as not sensibly to contribute to the retardations of signals, or in any other way affect the practical working of submarine lines, is a fundamental assumption made by W. Thomson in his mathematical theory of the submarine telegraph, and justified by the following considerations :—(1), That any agency depending on electric inertia must give rise to retardations and certain other effects, in simple proportion to the lengths of line used; and therefore, that if any such effects bear a considerable part in the remarkable phenomena observed in signalling through submarine lines of 300 miles and upwards in length, they must, in lines of shorter length, give results very different from those actually observed.   (2), That the character of the phenomena actually observed in submarine lines presents no feature attributable to electric inertia.   (3), That a mathematical investigation, not yet published, shewed that mutual electro-magnetic induction between the different conductors either of an ordinary multiple wire cable (in which the gutta percha coats of the different wires are each moistened all around by the sea water), or of a cable with two or more wires insulated in one continuous mass of gutta percha, and consequently electro-magnetic "inertia," in a single submarine conductor, cannot be sensible in comparison to what he called "peristaltic induction*," if the length of the line be more than one hundred miles; although in lines of ten miles or less, effects of the former class, being in proportion to the length of the line, may actually preponderate over effects of the latter, which are in proportion to the square of the length of the line.

The theory founded upon this assumption, excluding the second of the three properties of electric action mentioned above, must of course rest on the first and third; and its fundamental

* Induction of charge by electrostatic force. [Art. LXXV. above.]

principles are therefore the law of charge and the law of conduction. The retardations which it shews (depending on slowness
of viscosity, not slowness of "inertia") follow the law which
Fourier long ago discovered in his beautiful mathematical theory
of the conduction of heat through a long bar (the "linear propagation of heat"). Thus an instantaneous application of electro-
motive force at one end of an indefinitely extended line, gives
rise to a long gradual swell, and still more gradual subsidence,
of electric current through any distant part of the conductor
the instant when the maximum strength of this current, or its
maximum rate of increase, or its maximum rate of diminution,
or when any stated proportion of the maximum is reached by
the rising or falling flow at any point of the line, is later than
the time of the initial impulse by an interval increasing in proportion to the square of the distance from the origin. The beginning of the current is instantaneous all along the line, according to this theory;—is in reality delayed only by electric inertia,
and that not at all sensibly, and is practically observable after a
smaller and smaller interval, the more sensitive the instrument
employed to detect it.

Thus, in Queenstown Harbour, in 1857, the ordinary telegraph
galvanometers and relays employed in observing the transmission
of signals through the 2,500 miles of cable on board the two ships,
gave their indications after a retardation of $1\frac{3}{4}$ seconds from the
instant of the application of an electro-magnetic impulse at the
remote end. At Keyham, in 1858, before the cable was again
taken to sea, a quicker and more sensitive instrument—Thomson's
mirror galvanometer—gave a sensible indication of the rising
current at one end of 3,000 miles of cable about a second after the
application of a Daniell's battery at the other.

Observations with the same instrument, or with different
instruments of the same degree of sensibility, and with coils
presenting no sensible resistances in comparison with the whole
resistance in the line, would, if the insulation of the line were
perfect, show retardations proportional to the squares of the
distances travelled over by the impulse, provided the battery
power is varied in simple proportion to the distance. In other
words, the "velocity" of propagation might be said to be inversely
proportional to the distance travelled. A rigorous verification

of this law cannot be obtained in practice, because perfect insulation is unattainable; but one striking fact, alluded to above,
is clearly explained by it—that, on the whole, the greater the
length of line used, the less has been found the apparent
velocity.

With reference to the velocity of propagation of regularly
continued pèriodic impulses, whether from battery or from induction coils, applied at one end of a long submarine wire, the
mathematical theory has given results identical with those which
Fourier found in his investigation of the propagation of the
summer heat and winter cold into the earth; and it thus appears
that, in a telegraphic line of indefinite length, with a regular
harmonic variation of potential applied at one point—(1), The
retardations of maxima, of zeros, and of minima of potential and
of current, are in simple proportion to the distances along the
line.   (2), The magnitudes of the effect diminish in geometrical
progression, at equal intervals of greater and greater distance
along the line.   (3), The velocity of propagation of the phases
mentioned in No. 1 is inversely proportional to the square root of
the periodic time.

References.   Faraday, Lecture to Royal Institution, January
20, 1854; *Journal* of the Institution, and *Philosophical Magazine;*
W. Thomson, *Proceedings of the Royal Society* (regularly re-
published a few months after each date, in the *Philosophical
Magazine*), "On the Theory of the Electric Telegraph," May, 1855
[Art. LXXIII. above]; "On Mutual Peristaltic Induction between the
wires of a multiple electric cable," May, 1856 [Art. LXXV. above];
Letter to the *Athenæum*, October, 1856 [Art. LXXVI. above];
Kirchoff, Poggendorf's *Annalen*, Band C., p. 193, Band CII.,
p. 529 [and, (added July 29, 1883,) Maxwell's *Electricity and
Magnetism*, Vol. II. chap. XX., "Electro-magnetic Theory of
Light"].

ART. LXXXII. EXTRACT FROM ARTICLE "TELEGRAPH" OF
NICHOL'S "CYCLOPÆDIA." [Second Edition. 1860.]

IN considering the question of the practicability of completing
the Atlantic telegraphic cable, an unexpected difficulty had presented itself. In working the telegraph from Harwich to the
Hague, it was perceived that the signals were given more slowly
and less sharply defined than usual. Professor Faraday, having
undertaken to investigate the cause, discovered that the gutta
percha acted as a Leyden jar, the conducting wire serving for
the interior coating, and the water, or the enclosing wire, for
the exterior coating. By using Mr. Bain's process, the currents
were made to write their own history, and were found to be
retarded both in the commencement and in the duration of their
effect, so that what ought to have been a dot was converted into
a line, faintly marked at each end.

In the following year, 1855, Mr Whitehouse shewed the same
results in a higher degree, by using a wire 1,125 miles in length.
He found that the current produced by an instantaneous electro-
motive force was detained in the wire so as to occupy more than a
second and a-half in recording itself.—Professor W. Thomson
had computed that the retardation would be directly proportional
to the square of the length of wire, and inversely to the area of
its transverse section for a given proportion between the area
of the copper and the area of the gutta percha section. Thus,
since the wire from Newfoundland to Ireland would be nearly
double the length of that which Mr Whitehouse employed, a
signal would occupy four times as long, or more than six seconds,
supposing the wire and its coating to be of the same sectional area.

Mr Whitehouse argued that if this law were true, the telegraph to America would be practically useless; and at the meeting of the British Association, in 1856, he described experiments from which he concluded that a double length of wire produced little more than a double retardation, instead of fourfold. Thomson, in reply (*Athenæum*, Nov. 1, 1856) [Art. LXXVI. above], pointed out that this conclusion was altogether unsupported by the facts adduced, and that "the law of squares" was untouched by Whitehouse's investigation. It has since received a very decided experimental confirmation in the rates of ordinary Morse signalling which have been found practicable through different lengths of the Atlantic cable before submergence, and by Mr Jenkin's recent experiments (communicated to the British Association, Aberdeen, 1859), on the Red Sea cable—the best experiments yet made on any submarine telegraph, so far as illustration of the mathematical theory is concerned. In trials through portions of the Atlantic cable lying at Keyham during the winter and spring, 1857-8, when lengths of twelve or thirteen hundred miles were exceeded, the rates attained proved to be nearly in the inverse ratios of the squares of the lengths, as Thomson had anticipated in the year 1854; and through 2,500 miles or upwards no greater speed than one word a minute could be reached in the transmission of messages if received and recorded by instruments of the common construction. By the use of a new class of instruments, however, much more rapid and sure signalling was effected. After submergence, messages were transmitted between Newfoundland and Ireland at the rate of from two to two and a-half words per minute, and received with perfect distinctness on Thomson's mirror galvanometer, a wrong or a doubtful letter scarcely ever occurring, even when the cable was in a condition of so defective insulation that the ordinary recording instruments, which had been prepared for the Company, failed to give any intelligible signals whatever. There can be no doubt but that a considerably higher speed would have been attained if the cable had not entirely failed before arrangements could be made to take advantage of the indications of the mathematical theory as to the best mode of *sending* as well as of receiving messages through a submarine line of so great length.

Several short papers on the mathematical theory of these phenomena and its application to the solution of practical problems*, have been contributed by Professor W. Thomson to the *Proceedings of the Royal Society* (May 1855, May 1856, Dec. 1856), the *Philosophical Magazine* (vol. July to Dec. 1855), the *British Association Report* (Glasgow, 1855), and the *Athenæum* (Nov. 1, 1856) [Arts. LXXIII. to LXXVII. above]. In the first of these [Art. LXXIII. above] the equations of electric conduction in a submerged wire were investigated, and the integrals, adapted to the expression of the most marked features of the phenomena which had attracted attention to the subject, were given. In the second (communicated to the *Philosophical Magazine*, June, 1854, and published about a year later) [Art. LXXIV. above] the electrostatic capacity of any portion of a submerged wire was investigated. In the third (*Proceedings of the Royal Society*, May, 1856) [Art. LXXV. above] the equations of electric conduction through any number of wires insulated from one another in one mass of gutta percha, under a common metallic sheath, were investigated, and the proper modes of integration for the solution of practical problems were indicated. Among other conclusions, one of practical value was pointed out from this investigation—that, contrary to the expectations of some of the most eminent practical men, as exhibited in patented projects, and supposed to be verified by elaborate experiments, no diminution of inductive embarrassment could be obtained by the use of a complete metallic circuit of two wires separately insulated, beside one another in one mass of gutta percha†; a conclusion which is now generally admitted. In a letter to the *Athenæum* (No. for November 1, 1856) [Art. LXXVI. above], replying to Mr Whitehouse, and in subsequent communications to the Royal Society, various practical conclusions regarding the rates of signalling attainable through air and

---

* [See also Art. LXXXV. below.]

† An Atlantic Telegraph had been projected, and the construction of a Mediterranean telegraph was on the point of being commenced, on this plan as a patented invention, in the year 1857, when the attention of the manufacturers was urgently called to the mathematical conclusion stated in the text; and reasons were given for not trusting to the supposed experimental evidence which had led them to the contrary conclusion. The result was, that they gave up their plan of a double wire telegraph, and made their Mediterranean and each subsequent cable with a single insulated conductor. It may be remarked that a metallic circuit of two well insulated wires (whether in one insulating mass of gutta percha, or with water and wet hemp between them, or in two separate cables) has the advantage of being quite free from the disturbance of earth currents, although no such arrangement can diminish *inductive embarrassment*.

submarine lines of different lengths are stated, and a new
method of working, founded on a theoretical investigation of the
operations best adapted for giving a highly condensed single
electric pulse at the remote end of a long line of submerged
wire, such as that by which it is proposed to establish electric
communication between Ireland and Newfoundland, is indicated.
In the first communication on the mathematical theory, consisting chiefly of two letters written to Professor Stokes in Oct.
and Nov. 1854, and afterwards published (May, 1855,) in the
Proceedings of the Royal Society [Art. LXXIII. above], two general
laws were investigated. The first of these, which has been
called "the Law of Squares," is this:—The time required to
charge to a stated proportion of the ultimate electrification
producible by a given battery power, or to discharge a stated
proportion of a given electrification, is proportional to the squares
of the lengths in different cables of the same lateral dimensions. As a particular case, it was stated that the retardation,
from the instant of applying an electro-motive force at one
end, until a stated proportion of the maximum effect is experienced at the other end, is proportional to the square of the
length of the cable.—The second was: That if at one end of an
infinitely long submerged wire be applied an electro-motive force
regularly changing from positive to negative symmetrically in
equal successive intervals of time, electrical waves will be propagated along the wire at a rate which tends to perfect uniformity
the greater the distance from the operating end, and with amplitudes rapidly decreasing, according to a law which tends to a
geometrical progression at greater and greater distances in arithmetical progression. It was remarked that the law of this phenomenon is identical with that which Fourier, in one of the most
admirable of all the beautiful applications he made of his mathematical theory of heat, found as the law of propagation of the
summer heat and winter cold to different depths below the
surface of the earth.—Mr Whitehouse's experiments, referred to
above, afford many interesting illustrations of particular features
of this law, and his supposed conclusion against the law of squares,
is in reality a partial discovery, by experiment, of the uniform
velocity which the mathematical theory had indicated as early as
the year 1854.

ART. LXXXIII. ON THE TRUE AND FALSE DISCHARGE OF A
COILED ELECTRIC CABLE. BY PROF. THOMSON AND MR
FLEEMING JENKIN.

[From the *Phil. Mag.*, Sept. 1861.]

IN an article in the last May Number of this Magazine, "On
the Galvanic Polarization of Buried Metal Plates," translated
from Poggendorff's *Annalen*, No. 10, 1860, Dr Karl describes
certain interesting experiments on the electro-polarization produced between two large zinc plates buried in the garden of the
Observatory of Munich, by opposing and by augmenting the
natural earth-current between them by the application of a
single element of Daniell's; and concludes with the following
remark :—

"The above experiments disclose nothing at variance with the
"known laws of galvanism; but it nevertheless appeared to me ad-
"visable to make them known, as they afford a simple explanation
"of certain phenomena which Professor Thomson has described
"(Report of the Twenty-ninth Meeting of the British Association,
"Aberdeen, 1859, Trans. of Sections, p. 26) [Art. LXXX. above], and
"which he seems to attribute to entirely different causes."

In the report of Prof. Thomson's communication to the British Association here referred to, it is stated that (after mentioning certain experiments by Mr F. Jenkin on submarine cables
coiled in the manufactory of Messrs. Newall and Co., Birkenhead,
in which one end of the battery used, and one end of the cable
experimented on, in each case was kept in connexion with the
earth while the other end of the cable, after having been for a
time in connexion with the insulated pole of the battery, was

suddenly removed from the battery and put in connexion with the earth through the coil of a galvanometer) Prof. Thomson and Mr Fleeming Jenkin remarked that the deflections recorded in these experiments were in the contrary direction to that which

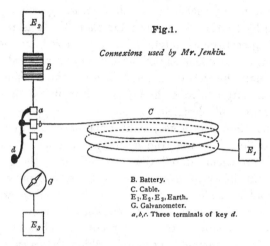

Fig.1.

Connexions used by Mr. Jenkin.

B. Battery.
C. Cable.
E_1, E_2, E_3, Earth.
G. Galvanometer.
a, b, c. Three terminals of key d.

the true discharge of the cable would give; and at Prof. Thomson's request Mr Jenkin repeated the experiments, watching carefully for indications of reverse currents to those previously noted. It was thus found that the first effect of pressing down the key [to throw the cable from battery to earth through galvanometer] was to give the galvanometer a deflection in the direction corresponding to the true discharge current; and that this was quickly followed by a reverse current generally greater in degree, which gave a deflection corresponding to a current in the same direction as that of the original flow through the cable.

Professor Thomson explained this second current, or false discharge, as it has since been sometimes called, by attributing it to mutual electro-magnetic induction between different parts of the coil, and anticipated that no such reversal could ever be found in a submerged cable. The effect of this induction is to produce in those parts of the coil first influenced by the motion of the key, a tendency for the electricity to flow in the same direction as that of the decreasing current flowing through the remoter parts of the coil. Thus, after the first violence of the back flow through the key and galvanometer, the remote parts of the cable begin, by their electro-magnetic induction on the

near parts, to draw electricity back from the earth through the
galvanometer into the cable again, and the current is once more
in one and the same direction throughout the cable.

The phenomena thus described and explained are entirely
different from any that could result from the galvanic polarization
supposed by Dr Karl to account for them*. It is true that the
discharging earth-plate might become polarized by the discharge
in certain cases sufficiently to cause a slight reversal in the current through the galvanometer coil, after the subsidence of the
violent discharge current through it. But in no case could the
whole quantity of electricity flowing in this supposed polarization
current be more than a very small fraction of the quantity which
previously flowed in the true discharge current, of which it is a
feeble electro-chemical reflexion. Its effect on the galvanometer
needle must in every case be as nothing in comparison to the
great impulsive deflection produced by the true discharge current;
and there is no combination of circumstances, as to size of the
earth-plates, amount of the battery power, and rapidity or sensibility of the galvanometer needle, in which the cause supposed by
Dr Karl could possibly be adequate to explain the phenomena
described in Prof. Thomson's communication.

In point of fact, all effects of polarization of the earth-plates
were extremely small in comparison with the main currents observed, which in the experiments on cables with one end kept
to earth, consisted of (1) the constant *through-current*, produced
by a battery of 72 elements of Daniell's in series; (2) the true

* They are also different from any effects which could result from polarization
of the plate connecting the far end of the cable with earth; a cause suggested by
Prof. Wheatstone in a report published by the Committee appointed by the Board
of Trade to inquire into the Construction of Submarine Cables. In support of
his opinion, Prof. Wheatstone quotes some experiments in which he could observe
only the well-known effects due to polarization, which on the short pieces of wire
at his command quite overpowered both the true and false discharge. The current
from the polarized end of a cable is always in the direction of the true discharge
when the battery has been long enough applied: it is observed on both straight
and coiled cables, and is capriciously variable. The details given in the present
paper shew that the currents due to electro-magnetic induction, called false
discharge currents, are on the contrary always in the opposite direction to that
of the true discharge; that they can only be observed on coiled cables; and that
they are in each case sensibly constant. The galvanometer used by Mr Jenkin
would not have been deflected half a degree by the current from a polarized
earth-plate at the end of cables from 300 to 500 knots in length.

discharge through the galvanometer to be observed instantly after breaking the battery connexion of the end of the cable to which the battery was applied, and making instead a connexion, through the galvanometer coil, between the same end of the cable and the earth; and (3) the "false discharge," so called because it must have been often mistaken for the true discharge, which almost necessarily escapes notice altogether when short lengths of coiled cable are tested with slow galvanometer needles. The *through-current* (1) was measured at the beginning of the discharge experiments by introducing the galvanometer into the circuit of cable and battery. Neither the whole amount of the true and false discharges, nor the rapidly varying strength of the current from instant to instant, could be distinctly observed, because the period of vibration of the galvanometer needle, being about $4\frac{1}{2}$ seconds each way, was neither incomparably greater nor incomparably smaller than the duration of the current in either direction. Thus the back-flow, or true discharge, which was of comparatively short duration, first gave the needle an impulse to the left (let us suppose); but before its natural swing, from even an instantaneous impulse, could have allowed it to begin to return, it was caught by the reverse current of false discharge and turned and thrown to the other side of zero through an angle to the right, which, except in the cases of the longest lengths of cable experimented on, was much greater than the angle of the first deflection to the left. It is obvious from what has been stated, that the durations of these deflections of the needle on the two sides do not even approximately coincide with the times during which the current flowed in the directions of the true and false discharges respectively, but that they depend in a complicated manner on the inertia of the needle and the varying forces to which it is subjected. The general character of the phenomena will be made sufficiently clear by the following examples, which are quoted from letters of Mr Jenkin to Prof. Thomson, of dates April 9 and April 22, 1859.

TABLE I.

Lengths of cable in nautical miles *; the first being for the Dardanelles, and the other three, of a different gauge, for the Alexandria and Candia telegraph.	Remote end of cable kept insulated.	Remote end of cable kept to earth.
	First throw of needle.	First observed throw of needle.
123	$1\overset{\circ}{2}$ left	$3\overset{\circ}{2}$ right
137¾	15½ ,,	37 ,,
261½	28½ ,,	31 ,,
399¼	41½ ,,	21 ,,

To explain the cause of the deflections to the right recorded in the last column of this Table, the following observations were made, with care that the first motion of the needle in either direction, however slight or rapid, should not escape notice.

TABLE II.—455 nautical miles of Alexandria and Candia Cable.

Remote end of cable kept.	First throw of needle.	Recoil of second throw.	Excess of recoil above first throw.
1. To earth direct	$2\overset{\circ}{\frac{1}{2}}$ right	$24\overset{\circ}{\frac{1}{2}}$ left	$2\overset{\circ}{2}$
2. To earth through 50 German miles resistance units†	5 ,,	22 ,,	17
3. To earth through 50 ,,	11½ ,,	18½ ,,	7
4. To earth through 90 ,,	16½ ,,	21 ,,	4½
5. Insulated	44½ ,,	not observed	
6. To earth direct, and key " pressed very sharp home"	3¼ ,,	24 ,,	21¾

* A nautical or geographical mile, or a knot as it is generally called in nautical language, is taken as 6087 feet.

† The resistance of this unit was found by experiment to be equal to about 190×10^6 British absolute units of feet per second, or to 6¼ nautical miles of the Alexandria and Candia cable, or to 4·39 of the Dardanelles, or to 7·44 of the Red Sea.

If the whole duration of current, with or without reversal, through the galvanometer coil had been infinitely small in comparison with the natural time of oscillation of the needle (which, reckoned in one direction, was about nine seconds), the recoils would have been sensibly equal to the first throws in the contrary direction, being only less by the effect of resistance of the air, &c. to the motion of the needle. Hence the numbers in the last column of the preceding Table prove that at some interval of time, not incomparably less than nine seconds, after the first motion of the needle, there was a current through the galvanometer coil opposite in direction to that which produced the first or *right* deflection, in each case except No. 5, or that in which the remote end of the cable was insulated. It may be safely assumed that the conductors used in cases 2, 3 and 4 to give the stated resistances between the remote end of the cable and the earth, exercise no sensible electro-magnetic influence, and held no sensible charge, in the actual circumstances ; and it is interesting to see how the greater the resistance thus introduced, that is to say the more nearly is the remote end insulated, the greater is the first throw (due, as explained above, to true discharge), and the less is the excess of the recoil above it.

This excess, shewn in the last column of the Table, exhibits the effect of the electro-magnetic induction from coil to coil which stops short the true discharge, and produces after it a reverse current constituting the "false discharge." The following experiments, performed by Mr Jenkin on the 19th of April, 1859, on different lengths of the Red Sea cable, illustrate the relations between true and false discharge.

TABLE III.

Lengths of Red Sea cable.	Remote length of end used, kept insulated. Discharge from electrification of 36 cells.			Remote end of length used, kept to earth. True and false discharge from electrification and current of 72 cells.		
	First throw.	Recoil.	Excess of first throw above recoil.	First throw.	Recoil.	Excess of recoil above first throw.
312 nautical miles......	20° left	19° right	1°	1¼° left	18° right	16¾°
546 ,,	29½ ,,	27 ,,	2½	5¾ ,,	15 ,,	9¼
858 ,,	35 ,,	14 ,,	21	17 ,,	22 ,,	5
Col. 1.	Col. 2.	Col. 3.	Col. 4.	Col. 5.	Col. 6.	Col. 7.
	True discharge.	Inertia of needle.	Effect of duration of discharge.	True discharge.	"False discharge," and inertia.	"False discharge," or effect of electro-magnetic induction.

The great increase of the numbers in column 4, for the longer portions of cable, illustrates the fact first demonstrated by Prof. Thomson in 1854[*], that, when undisturbed by electro-magnetic induction, the discharge of a cable takes place at a rate inversely proportional to the square of the length. The duration of the discharge, which, when the remote end is kept insulated, is probably much increased by electro-magnetic induction, must be very considerable in the case of the 858 miles length, to produce so great a diminution as 21° in the recoil, from a throw of 35°, on a needle whose period of vibration was 4½ seconds. The diminution of 1° from the throw of 20°, as observed in the case of the 312 miles length, may be to some considerable proportion of its amount due to resistance of the air, although, as this is probably scarcely sensible on a single swing of the needle, it may be supposed that it is chiefly the effect of the duration of the discharge current. From column 7 it is clear that nearly all trace of the electro-magnetic influence would be lost sight of in comparison with the greater effect of true discharge, in the method of experimenting that was followed, if applied to lengths exceeding 1000 knots, in a coil or coils of similar dimensions to those actually used; while for the 546 knots, and shorter lengths, the effect of electro-magnetic induction is greater than that of the true discharge. It is remarkable that the effect of

---

[*] *Proceedings of the Royal Society*, May, 1855; and *Phil. Mag.* Feb. 1856 [Art. LXXIII. above].

electro-magnetic induction is absolutely greater for the shortest of the three lengths. These relations between the different lengths must of course, according to the explanation we have given, depend on the plan of coiling, whether in one coil or in several coils, and on the dimensions of the coil or coils, as well as on the dimensions of the conductor, the gutta percha, and the outer iron sheath of the cable. The magnetic properties of the iron sheath must greatly influence the false discharge, and it would be interesting to compare the discharge from a plain gutta percha covered wire coiled under water with that from an iron sheathed cable.

The following set of experiments, the last which we at present adduce, illustrate the influence of less or greater intervals of time during which the near end of the cable remains insulated, after removal from the battery but before application to earth through the galvanometer coil.

TABLE IV.—455 nautical miles of Alexandria and Candia Cable, remote end kept to earth.   Battery of 72 cells Daniell's.

Experiment.		Throw of needle by true dis-charge.	Recoil, if any, and throw by false discharge.
No. 1	Key struck down...................	3̊ left	27̊ right
2	Key pressed down as usual	2¾ ,,	26 ,,
3	Key pressed very gently.........	2½ ,,	20½ ,,
4	Key held 5 seconds half-way...	0 ,,	14 ,,
5	,, 10 ,, ,,	0 ,,	17 ,,
6	,, 15 ,, ,,	0 ,,	4 ,,

In order to detect whether there might not have been "a slight hesitation in these three last instances, a much more delicate instrument was taken, but no such hesitation could be detected." These results are very remarkable, especially as regards the duration of the electro-magnetic influence. If the conductor of the cable were circumstanced like that of a common electromagnet, and had no sensible electrostatic capacity, the "mechanical value* of the current in it" at the instant of the connexion

---

* See a paper "On Transient Electric Currents," by Prof. W. Thomson, *Phil. Mag.* June, 1853 [Art. LXII. Vol. I. above.], where it is shown that, like the mechanical value of the motion of a moving body, which is equal to half the square of its velocity, multiplied by its mass, the mechanical value of a current

between its near end and the battery being broken, would be
spent in a spark, or electric arc of sensible duration between the
separated metal surfaces. But in the cable, the electrostatic capacity of the near portions of the conductor has an effect analogous
to that of Fizeau's condenser in the Ruhmkorff coils ; and there
was little or no spark (none was observed, although it was looked
for, in the key) on breaking the battery circuit, and consequently, as nearly as may be, the whole mechanical value of the
current left by the battery must have been expended in the development of heat in the conductor itself, and by induced currents
in the iron of the sheath; and we need therefore not wonder at
the great length of time during which electric motion remains in
the cable.

The first column of results for experiments Nos. 1, 2, and 3,
and the two columns for Nos. 4, 5, and 6, shew that the continued flow of the main current through the cable, after the near
end is removed from the battery and kept insulated, is to reduce
its potential gradually from that of the battery (which for the
moment we may call positive), through zero, to negative, in some
time less than five seconds, and to keep it negative ever after, if
it is kept insulated, as long as any trace of electro-dynamic
action remains*. It is probable that, at the same time, there
may be oscillations of current backwards and forwards again†,
and of potential to negative, and positive again, in some parts,
especially towards the middle, of the cable. The mathematical
theory of the whole action is very easily reduced to equations:
but anything like a complete practical analysis of these equations

at any instant, in a coiled conductor, depending on electro-magnetic induction,
is equal to half the square of the strength of the current through it, multiplied
by a constant which the author defined as the "electro-dynamic capacity of the
conductor," and which he shewed how to calculate according to the form and
dimensions of the coil. Additional explanations and illustrations will be found
in Nichol's *Cyclopædia of Physical Science*, second edition, 1859, under the heads
"Magnetism—Dynamical Relations of" [Art. LXI. Vol. I. above], and "Electricity—
—Velocity of" [Art. LXXXI. above].

    * After what has been said in the text above, it is scarcely necessary to point
out that this effect is both opposed to, and much greater than, anything producible
by polarization of the earth-plates.

    † As in the oscillatory discharge of a Leyden phial, investigated mathematically
by Prof. Thomson ("Transient Electric Currents," *Phil. Mag.* June, 1853 [Art.
LXII. Vol. I. above], and actually observed by Feddersen, in his beautiful photo-
graphic investigation of the electric spark (Poggendorff's *Ann.* Vol. CVIII. p. 497,
probably year 1860; also second paper, year 1861).

presents what may be safely called insuperable difficulties, because of the mutual electro-magnetic influence of the different parts of the cable with differently varying current through them. These peculiar difficulties do not, theoretically viewed, present any specially interesting features; and the problem is of little practical importance when once practical electricians are warned to avoid being misled by electro-magnetic induction, in testing by discharge during either the manufacture, the submergence or lifting of a cable, and not to *under-estimate* the rate of signalling through a long submarine cable to be attained when it is laid, from trials through the same cable in coils, when electro-magnetic induction must embarrass the signalling more or less according to the dimensions and disposition of the coils, and probably does so in some cases to such an extent as to necessitate a considerably slower rate of working than will be found practicable after the cable is laid.

The theoretical conclusion that the "false discharge" would not be observed in submerged cables, has been recently verified by Mr Jenkin on various lengths of Bona cable up to 100 miles, which he was engaged in recovering, and which, under careful tests, never gave the slightest indication of "false discharge," although, even when the remote end had completely lost insulation, they gave not only polarization effects*, but also, in the same direction as these, but distinguishable from them, indications of true discharge.   But, in fact, a fortnight before the theoretical conclusion was published by Prof. Thomson at the Aberdeen meeting, a most remarkable and decisive experimental demonstration of it was published by Mr Webb, Engineer to the Electric and International Telegraph Company, who had independently discovered the phenomena which form the subject of this paper, and given substantially the same explanation as that which we now maintain.   If there could be a doubt as to the electro-magnetic theory, the following extract from a letter of Mr Webb's, published in *The Engineer* of August 26, 1859, is decisive :—

"It is, however, on making contact at *F* with earth [that is to say, putting what we have called the near end of the cable to earth] that the greatest and most singular difference occurs

---

* Of the same nature as those observed by Prof. Wheatstone on his short cables.

[between straight and coiled cables]. It will then be seen that the needle at *A* [that is to say, the needle of a galvanometer in cir-

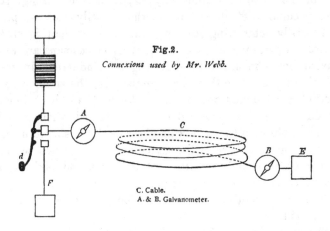

Fig.2.

Connexions used by Mr. Webb.

C. Cable.
A.& B. Galvanometer.

cuit between key and cable instead of between key and earth, as in our experiments], instead of being reversed will continue deflected in the original direction, and both needles will very gradually resume the perpendicular."

"There is a most marked difference between the effect produced between a coiled and a straight cable. The return current appears obliterated, or rather it is overpowered by the effects of the inductive action which takes place from coil to coil. The deflection thus produced is much greater than that produced by the return current. I have had perhaps peculiar facilities for observing this striking phenomenon. Whilst picking up a cable at sea, I frequently test the length I am operating on for return current; *and as the cable becomes coiled into the ship the deflection of the needle, when testing for return current, becomes reversed.*

"It is also my practice to cut the cable at certain distances as it is picked up, and then test such sections separately. On these occasions, sections which, when one end is insulated, will give a charge and discharge of 5°, will when that end is to earth, give a current at the battery end after contact of 90°, but in the reverse direction to that in which the discharge or return current would be if the cable were laid out straight."

ART. LXXXIV. ON THE FORCES CONCERNED IN THE LAYING
AND LIFTING OF DEEP-SEA CABLES.

[From the *Proc. Roy. Soc.* Dec. 1865.]

THE forces concerned in the laying and lifting of deep submarine cables attracted much public attention in the year 1857-58.

An experimental trip to the Bay of Biscay in May, 1858, proved
the possibility, not only of safely laying such a rope as the old
Atlantic cable in very deep water, but of lifting it from the bottom
without fracture. The speaker had witnessed the almost incredible
feat of lifting up a considerable length of that slight and seemingly
fragile thread from a depth of nearly $2\frac{1}{2}$ nautical miles*. The
cable had actually brought with it safely to the surface, from the
bottom, a splice with a large weighted frame attached to it, to
prevent untwisting between the two ships, from which two portions
of cable with opposite twists had been laid. The actual laying of
the cable a few months later, from mid-ocean to Valencia on
one side, and Trinity Bay, Newfoundland, on the other, regarded
merely as a mechanical achievement, took by surprise some of the
most celebrated engineers of the day, who had not concealed their
opinion, that the Atlantic Telegraph Company had undertaken an
impossible problem. As a mechanical achievement it was completely successful; and the electric failure, after several hundred
messages (comprising upwards of 4359 words) had been transmitted
between Valencia and Newfoundland, was owing to electric faults
existing in the cable before it went to sea. Such faults cannot
escape detection, in the course of the manufacture, under the

---

* Throughout the following statements, the word mile will be used to denote
(not that most meaningless of modern measures, the British statute mile but)
the nautical mile, or the length of a minute of latitude, in mean latitudes, which
is 6073 feet. For approximate statements, rough estimates, &c., it may be taken
as 6000 feet, or 1000 fathoms.

improved electric testing since brought into practice, and the causes which led to the failure of the first Atlantic cable no longer exist as dangers in submarine telegraphic enterprise. But the possibility of damage being done to the insulation of the electric conductor before it leaves the ship (illustrated by the occurrences which led to the temporary loss of the 1865 cable), implies a danger which can only be thoroughly guarded against by being ready at any moment to back the ship and check the egress of the cable, and to hold on for some time, or to haul back some length according to the results of electric testing.

The forces concerned in these operations, and the mechanical arrangements by which they are applied and directed, constitute one chief part of the present address; the remainder is devoted to explanations as to the problem of lifting the west end of the 1200 miles of cable laid last summer, from Valencia westwards, and now lying in perfect electric condition (in the very safest place in which a submarine cable can be kept), and ready to do its work, as soon as it is connected with Newfoundland, by the 600 miles required to complete the line.

Forces concerned in the Submergence of a Cable.

In a paper published in the *Engineer* Journal for October 16th, 1857, [of which the substance is given below as Appendix II. to the present Article,] the speaker had given the differential equations of the catenary formed by a submarine cable between the ship and the bottom, during the submergence, under the influence of gravity and fluid friction and pressure; and he had pointed out that the curve becomes a straight line in the case of no tension at the bottom. As this is always the case in deep-sea cable laying, he made no further reference to the general problem in the present address.

When a cable is laid at uniform speed, on a level bottom, quite straight, but without tension, it forms an inclined straight line, from the point where it enters the water, to the bottom, and each point of it clearly moves uniformly in a straight line towards the position on the bottom that it ultimately occupies*. That is to say, each particle of the cable moves uniformly along the base of an isosceles triangle, of which the two equal sides are the inclined

* Precisely the movement of a battalion in line changing front.

portion of the cable between it and the bottom, and the line along the bottom which this portion of the cable covers when laid. When the cable is paid out from the ship at a rate exceeding that of the ship's progress, the velocity and direction of the motion of any particle of it through the water are to be found by compounding a velocity along the inclined side, equal to this excess, with the velocity already determined, along the base of the isosceles triangle.

The angle between the equal sides of the isosceles triangle, that is to say, the inclination which the cable takes in the water, is determined by the condition, that the transverse component of the cable's weight in water is equal to the transverse component of the resistance of the water to its motion. Its tension where it enters the water is equal to the longitudinal component of the weight (or, which is the same, the whole weight of a length of cable hanging vertically down to the bottom), diminished by the longitudinal component of the fluid resistance. In the laying of the Atlantic cable, when the depth was two miles, the rate of the ship six miles an hour, and the rate of paying out of the cable seven miles an hour, the resistance to the egress of the cable, accurately measured by a dynamometer, was only 14 cwt. But it must have been as much as 28 cwt., or the weight of two miles of the cable hanging vertically down in water, were it not for the frictional resistance of the water against the cable slipping, as it were, down an inclined plane from the ship to the bottom, which therefore must have borne the difference, or 14 cwt. Accurate observations are wanting as to the angle at which the cable entered the water; but from measurements of angles at the stern of the ship, and a dynamical estimate (from the measured strain) of what the curvature must have been between the ship and the water, I find that its inclination in the water, when the ship's speed was nearly $6\frac{1}{2}$ miles per hour, must have been about $6\frac{3}{4}°$, that is to say, the incline was about 1 in $8\frac{1}{2}$. Thus the length of cable, from the ship to the bottom, when the water was 2 miles deep, must have been about 17 miles.

The whole amount (14 cwt.) of fluid resistance to the motion of this length of cable through it, is therefore about 81 of a cwt. per mile. The longitudinal component velocity of the cable through the water, to which this resistance was due, may be taken, with but very small error, as simply the excess of the speed of paying out above the speed of the ship, or about 1 mile an hour. Hence, to haul up a piece of the cable vertically through the water, at the rate of 1 mile an hour, would require less than 1 cwt. for over-coming fluid friction, per mile length of the cable, over and above its weight in water. Thus fluid friction, which for the laying of a cable performs so valuable a part in easing the strain with which it is paid out, offers no serious obstruction, indeed, scarcely any sensible obstruction, to the reverse process of hauling back, if done at only one mile an hour, or any slower speed.

As to the transverse component of the fluid friction, it is to be remarked that, although not directly assisting to reduce the egress strain, it indirectly contributes to this result; for it is the transverse friction that causes the gentleness of the slope, giving the sufficient length of 17 miles of cable slipping down through the water, on which the longitudinal friction operates, to reduce the egress strain to the very safe limit found in the recent expe-dition. In estimating its amount, even if the slope were as much as 1 in 5, we should commit only an insignificant error, if we supposed it to be simply equal to the weight of the cable in water or about 14 cwt. per mile for the 1865 Atlantic cable. The transverse component velocity to which this is due may be estimated with but insignificant error, by taking it as the velocity of a body moving directly to the bottom in the time occupied in laying a length of cable equal to the 17 miles of oblique line from the ship to the bottom. Therefore, it must have been about 2 miles in $17 \div 6\frac{1}{2} = 2\cdot 61$ hours, or $\cdot 8$ of a mile per hour. It is not probable that the actual motion of the cable lengthwise through the water can affect this result much. Thus, the *velocity of settling* of a horizontal piece of the cable (or velocity of sinking through the water, with weight just borne by fluid friction) would appear to be about $\cdot 8$ of a mile per hour. This may be contrasted with longitudinal friction by remembering that, according to the previ-ous result, a longitudinal motion through the water at the rate of 1 mile per hour is resisted by only $\frac{1}{17}$th of the weight of the portion of cable so moving.

These conclusions justify remarkably the choice that was made of materials and dimensions for the 1865 cable. A more compact cable (one for instance with less gutta percha, less or no hemp round the iron wires, and somewhat more iron), even if of equal strength and

equal weight per mile in water, would have experienced less transverse resistance to motion through the water, and therefore would
have run down a much steeper slope to the bottom.   Thus, even
with the same longitudinal friction per mile, it would have been
less resisted on the shorter length ; but even on the same length
it would have experienced much less longitudinal friction, because
of its smaller circumference.   Also, it is important to remark that
the roughness of the outer hemp covering undoubtedly did very
much to ease the egress strain, as it must have increased the fluid
friction greatly beyond what would have acted on a smooth gutta
percha surface, or even on the surface of smooth iron wires, presented by the more common form of submarine cables.

The speaker shewed models illustrating the paying-out machine
used on the Atlantic expeditions of 1858 and 1865.   He stated
that nothing could well be imagined more perfect than the action
of the machine of 1865 in paying out the 1200 miles of cable
then laid, and that if it were only to be used for *paying out*, no
change either in general plan or in detail seemed desirable, except
the substitution of a softer material for the "jockey pulleys," by
which the cable in entering the machine has the small amount of
resistance applied to it which it requires to keep it from slipping
round the main drum.   The rate of egress of the cable was kept
always under perfect control by a weighted friction brake of Appold's
construction (which had proved its good quality in the 1858 Atlantic expedition) applied to a second drum carried on the same shaft
with the main drum.   When the weights were removed from the
brake (which could be done almost instantaneously by means of a
simple mechanism), the resistance to the egress of the cable, produced by "jockey pulleys," and the friction at the bearings of the
shaft carrying the main drum, &c., was about $2\frac{1}{2}$ cwt.

*Procedure to Repair the Cable in case of the appearance of an
Electric Fault during the Laying.*

In the event of a fault being indicated by the electric test at
any time during the paying out, the safe and proper course to
be followed in future (as proved by the recent experience), if the
cable is of the same construction as the present Atlantic cable,
is instantly, on order given from an authorised officer in the electric
room, to stop and reverse the ship's engines, and to put on the

greatest *safe* weight on the paying-out break. Thus in the course
of a very short time the egress of the cable may be stopped, and,
if the weather is moderate, the ship may be kept, by proper use of
paddles, screw, and rudder, nearly enough in the proper position
for hours to allow the cable to hang down almost vertically, with
little more strain than the weight of the length of it between the
ship and the bottom.

The best electric testing that has been practised or even planned
cannot shew within a mile the position of a fault consisting of a
slight loss of insulation, unless both ends of the cable are at hand.
Whatever its character may be, unless the electric tests demonstrate its position to be remote from the outgoing part, the only
thing that can be done to find whether it is just on board or just
overboard, is to cut the cable as near the outgoing part as the
mechanical circumstances allow to be safely done. The electric
test immediately transferred to the fresh-cut seaward end shows
instantly if the line is perfect between it and the shore. A few
minutes more, and the electric tests applied to the *two ends* of the
remainder on board, will, in skilful hands, with a proper plan of
working, shew very closely the position of the fault, *whatever its
character may be.* The engineers will thus immediately be able to
make proper arrangements for resplicing and paying out good
cable, and for cutting out the fault from the bad part.

But if the fault is between the land end and the fresh-cut seaward end on board ship, proper simultaneous electric tests on board
ship and on shore (not hitherto practised, but easy and sure if pro-
perly planned) must be used to discover whether the fault lies so
near the ship that the right thing is to haul back the cable until it
is got on board. If it is so, then steam power must be applied to
reverse the paying-out machine, and, by careful watching of the
dynamometer, and controlling the power accordingly (hauling in
slowly, stopping, or veering out a little, but never letting the dynamometer go above 60 or 65 cwt.), the cable (which can bear 7
tons) will not break, and the fault will be got on board more
surely, and possibly sooner, than a "sulky" salmon of 30 lbs. can be
landed by an expert angler with a line and rod that could not
bear 10 lbs. The speaker remarked that he was entitled to make
such assertions with confidence now, because the experience of the
late expedition had not only verified the estimates of the scientific

committee and of the contractors as to the strength of the cable, its weight in water (whether deep or shallow), and its mechanical manageability, but it had proved that in moderate weather the Great Eastern could, by skilful seamanship, be kept in position and moved in the manner required. She had actually been so for thirty-eight hours, and eighteen hours during the operations involved in the hauling back and cutting out the first and second faults, and reuniting the cable, and during seven hours of hauling in, in the attempt to repair the third fault.

Should the simultaneous electric testing on board and on shore prove the fault to be 50 or 100 or more miles from the ship, it would depend on the character of the fault, the season of the year, and the means and appliances on board, whether it would be better to complete the line, and afterwards, if necessary, cut out the fault and repair, or to go back at once and cut out the fault before attempting to complete the line. Even the worst of these contingencies would not be fatal to the undertaking with such a cable as the present one. But all experience of cable-laying shews that almost certainly the fault would either be found on board, or but a very short distance overboard, and would be reached and cut out with scarcely any risk, if really prompt measures, as above described, are taken at the instant of the appearance of a fault, to stop as soon as possible with safety the further egress of the cable.

The most striking part of the Atlantic undertaking proposed for 1866, is that by which the 1200 miles of excellent cable laid in 1865 is to be utilised by completing the line to Newfoundland.

That a cable lying on the bottom in water two miles deep can be caught by a grapnel and raised several hundred fathoms above the bottom, was amply proved by the eight days' work which followed the breakage of the cable on the 3d of August last. Three times out of four that the grapnel was let down, it caught the cable, on each occasion after a few hours of dragging, and with only 300 or 400 fathoms more of rope than the 2100 required to reach the bottom by the shortest course. The time when the grapnel did not hook the cable it came up with one of its flukes caught round by its chain; and the grapnel, the short length of chain next it, and about 200 fathoms of the wire-rope, were proved to have been dragged along the bottom, by being found when brought on board to have interstices filled with soft light gray ooze (of which the speaker shewed a specimen to the Royal Society). These results are quite in accordance with the dynamical theory indicated above (see Appendix II.), according to which a length of such rope as the electric cable, hanging down with no weight at its lower end, and held by a ship moving through the water at half a mile an hour, would slope down to the bottom at an angle from the vertical of only 22°; and the much heavier and denser wire-rope that was used for the grappling would go down at the same angle with a considerably more rapid motion of the ship, or at a much steeper slope with the same rate of motion of the ship.

The only remaining question is: How is the cable to be brought to the surface when hooked? The operations of last August failed from the available rope, tackle, and hauling machine not being strong enough for this very unexpected work. On no occasion was the electric cable broken*. With strong enough tackle, and a hauling machine, both strong enough, and under perfect control, the lifting of a submarine cable, as good in mechanical quality as the Atlantic cable of 1865, by a grapnel or grapnels, from the bottom at a depth of two miles, is certainly practicable. If one attempt fails, another will succeed; and there is every reason, from dynamics as well as from the 1865 experience, to believe that in any moderate weather the feat is to be accomplished with little delay, and with very few if any failing attempts.

The several plans of proceeding that have been proposed are of two classes—those in which, by three or more ships, it is proposed to bring a point of the cable to the surface without breaking it at all; and those in which it is to be cut or broken, and a point of the cable somewhat eastward from the break is to be brought to the surface.

---

* The strongest rope available was a quantity of rope of iron wire and hemp spun together, able to bear 14 tons, which was prepared merely as *buoy-rope* (to provide for the contingency of being obliged, by stress of weather or other cause, to cut and leave the cable in deep or shallow water), and was accordingly all in 100 fathoms-lengths, joined by shackles with swivels. The wire and hemp rope itself never broke, but on two of the three occasions a swivel gave way. On the last occasion, about 900 fathoms of Manilla rope had to be used for the upper part, there not being enough of the wire buoy-rope left; and when 700 fathoms of it had been got in, it broke on board beside a shackle, and the remaining 200 fathoms of the Manilla, with 1540 fathoms of wire-rope and the grapnel, and the electric cable which it had hooked, were all lost for the year 1865.

With reference to either class, it is to be remarked that, by lifting simultaneously by several grapnels so constructed as to hold the cable without slipping along it or cutting it, it is possible to bring a point of the cable to the surface without subjecting it to any strain amounting to the weight of a length of cable equal to the depth of the water. But so many simultaneous grapplings by ships crossing the line of cable at considerable distances from one another would be required, that this possibility is scarcely to be reckoned on practically, without cutting or breaking the cable at a point westward of the points raised by the grapnels. On the other hand, with but three ships the cable might, no doubt, be brought to the surface at any point along the line, without cutting it, and without subjecting it at any point to *much* more strain than the weight corresponding to the vertical depth, as is easily seen when it is considered that the cable was laid generally with from 10 to 15 per cent. of slack. And if the cable is cut at some point not far westward of the westernmost of the grapnels, there can be no doubt but it could be lifted with great ease by three grapnels hauled up simultaneously by three ships. The catenaries concerned in these operations were illustrated by a chain with 15 per cent. of slack hauled up simultaneously at three points.

The plan which seemed to the speaker surest and simplest is to cut the cable at any chosen point, far enough eastward of the present broken end to be clear of entanglement of lost buoy-rope, grapnels, and the loose end of the electric cable itself; and then, or as soon as possible after, to grapple and lift at a point about three miles farther eastward. This could be well and safely done by two ships, one of them with a cutting grapnel, and the other (the Great Eastern herself) with a holding grapnel. The latter, on hooking, should haul up cautiously, never going beyond a safe strain, as shown by the dynamometer. The other, when assured that the Great Eastern has the cable, should haul up, at first cautiously, but ultimately, when the cable is got well off the bottom by the Great Eastern, the western ship should move slowly eastwards, and haul up with force enough to cut or break the cable. This leaves three miles of free cable on the western side of the Great Eastern's grapnel, which will yield freely eastwards (even if partly lying along the bottom at first), and allow the Great Eastern to haul up and work slowly eastwards, so as to keep its grappling rope, and therefore ultimately the portions of electric cable hanging down on the two sides of its grapnel, as nearly vertical as is necessary to make sure work of getting the cable on board. This plan was illustrated by lifting, by aid of two grapnels, a very fragile chain (a common brass chain in short lengths, joined by links of fine cotton thread) from the floor of the Royal Society. It was also pointed out that it can be executed by one ship alone, with only a little delay, but with scarcely any risk of failure. Thus, by first hooking the cable by a holding grapnel, and hauling it up 200 or 300 fathoms from the bottom, it may be left there hanging by the grapnel-rope on a buoy, while the ship proceeds three miles westwards, cuts the cable there, and returns to the buoy. Then, it is an easy matter, in any moderate weather, to haul up safely and get the cable on board.

The use of the dynamometer in dredging was explained; and the forces operating on the ship, the conditions of weather, and the means of keeping the ship in proper position during the process of slowly hauling in a cable, even if it were of strength quite insufficient to act, when nearly vertical, with any sensible force on the ship, were discussed at some length. The manageability of the Great Eastern, in skilful hands, had been proved to be very much better than could have been expected, and to be sufficient for the requirements in moderate weather. She has both screw and paddles—an advantage possessed by no other steamer in existence. By driving the screw at full power ahead, and backing the paddles, to prevent the ship from moving ahead, or (should the screw overpower the paddles), by driving the paddles full power astern, and driving at the same time the screw ahead with power enough to prevent the ship from going astern, "steerage way" is *created* by the lash of water from the screw against the rudder; and thus the Great Eastern may be effectually steered without going ahead. Thus she is in calm or moderate weather, almost as manageable as a small tug steamer, with reversing paddles, or as a rowing boat. She can be made still more manageable than she proved to be in 1865, by arranging to disconnect either paddle at any moment; which, the speaker was informed by Mr Canning, may easily be done.

The speaker referred to a letter he had received from Mr Canning, chief engineer of the Telegraph Construction and Maintenance Company, informing him that it is intended to use three ships, and to be provided both with cutting and with holding grapnels, and expressing great confidence as to the success of the attempt.   In this confidence the speaker believed every practical man who witnessed the Atlantic operations of 1865 shared, as did also, to his knowledge, other engineers who were not present on that expedition, but who were well acquainted with the practice of cable-laying and mending in various seas, especially in the Mediterranean. The more he thought of it himself, both from what he had witnessed on board the Great Eastern, and from attempts to estimate on dynamical principles the forces concerned, the more confident he felt that the contractors would succeed next summer in utilising the cable partly laid in 1865, and completing it into an electrically perfect telegraphic line between Valencia and Newfoundland.

Appendix I.

Descriptions of the Atlantic Cables of 1858 and 1865.

(Distance from Ireland to Newfoundland, 1670 nautical miles.)

Old Atlantic Cable, 1858.

Conductor.—A copper strand, consisting of seven wires (six laid round one), and weighing 107 lbs. per nautical mile.

Insulator.—Gutta percha laid on in three coverings, and weighing 261 lbs. per knot.

External Protection.—Eighteen strands of charcoal iron wire, each strand composed of seven wires (six laid round one), laid spirally round the core, which latter was previously padded with a serving of hemp saturated with a tar mixture.   The separate wires were each 22 gauge; the strand complete was No. 14 gauge.

Circumference of Finished Cable, 2 inches.

Weight in Air, 20 cwt. per nautical mile.

Weight in Water, 13·4 cwt. per nautical mile.

Breaking Strain, 3 tons 5 cwt., or equal to 4·85 times the cable's weight in water per mile.   Hence the cable would bear its own weight in nearly five miles depth of water, or 2·05 times the—

Deepest Water to be encountered, 2400 fathoms, being less than 2½ nautical miles.

Length of Cable Shipped, 2174 nautical miles.

New Atlantic Cable, 1865.

Conductor.—Copper strand consisting of seven wires (six laid round one), and weighing 300 lbs. per nautical mile, embedded for solidity in Chatterton's compound. Diameter of single wire ·048 = ordinary 18 gauge. Gauge of strand ·144=ordinary No. 10 gauge.

Insulation.—Gutta percha, four layers of which are laid on alternately with four thin layers of Chatterton's compound. The weight of the entire insulation 400 lbs. per nautical mile. Diameter of core ·464 of an inch ; circumference of core 1·46 inches.

External Protection.—Ten solid wires of diameter ·095 (No. 13 gauge) drawn from Webster and Horsfall's homogeneous iron, each wire surrounded separately with five strands of Manilla yarn, saturated with a preservative compound, and the whole laid spirally round the core, which latter is padded with ordinary hemp, saturated with preservative mixture.

Circumference of Finished Cable, 3·534 inches.

Weight in Air, 35 cwt. 3 qrs. per nautical mile.

Weight in Water, 14 cwt. per nautical mile.

Breaking Strain, 7 tons 15 cwt., or equal to eleven times the cable's weight in water per mile. Hence the cable will bear its own weight in eleven miles depth of water, or 4·64 times the—

Deepest Water to be encountered, 2400 fathoms, or less than 2½ nautical miles.

Length of Cable Shipped, 2300 nautical miles.

Appendix II.

Let W be the weight of the cable per unit of its length in water, T the force with which the cable is held back at the point where it reaches the water (which may be practically regarded as equal to the force with which its egress from the ship is resisted by the paying-out machinery, the difference amounting only to the weight in air of a piece of cable equal in length to the height of the stern pulley above the water); P and Q the transverse and longitudinal components of the force of frictional resistance experienced by the cable in passing through the water from surface to bottom; i the inclination of its line to the horizon; D the depth of the water.

The whole length of cable from surface to bottom will be $\dfrac{D}{\sin i}$; and the transverse and longitudinal components of the weight of this portion are therefore $\dfrac{WD}{\sin i} \cos i$, and WD respectively. These are balanced by $P \dfrac{D}{\sin i}$ and $T + Q \dfrac{D}{\sin i}$.

Hence

$$P = W \cos i, \quad Q = \left(W - \frac{T}{D} \right) \sin i \dots\dots\dots\dots\dots(1).$$

To find the corresponding components of the velocity of the cable through the water, which we shall denote by p and q, we have only to remark that the actual velocity of any portion of the cable in the water may be regarded as the resultant of two velocities,—one equal and parallel to that of the ship forwards, and the other equal to that of the paying-out, obliquely downwards along the line of the cable (since if the cable were not paid out, but simply dragged, while by any means kept in a straight line at any constant inclination, its motion would be simply that of the ship). Hence, if v be the ship's velocity, and u the velocity at which the cable is paid out from the ship, we have

$$p = v \sin i, \quad q = u - v \cos i \dots\dots\dots\dots\dots(2).$$

Now, as probably an approximate, and therefore practically useful, hypothesis, we may suppose each component of fluid friction to depend solely on the corresponding component of the fluid velocity, and to be proportional to its square. Thus we may take

$$P = W \frac{p^2}{\mathfrak{p}^2}, \quad Q = W \frac{q^2}{\mathfrak{q}^2} \dots\dots\dots\dots\dots(3),$$

where $\mathfrak{p}$ and $\mathfrak{q}$ denote the velocities, transverse and longitudinal, which would give frictions amounting to the weight of the cable; or, as we may call them, the transverse and longitudinal *settling velocities.* We may use these equations merely as introducing a convenient piece of notation for the components of fluid friction, without assuming any hypothesis, if we regard $\mathfrak{p}$ and $\mathfrak{q}$ as each some unknown function of p and q. It is probable that $\mathfrak{p}$ depends to some degree on q, although chiefly on p; and *vice versa*, $\mathfrak{q}$ to some degree on p, but chiefly on q. It is almost certain, however, from experiments such as those described in "Beaufoy's Nautical Experiments," that $\mathfrak{p}$ and $\mathfrak{q}$ are each *very nearly* constant for all practical velocities.

Eliminating p and q between (1), (2), and (3), we have

$$W \cos i = W \left(\frac{v \sin i}{\mathfrak{p}}\right)^2,$$

which gives

$$\mathfrak{p} = \frac{v \sin i}{\sqrt{\cos i}} \quad\quad\quad\quad (4),$$

and

$$(WD - T) \sin i = WD \left(\frac{u - v \cos i}{\mathfrak{q}}\right)^2 \quad\quad (5),$$

which gives

$$\mathfrak{q} = (u - v \cos i) \sqrt{\frac{WD}{(WD - T) \sin i}} \quad\quad (6).$$

These formulæ apply to every case of uniform towing of a rope under water, or hauling in, or paying out, whether the lower end reaches the bottom or not, provided always the lower end is free from tension; but if it is not on the bottom, D must denote its vertical depth at any moment, instead of the whole depth of the sea. To apply to the case of merely towing, we must put $u = 0$; or, to apply to hauling in, we must suppose u negative.

It is to be remarked that the inclination assumed by the cable under water does not depend on its longitudinal slip through the water (since we assume this not to influence the transverse component of fluid friction), and that, according to equation (4), it is simply determined by the ratio of the ship's speed to the transverse "settling velocity" of the cable.

The following table shews the ratio of the ship's speed to the "transverse settling velocity" of the cable for various degrees of inclination of the cable to the horizon :—

Inclination of Cable to Horizon.	Ratio of Ship's Speed to "*transverse settling velocity*" of Cable.	Inclination of Cable to Horizon.	Ratio of Ship's Speed to "*transverse settling velocity*" of Cable.
i	$\dfrac{v}{\mathfrak{p}} = \dfrac{\sqrt{\cos i}}{\sin i}$	i	$\dfrac{v}{\mathfrak{p}} = \dfrac{\sqrt{\cos i}}{\sin i}$
5^0	11·4518	45^0	1·1892
		50	1·0466
$6^0\ 45'$	8·4784	$51^0\ 50'$	1·0000
		55	·9232
10	5·7149	60	·8165
15	3·7973	65	·7173
20	2·8343	70	·6224
25	2·2013	75	·5267
30	1·8612	80	·4231
35	1·5779	85	·0875
40	1·3616		

angle whose sine is $\frac{1}{8\frac{1}{4}}$ } $6^0\ 45'$

If the inclination of the cable had been exactly 6° 45′ when the speed of the Great Eastern was exactly 6½ miles per hour, the value of $\mathfrak{p}$ for the Atlantic cable of 1865 would be exactly 6½÷8·478, or ·765 of a mile per hour.

[From *Transactions of the Institution of Engineers and Shipbuilders in Scotland,* 18th March, 1873.]

ART. LXXXV.—ON SIGNALLING THROUGH SUBMARINE CABLES.

THE Lecturer began by explaining the terms "electrostatic capacity" as applied to a submarine telegraph cable, and, referring to the phenomena of the transmission of an electric impulse through such a cable, he described how that quality effected the speed of signalling, by giving rise to "Retardation." These points were illustrated by signals sent through a model cable—an ingenious apparatus devised and constructed by Mr Cromwell Varley, and consisting of a combination of his condensers with resistance-coils, embodying in an instrument convenient for experiment, the conditions under which electric signals are transmitted through a submarine cable.

The instrument first used for receiving signals through a long submarine cable (the short-lived 1858 Atlantic cable) was the Mirror Galvanometer, which consisted of a small mirror with four light magnets attached to its back (weighing, in all, less than half-a-grain), suspended by means of a single silk fibre, in a proper position within the hollow of a bobbin of fine wire: a suitable controlling magnet being placed adjacent to the apparatus. The action of this instrument is as follows. On the passage of a current of electricity through the fine wire coil, the suspended magnets with the mirror attached, tend to take up a position at right angles to the plane of the coil, and are deflected to one side or the other according as the current is in one direction or the other.

Of various other forms of *receiving* instruments which he had devised, the lecturer referred specially to the Spark Recorder, both on account of the principles involved in its construction, and because it in some respects foreshadowed the more perfect instrument, the Siphon Recorder, which he introduced some years later. The action of the spark recorder was as follows. An indicator, suitably

supported, was caused to take a to and fro motion, by means of
the electro-magnetic actions due to the electric currents constituting the signals. This indicator was connected to a Ruhmkorff coil or other equivalent apparatus, designed to cause a
continual succession of sparks to pass between the indicator, and
a metal plate situated beneath it and having a plane surface
parallel to its line of motion. Over the surface of this plate and
between it and the indicator, there was passed, at a regularly
uniform speed in a direction perpendicular to the line of motion
of the indicator, a material capable of being acted on physically
by the sparks, either through their chemical action, their heat, or
their perforating force. The record of the signals given by this
instrument was an undulating line of fine perforations or spots,
and the character and succession of the undulations were used
to interpret the signals desired to be sent.

The latest form of *receiving* instrument for long submarine
cables, is that of the Siphon Recorder, for which the lecturer
obtained his first patent in 1867. Within the three succeeding
years he effected great improvements on it, and the instrument
has, since that date, been exclusively employed in working most
of the more important submarine cables of the world—indeed
all except those on which the Mirror-Galvanometer method is still
in use.

In the siphon recorder (a view of which is shewn in Fig. 1),
the indicator consists of a light rectangular signal-coil of fine
wire, suspended between the poles of a powerful electro-magnet, so as to be free to move about its longer axis which is
vertical, and so joined up that the electric currents constituting
the signals through the cable, pass through it. A fine glass siphon-
tube is suitably suspended, so as to have only one degree of
freedom to move, and is connected to the signal-coil so as to move
with it. The short leg of the siphon-tube dips into an insulated
ink bottle, which permits of the ink contained by it being electrified, while the long leg is situated so that its open end is at a
very small distance from a brass table, placed with its surface
parallel to the plane in which the mouth of this leg moves, and
over which a slip of paper may be passed at a uniform rate as
in the Spark Recorder. The effect of electrifying the ink is
to cause it to be projected in very minute drops from the
open end of the siphon-tube, towards the brass table or on the

Fig. 1.

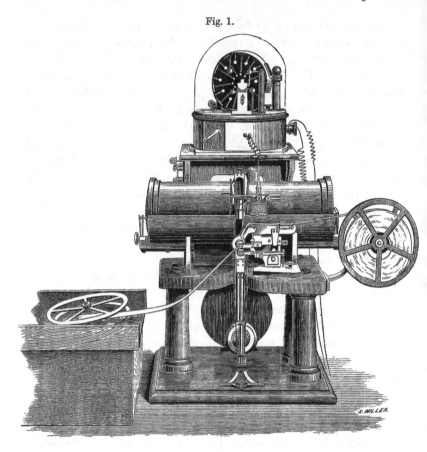

paper-slip passing over it. Thus when the signal-coil moves in obedience to the electric signal currents passed through it, the motion then communicated to the siphon, is recorded on the moving slip of paper by a wavy line of ink marks very close together. The interpretation of the signals is according to the Morse code; the dot and dash being represented by deflections of the line to one side or other of the centre line of the paper.

[Addition of December, 1883. Within the last five years I have brought into use a very much simpler form of siphon recorder. In this form of the instrument, instead of the electro-magnets, I use two bundles of long bar-magnets of square section and made up of square bars of glass-hard steel. The two bundles are supported vertically on a cast-iron socket, and on the upper end of each is fitted a soft iron shoe, so shaped as to concentrate the lines of force and thus produce a strong magnetic field in the space within which the signal-coil is suspended. I have made instruments of this kind to work both with and without electrification of the ink. Without electrification the instrument, as shewn in Fig. 2, is exceedingly simple and compact, and in this

Fig. 2.

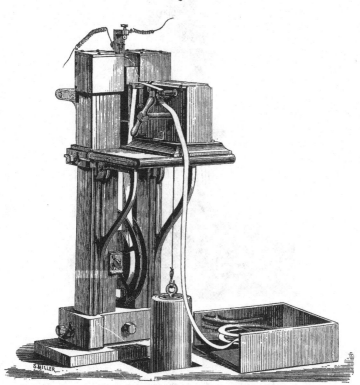

form is capable of doing good work on cables of lengths up to 500 or 600 miles. When constructed for electrification of the ink, as shewn in Fig. 3, it is of course available for much longer lengths of cable, but for cables such as the Atlantic cables, the original form of the siphon recorder is that still chiefly used. The strongest magnetic field hitherto obtained by permanent magnets (of glass-hard steel) is about 3000 C. G. S. With the electro-magnets of the original form of siphon recorder as in ordinary use a magnetic field of about or over 5000 C. G. S. is easily attained.

In Fig. 4 is shewn a *fac simile* of part of a message received and recorded by a Siphon Recorder, such as is shewn in Fig. 1, from one of the Eastern Telegraph Co.'s Cables of about 830 miles length.   W. T.]

Fig. 3.

Fig. 4.

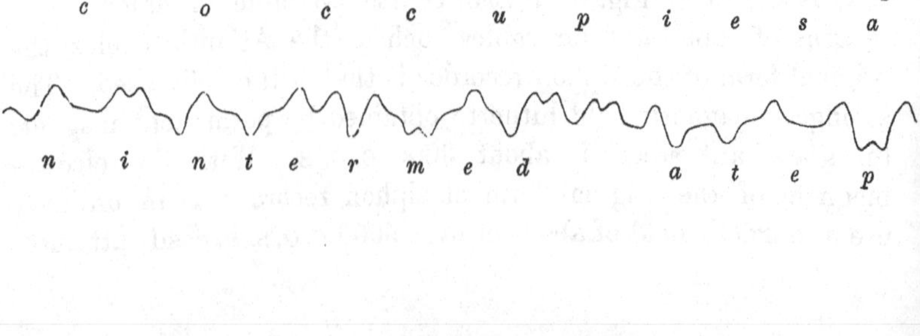

t   o   t   h   e   p   u   b   l   i

c   o   c   c   u   p   i   e   s   a

n   i   n   t   e   r   m   e   d   a   t   e   p

[From *Brit. Assoc. Rep.*, 1855. Part 2.]

ART. LXXXVI. ON THE EFFECTS OF MECHANICAL STRAIN ON
THE THERMO-ELECTRIC QUALITIES OF METALS.

HAVING found by experiment that iron and copper wires, when
stretched by forces insufficient to cause any permanent elongation,
had their thermo-electric qualities altered, but immediately fell
back to their primitive condition in this respect when the stretching forces were removed ; having remarked that these temporary
effects were in each case the reverse of the permanent thermo-
electric effects previously discovered by Magnus, as resulting from
permanent elongation of the wires, by drawing them through holes
in a draw-plate; and thinking it most probable that all these
effects depended on mechanical induction of the thermo-electric
qualities of a crystal in the metals operated upon; the author
undertook an experimental investigation of the thermo-electric
effects of mechanical strains, in which he intended to include
longitudinal extension, longitudinal compression, lateral compression, and lateral extension, and in each case to test both the
temporary effects of strains within the elastic limits of the substance, and the residual alterations in thermo-electric quality
manifested after the cessation of the constraining force, when this
has been so great as to give the substance a permanent set*. The
cycle of experiments has now been so nearly completed for both
the temporary and the permanent strains, as to allow the author
to conclude with certainty that the peculiar thermo-electric qualities induced in each case are those of a crystal. Thus, he finds
that iron bars, hardened by longitudinal compression, have the
reverse thermo-electric property to that discovered by Magnus in
iron wires hardened by drawing; and that iron wire, under lateral
compression, manifests the same thermo-electric property as the
author had discovered in iron wire while under a longitudinal
stretching force. The apparatus by which these results were

[* See Art. XCI., Part III., below.]

obtained was exhibited to the Section, and the mode of experimenting fully described. As regards iron, the general conclusion
is, that its thermo-electric quality, when under pressure in one
direction, deviates from that of the unstrained metal, towards
bismuth for currents in the direction of the strain, and towards
antimony for currents perpendicular to this direction; while for all
cases that have been examined, the residual thermo-electric effect
of a permanent strain is the reverse of the temporary thermo-
electric effect which subsists as long as the constraining force is
kept applied. Those of the other metals which have been as yet
examined, namely, Copper, Lead, Cadmium, Tin, Zinc, Brass, Steel,
and Platinum (specimens supplied as chemically pure by Messrs.
Matthey and Johnson being in general used), showed uniformly
the reverse effect to that of iron when similarly treated. The
effects of permanent lateral compression by hammering were
those which were chiefly tested for in this list of metals, and were
in almost every case of a very marked and unmistakeable kind.
Curious results were also obtained by carefully annealing portions
of wires which had been suddenly cooled, and leaving the remaining parts unannealed. Tin and Cadmium thus treated have, as yet,
given only doubtful results; Platinum has not been tried; Iron,
Steel, Copper, and Brass have given decided .indications, in which
the unannealed portions showed the same kind of thermo-electric
effect as had been found to be produced by permanent lateral
compression.

[From *Brit. Assoc. Rep.*, 1855. Part 2.]

ART. LXXXVII. ON THE USE OF OBSERVATIONS OF TERRESTRIAL
TEMPERATURE FOR THE INVESTIGATION OF ABSOLUTE DATES
IN GEOLOGY.

THE relative thermal conductivities of different substances have
been investigated by many experimenters; but the only absolute determinations yet made in this most important subject are
due to Professor James Forbes*, who has deduced the absolute
thermal conductivity of the trap rock of Calton Hill, of the sandstone of Craigleith Quarry, and of the sand below the soil of the
Experimental Gardens, from observations on terrestrial temperature, which were carried on for five years in these three localities
(all in the immediate neighbourhood of Edinburgh), by means of
thermometers constructed and laid, under his care, by the British
Association. The author of the present communication explained
briefly a method of reduction depending on elementary formulæ
of the theory of the conduction of heat given by the great French
mathematician Fourier, which proved to be more complete and
satisfactory than the method indicated by Poisson†, which had
been adopted by Professor Forbes. He applied it both to the series
of observations used by Professor Forbes, and to a continuation of
the observations on the trap rock of Calton Hill, which has been
carried on up to the present time at the Royal Observatory of
Edinburgh, and of which eleven years complete have been supplied to the author in manuscript, through the kindness of Professor Piazzi Smyth. The results, as regards thermal conductivities,
show that the determinations originally given by Professor Forbes
do not require very considerable corrections; and are satisfactory,
inasmuch as values derived from the diminution of the extent of
variation of the temperature for the deep thermometers agree very
closely with those derived from the retardation of the periods of

* "Account of some Experiments on the Temperature of the Earth near
Edinburgh," *Trans. Roy. Soc. Edinb.*, Vol. XVI. Part 2.
† Poisson's "Théorie Mathématique de la Chaleur," Chap. XII.

summer heat and winter cold at the different depths. They show
very decidedly a somewhat greater conductivity of the trap rock
at the greater depths (from twelve to twenty-four feet) than between the three feet deep and the six feet, or between the six feet
and the twelve feet thermometers, but do not establish any such
variation in the properties of the sandstone, and of the sand of
the two other localities. A comparison of the mean temperatures
of the four thermometers, for the whole sixteen years' observation,
shows an increase of indicated temperature in going downwards in
Calton Hill, which apparently is much more rapid between the
upper than between the lower thermometers; so much so, as not
to be referable to the greater conductivity of the rock in the lower
position. The author remarked, that, to make the observations
available for giving with accuracy the mean absolute temperatures
at the different depths, it would be necessary to have the thermometers taken up and re-compared with a standard thermometer.
It is most probable that the zero-points of all the thermometers
have risen considerably since they were first laid, because the apparent mean temperatures, as shown by the thermometers, are
much higher of late than they were at first. Thus, for the period
of five years examined by Professor Forbes, and for the succeeding
period of eleven years, the means at the different depths are as
follows :—

Trap Rock of Calton Hill.

	3 feet deep.	6 feet deep.	12 feet deep.	24 feet deep.
Period 1837 to 1842	45·49	45·86	46·36	46·87
,, 1843 to 1854	46·512	46·751	47·035	47·349

Notwithstanding the cause of uncertainty which has been
alluded to, these results make it highly probable that the augmentation of mean temperature from three feet to twenty-four feet
below the surface, apparently $1°·38$ Fahr. in the first period and
$84°$ in the second period, must be really more than half a degree,
or more than the greatest elevation of temperature that had been
observed, for a depth of twenty-one feet, in any other part of the
earth. The author was struck with this, and reflecting that probably the Edinburgh observations are the only ones that have
been made on the interior temperature of other igneous rocks
than granite, supposed it to indicate the comparatively modern

time at which the trap rock of Calton Hill has burst up in an incandescent fluid state. This conjecture, shortly after it occurred to him, was confirmed by the intelligence he received at Kreuznach, in Rhenish Prussia, that the temperature in the porphyry of that locality increases at the rate of from 2° to 3° Reaumur in 100 feet downwards, being more than double or triple the rate of augmentation which had been observed in numerous localities in England, France, and other parts of Europe, in granitic rocks and sedimentary strata, and found to be about 1° Fahr. of elevation of temperature in fifteen yards at the least or in twenty yards at the greatest, as Professor Phillips has shown in his Treatise on Geology, in Lardner's *Cyclopædia*, from careful observations made by himself and others. The author pointed out, that the mathematical theory of heat,—with data as to absolute conductivities of rocks, such as those supplied by Professor Forbes, and with the assistance of observation on the actual cooling of historic lava streams, such as the great outbreak from Etna which overthrew Catania in 1669, or of those of Vesuvius which may be seen in the incandescent state, and observed for temperature a few weeks oɪ months after the commencement of solidification,—may be applied to give estimates, within determined limits of accuracy, of the absolute dates of eruption of actual volcanic rocks of prehistoric periods of geology, from observations of temperature in bores made into the volcanic rocks themselves and the surrounding strata.

[From *Brit. Assoc. Rep.*, 1855. Part 2.   *Pogg. Ann.*, xcix. 1856.]

ART. LXXXVIII.   On the Electric Qualities of
Magnetized Iron.

THE well-known ordinary phenomena of magnetism prove that
there is a wonderful difference between the mutual physical
relations of the particles of a mass of iron according as it is
magnetized or in an unmagnetic condition.   Joule's important
discovery, that a bar of iron, when longitudinally magnetized,
experiences an increase of length, accompanied with such a diminution of its lateral dimensions as to leave its bulk unaltered, is
the first of a series by which it may be expected we shall learn
that all the physical properties of iron become altered when the
metal is magnetized, and that in general those qualities which
have relation to definite directions in the substance are differently
altered at different inclinations to the direction of magnetization.
In the present communication, the author described experiments
he had made—with assistance in defraying the expenses from the
Royal Society, out of the Government grant for scientific investigations—to determine the effects of magnetization on the thermo-
electric qualities, and on the electric conductivity, of iron.

The first result obtained was, that longitudinally magnetized
iron wire, in an electric circuit, differs thermo-electrically in the
same direction as antimony from unmagnetized iron.   This any
one may verify with the greatest ease by applying a spirit-lamp
to heat the middle of an iron wire or thin rod of iron a couple of
feet long, with a little magnetizing coil of copper wire (excited by
a cell or two of any ordinary galvanic battery) adapted to slide
freely on it, and so bring a magnetizing force to act on two or
three inches in any part of the length of the iron; and, when
the ends of the iron conductor are connected with the electrodes
of an astatic needle galvanometer of very moderate sensibility,
suddenly moving the coil from one side to the other of the flame
of the spirit-lamp.

The author next explained a series of experiments (not so easily described without the apparatus which was exhibited to the Section, or drawings of it), by which it was ascertained that magnetized iron, with electric currents crossing the lines of magnetization at right angles, differs from unmagnetized iron, thermoelectrically, in the same direction as *bismuth*, that is, in the opposite way to that previously found for iron magnetized along the line of current; and it was verified that an iron conductor, obliquely magnetized, and placed in a circuit of conducting matter, has a current excited through it when its two polar sides are maintained at different temperatures. The author also described and exhibited an experimental arrangement made, but not yet sufficiently tried, to test whether or not magnetized iron possesses a certain thermoelectric rotatory property which his theory of thermo-electricity in crystalline conductors had led him to believe might possibly exist in every substance possessing, either intrinsically or inductively, such a dipolar directional property as that of magnetism *

Regarding the thermo-electric properties of magnetized steel, the only experiments yet made, being on longitudinal magnetization, showed most decidedly the same kind of effect subsisting with the permanent magnetization, after the magnetizing agency is withdrawn, as had been found in iron while actually sustained in a state of magnetization by the electro-magnetic force.

The effects of magnetism on the conductivity of iron both for heat and electricity, in different directions with reference to the direction of magnetization, had been tested by different experimenters with no confirmed indications in the conduction of heat, and with only negative results regarding electric conductivity. The author of the present communication, feeling convinced that only tests of sufficient power are required to demonstrate real effects of magnetization on all physical properties of iron, tried to ascertain the particular nature of the conjectured effect in the case of electric conductivity; and at last, after many unsuccessful attempts, succeeded in establishing, that an iron conductor, sustained in a magnetic condition by a longitudinal magnetizing force, and brittle steel wires retaining longitudinal magnetism, resist the passage of electricity more, or, which is the same, possess less electric conductivity, than the same conductors when

---

* [See foot-notes, of date March 3, 1882, to §§ 163 and 176 of Art. xlviii. Vol. i. above, respecting Hall's great discovery; see also Article xci. below, Part iv. § 161.]

unmagnetic. It remains to be seen whether either iron or steel has, when magnetized, the electro-crystalline property of possessing different electric conductivities in different directions; and whether either has the possible rotatory property as regards conduction, which the intrinsically dipolar type of magnetization suggests *.

It is important to observe, that both the thermo-electric quality, and the effect on electrical conductivity induced in iron or steel, and sustained by the magnetizing force, are retained with the permanent magnetism in steel after the magnetizing force is removed, as Joule found to be the case with the alteration of dimensions, which he discovered as an effect of magnetism; while on the other hand, as the author showed in a previous communication to the Section [Art. LXXXVI. above], the thermo-electric quality he had discovered as an effect of mechanical strain, becomes reversed when the constraining force has been removed, if any permanent strain has been produced.

* See note, p. 179.

ART. LXXXIX. On the Thermo-electric Position of
Aluminium.

[*Brit. Assoc. Rep.*, 1855.   Part 2.]

THE author, through the kindness of Baron Liebig, having been
enabled to make experiments on a bar of aluminium with a view
to investigating its thermo-electric properties, found that it gave
currents when its ends were at different temperatures, and an inch
or two of its length was included in the circuit of a galvanometer
by means of wires of copper, of lead, of tin, or of platinum, bent
round it. These currents were in such directions as to show that
the Aluminium lies, in the thermo-electric series, on the side
towards bismuth, of Tin, Lead, Copper, and a certain platinum
wire (P_2); and, on the side towards Antimony, of another platinum
wire (P_3). They were in the same direction as regards the higher
and lower temperatures of the two junctions of the aluminium
with the other metal in each case, whether the whole bar was
heated so much by a spirit-lamp that it could scarcely be held in
the hand, or no part of it was heated above the temperature of the
air, and one end cooled by being covered with cotton kept moistened
with æther. Taking into account the results of previous experiments which the author had made on a number of different metals,
including three specimens of platinum wire (P_1, P_2, P_3), probably
differing from one another as to chemical purity, which he used as
thermo-electric standards, he concluded that at temperatures of
from 10° to 32° Cent., the following order subsists unchanged as
regards the thermo-electric properties of the metals mentioned:—
Bismuth, P_3, Aluminium, Tin, Lead, P_2, Copper, P_1, Zinc, Silver,
Cadmium, Iron. As he had found that a brass wire, on which he
experimented, is neutral to P_3 at − 10° Cent., and to P_2 at 38°, he
infers that at some temperature between − 10° and 38° Aluminium
must be neutral either to the brass or to P_3. He intends, as soon
as he can procure a few inches of aluminium wire to experiment
with, to determine this neutral point, and others which he infers
from the experiments already made, will probably be found at
some temperature not very low, between Aluminium and Tin,
and Aluminium and Lead; and to look for neutral points which
may possibly be found between Aluminium and P_3 and Aluminium
and P_2, at either high or low temperatures.

Art. XC. On the Origin and Transformations of
Motive Power.

[From the *Royal Institution Proc.*, ii., (Feb. 1856);

Chemist. iii., 1856.]

The speaker commenced by referring to the term *work done*, as
applied to the action of a force pressing against a body which yields,
and to the term *mechanical effect produced*, which may be either
applied to a resisting force overcome, or to matter set in motion.
Often the mechanical effect of work done consists in a combination
of those two classes of effects. It was pointed out that a careful
study of nature leads to no firmer conviction than that work cannot
be done without producing an indestructible equivalent of mechanical effect. Various familiar instances of an apparent loss of mechanical effect, as in the friction, impact, cutting, or bending of
solids, were alluded to, but especially that which is presented by a
fluid in motion. Although in hammering solids, or in forcing solids
to slide against one another, it may have been supposed that the
alterations which the solids experience from such processes constitute the effects mechanically equivalent to the work spent, no
such explanation can be contemplated for the case of work spent in
agitating a fluid. If water in a basin be stirred round and left
revolving, after a few minutes it may be observed to have lost all
sensible or otherwise discernible signs of motion. Yet it has not
communicated motion to other matter round it; and it appears as
if it has retained no effect whatever from the state of motion in
which it had been. It is not tolerable to suppose that its motion
can have come to nothing; and until fourteen years ago confession
of ignorance and expectation of light was all that philosophy taught
regarding the vast class of natural phenomena, of which the case
alluded to is an example. Mayer, in 1842, and Joule, in 1843,
asserted that heat is the equivalent obtained for work spent in
agitating a fluid, and both gave good reasons in support of their
assertion. Many observations have been cited to prove that heat is

not generated by the friction of fluids: but that heat is generated by
the friction of fluids has been established beyond all doubt, by the
powerful and refined tests applied by Joule in his experimental
investigation of the subject.

An instrument was exhibited, by means of which the temperature
of a small quantity of water, contained in a shallow circular case
provided with vanes in its top and bottom, and violently agitated
by a circular disc provided with similar vanes, and made to turn
rapidly round, could easily be raised in temperature several degrees
in a few minutes by the power of a man, and by means of which
steam power applied to turn the disc had raised the temperature of
the water by 30° in half an hour. The bearings of the shaft, to the
end of which the disc was attached, were entirely external; so
that there was no friction of solids under the water, and no way
of accounting for the heat developed except by the friction in the
fluid itself.

It was pointed out that the heat thus obtained is not *produced
from a source,* but is *generated;* and that what is called into existence by the work of a man's arm cannot be matter.

Davy's experiment, in which two pieces of ice were melted by
rubbing them together in an atmosphere below the freezing point,
was referred to as the first completed experimental demonstration
of the immateriality of heat, although not so simple a demonstration
as Joule's; and although Davy himself gives only defective reasoning
to establish the true conclusion which he draws from it. Rumford's
inquiry concerning the "Source of the Heat which is excited by
Friction" was referred to as only wanting an easy additional experiment—a comparison of the thermal effects of dissolving (in an
acid for instance), or of burning, the powder obtained by rubbing
together solids, with the thermal effects obtained by dissolving or
burning an equal weight of the same substance or substances in one
mass or in large fragments—to prove that the heat developed by
the friction is not *produced from the solids,* but is *called into existence between them.*   An unfortunate use of the word "capacity
for heat," which has been the occasion of much confusion ever since
the discovery of latent heat, and has frequently obstructed the
natural course of reasoning on thermal and thermo-dynamic
phenomena, appears to have led both Rumford and Davy to give
reasoning which no one could for a moment feel to be conclusive,
and to have prevented each from giving a demonstration which

would have established once and for ever the immateriality of heat.

Another case of apparent loss of work, well known to an audience in the Royal Institution—that in which a mass of copper is compelled to move in the neighbourhood of a magnet—was adduced; and an experiment was made to demonstrate that in it also heat appears as an effect of the work which has been spent. A copper ball, about an inch in diameter, was forced to rotate rapidly between the poles of a powerful electro-magnet. After about a minute it was found by a thermometer to have risen by 15° Fahr. After the rotation was continued for a few minutes more, and again stopped, the ball was found to be so hot that a piece of phosphorus applied to any point of its surface immediately took fire. It is clear that in this experiment the electric currents, discovered by Faraday to be induced in the copper in virtue of its motion in the neighbourhood of the magnet, generated the heat which became sensible. Joule first raised the question, Is any heat generated by an induced electric current in the locality of the inductive action? He not only made experiments which established an affirmative answer to that question, but he used the mode of generating heat by mechanical work established by those experiments, as a way of finding the numerical relation between units of heat and units of work, and so first arrived at a determination of the mechanical value of heat. At the same time (1843) he gave another determination founded on the friction of fluids in motion; and six years later he gave the best determination yet obtained, according to which it appears that 772 foot pounds of work, (that is 772 times the amount of work required to overcome a force equal to the weight of 1 lb. through a space of 1 foot,) is required to generate as much heat as will raise the temperature of a pound of water by one degree.

The reverse transformation of heat into mechanical work was next considered, and the working of a steam-engine was referred to as an illustration. An original model of Stirling's air-engine was shown in operation, developing motive power from heat supplied to it by a spirit lamp, by means of the alternate contractions and expansions of one mass of air. Thermo-electric currents, and common mechanical action produced by them, were referred to as illustrating another very distinct class of means by which the same transformation may be effected. It was pointed out that in each case, while heat is taken in by the material arrangement or machine, from the source of heat, heat is always given out in another locality, which is at a lower temperature than the locality at which heat is taken in. But it was remarked that the quantity of heat given out is not, (as Carnot pointed out it would be, if heat were a substance), the same as the quantity of heat taken in, but, as Joule insisted, less than the quantity taken in by an amount mechanically equivalent to the motive power developed. The modification of Carnot's theory to adapt it to this truth was alluded to ; and the great distinction which it leads to between reversible and not-reversible transformations of motive power was only mentioned.

To facilitate farther statements regarding transformations of motive power, certain terms, introduced to designate various forms under which it is manifested, were explained. Any piece of matter, or any group of bodies, however connected, which either is in motion, or can get into motion without external assistance, has what is called mechanical energy. The energy of motion may be called either " dynamical energy," or " actual energy." The energy of a material system at rest, in virtue of which it can get into motion, is called " potential energy," or, generally, motive power possessed among different pieces of matter, in virtue of their relative positions, is called "potential energy." To show the use of these terms, and explain the ideas of a *store of energy*, and of conversions and transformations of energy, various illustrations were adduced. A stone at a height, or an elevated reservoir of water, has potential energy. If the stone be let fall, its potential energy is converted into actual energy during its descent, exists entirely as the actual energy of its own motion at the instant before it strikes, and is transformed into heat at the moment of coming to rest on the ground. If the water flow down by a gradual channel, its potential energy is gradually converted into heat by fluid friction, and the fluid becomes warmer by a degree Fahr. for every 772 feet of the descent. There is potential energy, and there is dynamical energy, between the earth and the sun. There is most potential energy and least actual energy in July, when they are at their greatest distance asunder, and when their relative motion is slowest. There is least potential energy and most dynamical energy in January, when they are at their least distance, and when their relative motion is most rapid. The gain of dynamical energy from the one time to the other is equal to the loss of potential energy.

Potential energy of gravitation is possessed by every two pieces of matter at a distance from one another; but there is also potential energy in the mutual action of contiguous particles in a spring when bent, or in an elastic cord when stretched.

There is potential energy of electric force in any distribution of electricity, or among any group of electrified bodies. There is potential energy of magnetic force between the different parts of a steel magnet, or between different steel magnets, or between a magnet and a body of any substance of either paramagnetic or diamagnetic inductive capacity. There is potential energy of chemical force between any two substances which have what is called affinity for one another,—for instance, between fuel and oxygen, between food and oxygen, between zinc in a galvanic battery and oxygen. There is potential energy of chemical force among the different ingredients of gunpowder or gun cotton. There is potential energy of what may be called chemical force, among the particles of soft phosphorus, which is spent in the allotropic transformation into red phosphorus; and among the particles of prismatically crystallized sulphur, which is spent when the substance assumes the octahedral crystallization.

To make chemical combination take place without generating its equivalent of heat, all that is necessary is to resist the chemical force operating in the combination, and take up its effect in some other form of energy than heat. In a series of admirable researches on the agency of electricity in transformations of energy*, Joule showed that the chemical combinations taking place in a galvanic battery may be directed to produce a large, probably in some forms of battery an unlimited, proportion of their heat, not in the locality

* "On the Production of Heat by Voltaic Electricity," communicated to the Royal Society Dec. 17, 1840, (see Proceedings of that date,) and published *Phil. Mag.* Oct. 1841.

"On the Heat evolved by Metallic Conductors of Electricity, and in the cells of a battery during Electrolysis."—*Phil. Mag.* Oct. 1841.

"On the Electrical Origin of the Heat of Combustion."—*Phil. Mag.* March, 1843.

"On the Heat evolved during the Electrolysis of Water," *Proceedings of the Literary and Philosophical Society of Manchester*, 1843, Vol. VII., Part 3, Second Series.

"On the Calorific Effects of Magneto-Electricity, and on the Mechanical Value of Heat," communicated to the British Association (Cork), Aug. 1843, and published *Phil. Mag.* Oct. 1843.

"On the Intermittent Character of the Voltaic Current in certain cases of of combination, but in a metallic wire at any distance from that locality; or that they may be directed, not to generate that part of their heat at all, but instead, to raise weights by means of a rotating engine driven by the current. Thus if we allow zinc to combine with oxygen by the beautiful process which Grove has given in his battery, we find developed in a wire connecting the two poles the heat which would have appeared directly if the zinc had been burned in oxygen gas; or if we make the current drive a galvanic engine, we have, in weights raised, an equivalent of potential energy for the potential energy between zinc and oxygen spent in the combination.

The economic relations between the electric and the thermo-dynamic method of transformation from chemical affinity to available motive power were indicated, in accordance with the limited capability of heat to be transformed into potential energy, which the modification of Carnot's principle, previously alluded to, shows, and the unlimited performance of a galvanic engine in raising weights to the full equivalent of chemical force used, which Joule has established.

The transformation of motive power into light, which takes place when work is spent in an extremely concentrated generation of heat, was referred to. It was illustrated by the ignition of platinum wire by means of an electric current driven through it by the chemical force between zinc and oxygen in the galvanic battery; and by the ignition and volatilization of a silver wire by an electric current driven through it by the potential energy laid up in a Leyden battery, when charged by an electrical machine. The luminous heat generated in the last-mentioned case was the complement to a deficiency of heat of friction in the plate-glass and rubber of the machine, which a perfect determination, and comparison with the amount of work spent in turning the machine, would certainly have detected.

Electrolysis, and on the Intensity of various Voltaic arrangements."—*Phil. Mag.* Feb. 1844.

"On the Mechanical Powers of Electro-Magnetism, Steam, and Horses." By Joule and Scoresby.—*Phil. Mag.* June, 1846.

"On the Heat disengaged in Chemical Combination."—*Phil. Mag.* June, 1852.

"On the Economical Production of Mechanical Effect from Chemical Forces."—*Phil. Mag.* Jan. 1853.

[For all of the above-mentioned Papers except the seventh (which will appear in Vol. II.) see also "Joule's Scientific Papers" Vol. I.; published by the Physical Society of London, 1884.]

The application of mechanical principles to the mechanical actions of living creatures was pointed out. It appears certain from the most careful physiological researches, that a living animal has not the power of originating mechanical energy; and that all the work done by a living animal in the course of its life, and all the heat that has been emitted from it, together with the heat that would be obtained by burning the combustible matter which has been lost from its body during its life, and by burning its body after death, make up together an exact equivalent to the heat that would be obtained by burning as much food as it has used during its life, and an amount of fuel that would generate as much heat as its body if burned immediately after birth.

On the other hand, the dynamical energy of luminiferous vibrations was referred to as the mechanical power allotted to plants (not mushrooms or funguses, which can grow in the dark, are nourished by organic food like animals, and like animals absorb oxygen and exhale carbonic acid,) to enable them to draw carbon from carbonic acid, and hydrogen from water.

In conclusion, the sources available to man for the production of mechanical effect were examined and traced to the sun's heat and the rotation of the earth round its axis.

Published speculations* were referred to, by which it is shown to be possible that the motions of the earth and of the heavenly bodies, and the heat of the sun, may all be due to gravitation; or, *that the potential energy of gravitation may be in reality the ultimate created antecedent of all motion, heat, and light at present existing in the universe.*

---

* Prof. W. Thomson, "On the Mechanical Energies of the Solar System" (*Trans. Roy. Soc. Edinburgh*, April, 1854 [Art. LXVI. above]), and "On the Mechanical Antecedents of Motion, Heat, and Light" (*British Association Report*, Liverpool, 1854 [Art. LXIX. above]).

ART. XCI. ON THE ELECTRO-DYNAMIC QUALITIES OF METALS*.

[THE BAKERIAN LECTURE.]

[From *Transactions of the Royal Society*, Feb. 1856.]

1. AN electrified body may be regarded as a reservoir of potential energy, and any material combination in virtue of which bodies can receive charges of electricity is a source of motive power. The development of mechanical effect from the potential energy of electricity, or through electric means from any source of motive power, may take place in a great variety of ways. For instance, electro-statical attractions and repulsions may become direct moving forces (as in "Franklin's Spider"), to do work in the discharge of an electrified conductor or in the continued use of a continuous supply of electricity; or the forces of current electricity may, as in any kind of electro-magnetic engine, become working forces on bodies in motion; or the whole energy of the discharge may, as discovered by Joule, be converted into heat, which again may be transformed into other kinds of energy; or the heat evolved and absorbed by electricity, in a circuit of two different metals, at the places where it crosses the junctions from one metal to the other, being a thermal result of dynamic moment† when the junction at which heat is evolved is at a

---

* The author has to acknowledge much valuable assistance in the various experimental investigations described in this paper, from his assistant Mr McFarlane, and from Mr C. A. Smith, Mr R. Davidson, Mr F. Maclean, Mr John Murray, and other pupils in his laboratory.

† Either an evolution of heat at a temperature higher, or an absorption of heat at a temperature lower, than that of the atmosphere, may be taken advantage of to work an engine giving mechanical effect from heat; by using the atmosphere in one case as a recipient for discharged heat, in the other as the source of the heat taken in. Or an evolution of heat at any temperature and an absorption of heat at any lower temperature, may be taken advantage of for the same purpose, in a limited material system, neither taking heat from nor parting with heat to any external matter. Hence such a double thermal effect may be said to possess "dynamic moment." See the author's "Account of Carnot's Theory of the Motive Power of Heat," §§ 4 to 11, *Trans. Roy. Soc. Edinb.*, Jan. 2, 1849 [Art. XLI. Vol. I. above]; also his "Dynamical Theory of Heat," §§ 8, 13, 23 to 30, *Trans. Roy. Soc.*

higher temperature than the junction at which heat is absorbed, may be used in a thermo-dynamic engine. Again, a thermo-electric current is a dynamic result, derived from a definite absorption of heat in one locality and a definite evolution of heat in a locality of lower temperature.

2. Of these various kinds of action, all except the first mentioned, depend essentially on certain definite properties of matter in regard to which different metals have remarkably different qualities. Thus in electro-magnetic engines the electric conductivity of the coils through which the current passes, and the magnetic inductive capacity and retentiveness of the iron cores of the electro-magnets, are essentially involved; and as essentially, when permanent magnets are used, the magnetic properties of steel, loadstone, or other bodies possessing strong retentiveness for magnetism. In the simple conversion of any kind of energy into heat by means of electric currents in metals, their electric conductivities are essentially and solely concerned.

The inverse thermo-electric transformation of energy into an evolution and absorption of heat, at localities of different temperature, in quantities differing from one another *by the thermal equivalent* of the work spent in maintaining the current*, depends essentially on certain distinct properties of metals in regard to which their various qualities are shown by the differences of their positions in the thermo-electric series at different temperatures; and the accessory circumstances of such operations are influenced by the electric and thermal conductivities of the metals used. The same properties are involved in the direct thermo-electric transformation of energy in which electric currents, sustained by the communication of heat in a hot locality and the abstraction of a less quantity of heat in a locality lower in temperature, either produce any mechanical action, or are allowed to waste all their motive power in the frictional generation of heat†.

Edinb., March 17, 1851, and "Dynamical Theory of Heat, Part VI. Thermo-electric Currents," § 102, *Trans. Roy. Soc. Edinb.*, May 1, 1854. This latter series of articles under the general title "Dynamical Theory of Heat," has been republished in a succession of Numbers of the *Philosophical Magazine*, viz. §§ 1 to 80, Vol. July to Dec. 1852; §§ 81 to 96, Vol. Jan. to June, 1855; §§ 97 to 181, Vol. Jan. to June, 1856 [now constituting Art. XLVIII. Vol. I. above].

* See "Dynamical Theory of Heat, Part VI. Thermo-electric Currents," [Art. XLVIII. Vol. I. above], §§ 105, 110.

† ".... a current cannot pass through a homogeneous conductor without

3.   All properties, then, of electric and thermal conductivity, of magnetic inductive capacity and retentiveness, and of thermo-electric rank and its variations from one temperature to another, may be characterized as electro-dynamic; and the degrees to which these properties are possessed by different substances may be called their electro-dynamic qualities. Again, the variation which absolute magnetic inductive capacity, and magnecrystallic

generating heat in overcoming resistance. This effect, which we shall call the *frictional generation of heat*, has been discovered by Joule to be produced at a rate proportional to the square of the strength of the current; and, taking place equally with the current in one direction or the contrary, is obviously of an irreversible kind" (Dyn. Th. Heat, § 104). This definition was given merely to render circumlocution unnecessary in frequently referring to a mode of electric action which bore an obvious analogy to the action of a common fluid generating heat by friction among its particles as a dynamical equivalent to work spent upon it from without in forcing it to circulate in a tube, or otherwise keeping it in motion. It appears to me highly probable, however, that what I have, with reference only to recognized electric currents, defined as the frictional generation of heat, is precisely the mode of action by which all the heat is generated in every case when two solids are rubbed together. Certainly when two bad conductors of electricity are rubbed together, a portion of the heat of friction is generated in visible electric flashes; and a charged Leyden battery contains, in potential energy, a dynamic equivalent for a portion of the heat of friction between rubber and glass never made visible till the battery is discharged. As certainly a portion of the heat of friction between a metal and a bad conductor of electricity is *invisibly* generated by electric currents through a very minute depth of the metallic substance beside its rubbed surface. The first effect of chemical forces of affinity, as Joule has so powerfully demonstrated in a variety of cases, is to press electricity into motion; which motion may either subside into heat close to the locality of the combination (as [in ordinary combustion and as] when rough zinc is dropped into dilute sulphuric acid), or, reactively resisting the chemical combination, may transmit the work to a locality distant from the source, and may there either generate heat in a permanent metallic or other undecomposable conductor, or may, without any generation of heat at all, be wholly spent in effecting decompositions against chemical affinities infinitely less powerful than those from which it proceeds, or in raising weights. So it appears highly probable that the first effect of the force by which one solid is made to slide upon another is electricity set into a state of motion; that this electric motion subsides wholly into heat in most cases, either close to its origin and instantaneously, as when the solids are both of metal, or at sensible distances from the actual locality of friction and during appreciable intervals of time, as when the substance of one or both the bodies is of low conducting power for electricity; and that it only fails to produce the full equivalent in heat for the work spent in overcoming the friction, when the electric currents are partially diverted from closed circuits in the two bodies and in the space between them, and are conducted away to produce other effects in other localities. Still, no hypothesis need be implied by using the expression "the frictional generation of heat by an electric current," as defined in the passage quoted, and it is introduced into the present paper with no other justification than its convenience.

axial differences, experience with change of temperature may obviously be made the means of a transformation of heat into common mechanical energy, and we have thus a set of magneto-dynamic properties of matter which may almost in the present state of science be regarded as intrinsically electric, but which at all events (when we consider that the motions contemplated, taking place as they do under magnetic force, cannot but be accompanied by electric currents) may be fairly classed under the general designation of electro-dynamic. The variations of intrinsic magnetism, of magnetic inductive capacity, and of magnecrystallic properties, produced by variations of temperature, are therefore included among the electro-dynamic qualities of metals which I propose to investigate, although I have as yet made no progress in this branch of the subject.

PART I. ON THE ELECTRIC CONVECTION OF HEAT,

§§ 4 TO 77.

§§ 4—18. THEORETICAL INDICATIONS.

§§ 4—9. *Origin of the Investigation.*

4. In first attempting an application of the principles of the Dynamical Theory of Heat to show the mechanical relation between cause and effect in thermo-electric currents, I supposed the effects thermal and mechanical that can be produced by a thermo-electric current in any part of its circuit to be, as first suggested by Joule, due to the heat absorbed, according to Peltier's discovery, at the hot junction in virtue of the current crossing it, and I pointed out that the current crossing the cold junction must evolve a quantity of heat which, were this supposition true, would be less than that absorbed at the hot junction, by an amount precisely equivalent to all the effects, produced by the current in the rest of the circuit*.

---

* "Dynamical Theory of Heat," March 17, 1851 [Art. XLVIII. Vol. I. above], § 17.

5. Introducing Carnot's principle, as modified in the Dynamical Theory of Heat*, I found a relation between the quantities of heat absorbed or evolved by currents crossing metallic junctions at different temperatures ; which led immediately to a general expression for the electrical condition of a circuit of two metals with their junctions kept at any stated temperatures.

6. From this it appeared that the electromotive force should follow the same law of variation in every case, being expressed by a constant, (representing the thermo-electric difference between the two metals,) multiplied into an absolute function of the temperatures of their junctions, namely, the difference of their temperature on the absolute thermometric scale since proposed by Mr Joule and myself, and demonstrated by our experiments† to agree very approximately with their difference of temperature as indicated by an air-thermometer. Finding this conclusion contradictory to the statements made by experimenters, that the electromotive force does not vary with the temperature of the junctions according to the same law in circuits composed of different metals, I perceived that Peltier's discovery did not afford a sufficient explanation of the source whence a thermo-electric current derives its energy, but that electric currents must possess the previously undiscovered property of producing different thermal effects in passing from cold to hot and from hot to cold in the same metal, and must possess this property to different amounts in different metals.

7. Taking this new property of electric currents into account along with that discovered by Peltier, and introducing an application of Carnot's principle, I arrived at expressions for the relations between the heat absorbed and evolved in various parts of a circuit of any different metals, and between the electromotive force and the temperatures of the junctions, which appear to be

* This, the true form of Carnot's principle, was first published by Clausius in May 1850 (Poggendorf's *Annalen*). It had occurred to myself, and I had used it in discovering the true expression for the duty of a perfect thermo-dynamic engine shortly before that time. It was not, however, until the beginning of the year 1851 that I thought on a demonstration which would probably be admitted as conclusive in establishing the principle, and my investigation on the subject was only communicated in March 1851 to the Royal Society of Edinburgh. See *Trans. Roy. Soc. Ed.* of that date, "Dynamical Theory of Heat" [Art. XLVIII. Vol. I. above], § 14.

† "On the Thermal Effects of Fluids in Motion," *Transactions*, June, 1854 [Art. XLIX. Part 2, Vol. I. above].

in complete accordance with the facts. These investigations were communicated in December 1851, to the Royal Society of Edinburgh*.

8. Still simpler theoretical considerations (§§ 10—18 below) regarding the source of energy drawn upon in a thermo-electric current, make it certain that the phenomena of inversion discovered by Cumming could not exist, unless the metals presenting them had the property of experiencing, when unequally heated, unequal thermal effects from electric currents passing through them from hot to cold, and from cold to hot. Having satisfied myself, both by an examination of the evidence afforded by Becquerel's experiments (the original investigation on the subject by Cumming being at that time unknown to me), and by actual observation, in experiments of my own†, that the doubts which various writers had thrown on the existence of thermo-electric inversions were groundless, I concluded with certainty that the newly conceived thermal effect of electricity in unequally heated metals really exists. But the theory left it undecided what the absolute nature and amount of this effect may be, and only showed how, by observations on thermo-electric currents, its difference in different metals may be determined.

9. I therefore had recourse to direct experiment on the thermal effects of electric currents in unequally heated conductors, not to demonstrate the existence of the peculiar effect anticipated, but to ascertain its nature, with moreover a view of ultimately determining its absolute amount, in some particular metal or metals. Before proceeding to describe experiments, by which I have now discovered the quality of the new effect in several cases, I shall, without entering on the mathematical details of the theory, or the full application of Carnot's principle, repeat in a few words so much of my first communication on the subject to the Royal Society of Edinburgh, as to show the reasoning, founded on incontrovertible mechanical principles, which made me commence the experimental research with the certainty that the property looked for existed, whether I could find it or not.

---

* See *Proceedings* of that date, and *Philosophical Magazine*, June, 1852 [Art. XLVIII. Appendix, Note I. Vol. I. above].

† See below, Part II. §§ 79, 80, 81, 83, 84, &c.

§§ 10—15. *General inferences regarding the Electric Convection*
of Heat from Dynamical Principles.

10. Cumming has discovered that in many cases when one of
the junctions of a thermo-electric circuit of two metals is kept at
a fixed temperature, if that of the other be elevated gradually
from equality, an electro-motive force is produced, which first
increases to a maximum, then diminishes, vanishes for a certain
temperature of the junction, and acts in the contrary direction
with gradually increasing strength as the temperature is further
raised. It is clear that, at exactly that temperature of the hot
junction for which in any such case the electro-motive force is a
maximum, the two metals must be thermo-electrically neutral to
one another, and must present reverse thermo-electric relations
for temperatures below and above this point. Hence the *thermal
effect depending on the direction of a current crossing the junction
of two such metals* must be for temperatures above, the reverse
of what it is for temperatures below, the neutral point, and must
vanish when the metals are exactly at this temperature.

11. For although Peltier himself supposed the effect he had
discovered to depend on the conducting powers of the two metals
for heat, and remarked as an anomaly the case of bismuth and
copper, for which his supposition was violated, his own experiments show the truth to be, that in a circuit of two metals an
absorption of heat at the junction where the temperature is
higher, and an evolution of heat at the other, must be produced
by the thermo-electric current which is caused by the maintaining
of the difference of temperature between the junctions. That
this is universally true when the temperatures of the two junctions are on the same side of the neutral point, cannot, in the
present state of science as regards the theory of heat, be reason-
ably doubted.

12. If, therefore, a circuit of two metals have one junction
kept at the neutral point, and the other at some lower temperature, the current excited will cause the evolution of heat at
the cold junction, but neither absorption nor evolution of heat
at the hot junction; and in the rest of the circuit there will be
effects either purely thermal, or thermal and mechanical or chemical, according to the nature of the resistance against which the

electro-motive force is allowed to work. The source from which the electro-motive force derives its energy to produce these effects cannot be at the hot junction (§ 10), where heat is neither absorbed nor evolved, nor at the cold junction (§ 11), where heat is evolved, nor of course in any uniformly heated part in either metal, through all of which, provided the metal has no thermo-electric crystalline characteristic, there can be nothing but a frictional evolution of heat; that is, it is nowhere but in those portions of the circuit where the temperature varies between that of the cold and that of the hot junction. In those portions, therefore, there must be as much heat absorbed, in virtue of the current, as is equivalent to the aggregate mechanical value of the heat evolved at the cold junction, and all the effects, thermal, mechanical, and chemical, produced in the rest of the circuit.

13. If, for example, an electro-magnetic engine be introduced into the circuit, and be allowed to work at such a rate as to reduce, by its inductive reaction, the strength of the thermo-electric current to an infinitely small fraction of what it is when the engine is at rest, the heat absorbed in virtue of the current in the unequally heated parts of the two metals will be equal to the heat evolved at the cold junction, together with the thermal equivalent of the work done by the engine, and will be simply proportional to the strength of the current. On the other hand, if the engine be forced to work a little faster, so as to overbalance by an infinitely small amount the thermal electro-motive force, and cause a reverse current in the circuit, there must be heat evolved in virtue of this current in the unequally heated parts of the two metals to an amount equal to the heat absorbed at the cold junction, together with the thermal equivalent to the work done against electro-magnetic forces in the engine. It follows that in the unequally heated portions of the two metals, the current passing from cold to hot in one, and from hot to cold in the other, must produce a thermal effect, in simple proportion to its own strength, constituting on the whole an absorption of heat when the thermal electro-motive force is allowed to produce a current, and an evolution of heat when a current is forced by other means in the contrary direction.

14. Hence, for any two metals which are thermo-electrically neutral to one another at a certain temperature, and which possess

reverse thermo-electric properties for temperatures above and below the neutral point, we conclude the following propositions :—

(1) In one or other of the metals (and most probably in both) there must be a thermal effect due to the passage of electricity through a non-uniformly heated portion of it, which must be an absorption of heat or an evolution of heat, according to the direction of the current between the hot and cold parts, and proportional in amount to the whole quantity of electricity that passes in a stated time.

(2) The amount of this effect, with the same strength of current and the same difference of temperatures, must differ in the two metals to such an extent, that the effect of a current in passing from cold to hot in one metal, together with the effect of an equal current passing from a place equally hot to a place equally cold in the other, may amount to the absorption or evolution, the existence of which has been demonstrated.

15. The *reversible thermal effect** of electric currents in single metals of non-uniform temperature, which has been thus established, may obviously be called a Convection of Heat by electricity in motion. To avoid circumlocution, I shall express it that *the Vitreous Electricity carries heat with it*, when this convection is in the "nominal direction of the current." On the other hand, when the convection is against the "nominal direction of the current," it will be said that *the Resinous Electricity carries heat with it*.

§§ 16—18. *Dynamical Theory applied to draw, from thermo-electric data, inferences regarding the Electric Convection of Heat in Copper and in Iron.*

16. The application of the preceding theorem to the particular case of copper and iron is a consequence of Cumming's discovery, that, if one junction in a circuit of two arcs of those two metals be kept cold, and the other be heated gradually, a current at first sets from copper to iron through the hot junction

* See an article by the author, entitled "On a Universal Tendency in Nature to the Dissipation of Mechanical Energy" (*Proceedings Roy. Soc. Edinb.*, Feb. 16, 1852, and *Phil. Mag.*, Oct. 1852) [Art. LIX. Vol. I. above], where all natural operations are divided into two great classes, "reversible" and "irreversible." See foot-note on § 2 above, for an example of the second class.

with increasing strength; but begins to diminish after a certain temperature, which Becquerel found to be about 300° Cent., is exceeded; falls away to nothing when a red heat is attained; and sets in the reverse* direction when the elevation of temperature is pushed higher.

Some observations of Regnault's having appeared to indicate 240° Cent. as, more nearly than 300°, the temperature of the hot junction which gives the current its maximum strength, I concluded the following proposition :—

17. "When a thermo-electric current passes through a piece of iron from one end kept at about 240° Cent.†, to the other end kept cold, in a circuit of which the remainder is copper, including a long resistance wire of uniform temperature throughout, or an electro-magnetic engine raising weights, there is heat evolved at the cold junction of the copper and iron, and (no heat being either absorbed or evolved at the hot junction) there must be a quantity of heat absorbed on the whole in the rest of the circuit. When there is no engine raising weights, in the circuit, the sum of the quantities evolved, at the cold junction, and generated in the resistance wire,' is equal to the quantity absorbed on the whole in the other parts of the circuit. When there is an engine in the circuit, the sum of the heat evolved at the cold junction and the thermal equivalent of the weights raised, is equal to the

---

* Having myself experienced some difficulty in obtaining the reverse current in the manner described by M. Becquerel, in which one junction was heated in the flame of a spirit-lamp, while the other was kept at the atmospheric temperature, I found that it could be obtained so as to be observed with the greatest ease by means of a very ordinary galvanometer and an iron wire with copper wires twisted round its ends, by keeping the lower junction at a temperature considerably above that of the atmosphere, at 100° Cent. for instance; and I ascertained that when both junctions were kept at a very high temperature, in the flame of a spirit-lamp for instance, and one of them cooled a little below the temperature of the other, the current produced was the reverse of that which the same difference occasioned when both junctions were at ordinary temperatures. See Part II. below for further developments on this subject.

† I have since ascertained (see Part II. below), by keeping the ends of an iron wire, with copper wires from the galvanometer soldered to them, in separate vessels of hot oil, and determining different temperatures of the two which give no current, that the neutral point for the particular specimens of iron and copper which I used must be about 284° Cent. I should therefore, at present, substitute 284° for 240° in the proposition quoted in the text; without further research, however, it is impossible to pronounce upon the limits between which the neutral point of various specimens of copper and iron wires may be found to lie.

quantity of heat absorbed on the whole in all the circuit, except the cold junction *."

18.   Hence, if the reversible part of the effect of a current from hot to cold in iron is an evolution of heat, the corresponding effect in copper must be a greater evolution of heat.   But if, on the other hand, a cooling effect be produced by a current from hot to cold in iron, there must be either a less effect of the same kind, or a reverse effect, in copper.   It is left to experiment to determine which of the two hypotheses is true regarding iron; and should it turn out to be the latter, to ascertain which of the two remaining alternatives regarding copper must be concluded. With this object I commenced the experimental researches which I now proceed to describe.

§§ 19—77.   *Experimental Investigation of the Electrical Convection of Heat in Copper, in Iron, and in some other Metals.*

§§ 19, 20.   *Unsuccessful attempts, and first result.*

19.   I began, more than four years ago, by observing carefully the ignition produced in short wires of copper, iron, and platinum by electric currents alternately in the two directions, thinking that some of the effects described by various experimenters, as showing a superior heating power in the positive electrode, might possibly be dependent on the convective agency which I was endeavouring to discover.   But I never observed the slightest variation in the position of the incandescent part of the wire, with a sudden reversal of the current.   Sometimes the incandescence was assisted by a spirit-lamp flame applied to the middle part of the wire, and the ends were kept cool by wet threads. Sometimes in a long wire with a current through it not quite strong enough to keep it at a red heat, a small part was made incandescent by a slight application of heat as nearly as possible at one point, by a spirit-lamp flame.   Still there was never observed the slightest motion of the incandescent part, when the current was suddenly reversed, and I concluded that whatever had been observed in the way of different heating effects of the positive and negative electrodes, must have been owing to peculiar agencies of the current in passing between metal and rarefied air,

---

* *Proceedings of the Royal Society of Edinburgh,* Dec. 15, 1851, republished in *Phil. Mag.,* June, 1852 [Art. XLVIII. Appendix Note I. Vol. I. above].

or to some other cause than thermal convection in metals; and I saw that more powerful tests would be required to bring out the result I looked for.

20. I next made experiments on a conductor of bar iron bent into two equal upward vertical branches on each side of the horizontal part, which was kept immersed in a vessel of hot oil, while the upper ends of the vertical branches were kept cool by streams of cold water. Vessels of water were applied round the two vertical branches, as calorimetric arrangements to test heat evolved or absorbed in them by the agency of a current sent down one and up the other from a nitric acid battery of sixteen small iron cells, arranged as a single element.

The current was sent first for half an hour in one direction, then half an hour in the contrary direction; and so on, with a reversal every half-hour. The water round the two vertical branches was kept constantly stirred, and thermometers in fixed positions in them were observed at frequent intervals during the experiments, which were each continued for about two hours. A comparison of all the readings taken showed a rather higher mean temperature in the branch down which the current was passing than in the other; indicating, differentially, a *cooling* effect in the branch through which the current passes from the hot middle, and a heating effect in the other. This experiment appeared to show that "the resinous electricity" carries heat with it in an iron conductor; but the irregular variations of temperature in each thermometer were so much greater than the differential effect deduced, that I could not consider the conclusion satisfactorily established.

§§ 21—29. *Unsuccessful attempts with large bar conductors.*

21. There were difficulties connected with the arrangements of the calorimetric vessel, which made me judge that it would be better, instead of testing the average temperature of two portions of the conductor, each extending the whole way from the hot middle to the cold ends, to simply test the temperature of as nearly as possible one point midway between the hot and cold on each side; and it appeared that the heating could be more easily applied and better regulated by a source of heat at the middle of a straight horizontal conductor, than by the plan I had followed in the arrangement just described. I therefore got bars of copper

and iron, with holes to admit the bulbs of sensitive thermometers, made to the following dimensions:—

	Copper conductor. inches.	Iron conductor. inches.
Whole length	16	24
Breadth	1	2
Depth	$2\frac{1}{2}$	3
Depth of hollows.........	$2\frac{1}{10}$	$2\frac{1}{2}$
Diameter of hollows ...	$\frac{1}{4}$	$\frac{1}{5}$

These relative dimensions were chosen so that the conducting
powers of the two bars for electric currents, and consequently for
heat also, might be not very unequal.

A vessel of tin-plate, perforated to admit the bar through its
sides, was soldered round the middle of each conductor, and two
others so as to leave about 2 inches at the ends of the conductor
projecting beyond them. The parts of the conductors within
these vessels were about 3 inches long in the copper and $4\frac{1}{2}$ inches
in the iron, and the parts between the middle vessel and the
vessels at the two sides were $2\frac{1}{2}$ and 2 inches respectively. The
bores for the thermometer bulbs were exactly in the middle of
the last-mentioned parts. In experimenting on either conductor,
the central vessel was generally filled with oil or water, and kept
hot by a gas-lamp below it. Streams of cold water from the town
supply-pipes were kept flowing through the two lateral vessels.

22. To make these streams constant, whatever variations of
pressure might occur in the supply pipes, a cistern in a fixed
position above the conductor was kept full (overflowing), and the
coolers were supplied by pipes from this cistern. The supply
often failed for several minutes, and sometimes for much longer;
and after an experiment (Nov. 19, 1853) was nearly lost from
this cause, a plan was arranged to lift water up from a larger
cistern (into which the exit-streams from the coolers were discharged), and to pour it into the smaller cistern above, so as to
keep the stream constant in quantity (although not quite invari-
able in temperature) even when the proper supply failed.

23. A galvanic battery for exciting a current through these
conductors was prepared, consisting at first of four, and ultimately
of eight, large iron cells, each measuring internally 12 inches
deep, $10\frac{1}{2}$ inches broad, and $2\frac{1}{4}$ inches from side to side; eight
porous cells, each 12 inches deep, 10 inches broad, and 2 inches

from side to side; and eight zinc plates, each $9\frac{1}{2}$ inches by 10 inches. The iron cells were charged with a mixture of nitric acid two parts (bulk), sulphuric acid three parts, and water two. The porous cells were charged with dilute sulphuric acid. In each of the cells there were $1\frac{1}{3}$ square feet of zinc surface exposed to $2\frac{1}{2}$ square feet of iron, and the electro-motive force was not far from double that of a single cell of Daniell's.

24. After preliminary experiments in which, with oil in the central vessel kept hot by a gas-lamp, the temperatures were too unsteady to allow any results of value to be obtained, water was substituted for oil in the central vessel, and was kept boiling briskly by the gas, the place of the water evaporated being frequently supplied by small quantities of boiling water poured in, so that ebullition never ceased. The irregularities having been found to be much diminished, experiments were made in the following manner.

25. Four of the large iron cells, arranged as a single galvanic element, were used to excite the current. The experiment lasted about two hours, during which the current was sent through the conductor for twenty minutes at a time, alternately in the two directions, twice in each direction. Several minutes were spent in changing the direction of the current; the stiffness of the electrodes, and the clamps used for the connexions which had to be changed, rendering the process very troublesome. Readings of the thermometers were taken at intervals of five minutes during the flow of the current in both directions, as well as for some time before the current commenced and after it ceased.

26. The results of this experiment manifested, among great irregularities in the indications of the thermometers, a very decided differential variation between the two every time the direction of the current was changed; and appeared so promising, that a series of further experiments on the same copper conductor, and on the iron conductor similarly arranged, were immediately commenced, for the purpose of testing decisively the conclusion which had been indicated, and for discovering the corresponding effect in iron. To avoid the loss of time and the derangement in the position of the conductor by the shifting of the heavy clamps and stiff electrodes between its ends, in changing the direction of the current, a commutator, by which the

change could be effected nearly instantaneously, was constructed on the following plan.

27. Four square holes, each of 1 inch, in a square block of mahogany, were fitted with bottoms of thick copper slabs, passing

Fig. 1.

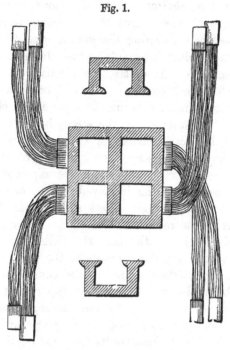

through the mahogany, and cemented with red lead so as to hold mercury, which was poured into each hole. The copper slabs projected outside to distances of about an inch, and each bore a bundle of 100 No. 18 copper wires soldered to it, two of which, connected with diagonally opposite copper slabs, served as battery electrodes, while the other two were clamped to the ends of the conductor to be tested. The four slabs have only to be connected by two conducting arcs parallel to one pair or to the other, of the sides of the square, to send the current one way or the other through the conductor; which was done by means of two heavy brass castings, as shown in the diagram (fig. 1). This commutator has been used in a considerable variety of experiments, and has been found very convenient. It gives the means of reversing almost instantaneously a very powerful current, without the necessity of bending any of the electrodes or deranging any part of the apparatus, and the conductors involved in it are so strong that it occasions very little resistance.

28. As the supposed differential effect had appeared not to be increased after the first five minutes of the flow of the current in either direction, shorter periods of various lengths were tried, and more frequent observations of the thermometers were made, for the purpose of discovering the gradual variation of the temperature in the conductor, towards its final distribution as affected by the current. Four more large iron cells were added to the battery, which made it consist in all of eight cells, arranged as a single galvanic element, exposing 20 square feet of iron to 10½ square feet of zinc surface. As the strength of current thus produced would be nearly double of that given in the previous experiment, any true effect of the kind sought would be augmented in the same ratio; and might be expected, both on this account and because of the improved system of observation, to become much more decided. These expectations, however, were not borne out by the results. The irregularities certainly became much diminished, but with these the differential effect on the thermometers, following the reversals of the current, either quite disappeared, or became very much less considerable than that which had been observed in the first experiment, and which I afterwards was led to attribute to some derangement in the position of the conductor occasioned by shifting the heavy clamps and stiff electrodes from between its two ends, causing the thermo meter bulbs to alter a little in their positions in the hollows.

29. Many experiments, both on the copper and iron conductor, were made, from October 1852 to March 1853, and the results of the observations (on each of the two principal thermometers either every half-minute, or every quarter-minute, during an experiment of about two hours) carefully reduced; with much labour at first when arbitrary scale thermometers were employed, but afterwards with far greater ease when centigrade thermometers, constructed for this investigation at the Kew Establishment, were received and brought into use.

In the months of September and October 1853 the investigation was taken up again. The thermometric observations which had been made in the previously completed experiments, were all reduced, on the plan of the Tables given (§§ 47 and 56) below for subsequent experiments, and, when thus tested, they appeared to contain some indications of the effect looked for. Several more sets of observations of the same kind were therefore executed, but with various modifications of details. Still no decided result could be obtained, and I concluded from all the experiments which had been made, that the anticipated effect must be too small to be discovered without either increasing the sensibility of the test or diminishing the irregularities. I therefore prepared new apparatus, by which the former, and as much as possible of the latter object, would be attained.

§§ 30—34. *Improvements and Modifications of Apparatus.*

30. Instead of increasing the power of the battery, which I reserved as a later resource, if necessary, or of increasing the length of the conductor between the heater and the coolers on each side, which, while it would increase in the same ratio the amount of the effect looked for, would increase in a duplicate ratio the time that would have to be given to allow it to reach a stated proportion of its limiting value, I had conductors made of about the same length as the others, but of considerably less section.

31. With a view to perfecting and testing the action of the heater and coolers, each conductor was made up of a number of

Fig. 2. Fig. 3.

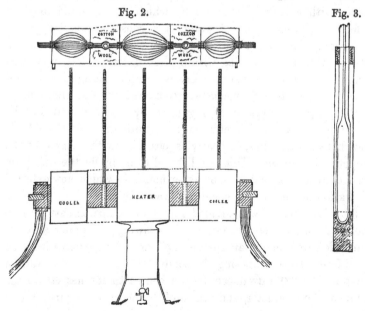

slips of flat sheet metal, bent and placed together, as shown in the accompanying diagram (fig. 2). The slips were held firmly together by a vice, while collars of sheet copper, separated from them by vulcanized india-rubber, were soldered round them in the places for the sides of the heater and coolers. Tin-plate vessels, as shown in the diagram, were then put together, and soldered to these collars. The interstices between the slips and the india-rubber, and the metal collars round the india-rubber, were stopped with red lead, and after some trouble were made water-tight. Thus the heater and coolers, without any metallic communication with the conductor, served the purpose of keeping the required supplies of hot and cold water round it in the proper places. The spaces for the thermometers were firmly stopped below with corks fitted to support the lower ends of the bulbs in perfectly fixed positions (fig. 3). Little collars of cork were put round the tubes just above the bulbs, and pushed down into the upper end of the hollow so as to hold the thermometers firmly and prevent all motion of their bulbs.

32. Various methods of heating the central part of the conductor were tried. First, as in previous experiments, water in the central vessel was kept boiling either by a gas-lamp under it, or by steam blown into it from a separate boiler; then a complex system with a boiling fountain, by which I attempted to get a perfectly uniform stream of water at a constant temperature, as little short of boiling as possible, to flow through the open spaces between the different slips within the central vessel, was used during several experiments. Lastly, water filling the central vessel was kept at a very constant temperature, near the boiling-point, by a gas-lamp below it, regulated by a person watching the indications of a thermometer with its bulb fixed in the middle space between the slips, as nearly as possible in the centre of the compound conductor. This last I found to be by far the best plan, and I used it in all subsequent experiments in which any external application of heat to the conductor was required.

33. Each cooler was divided into four compartments by partitions of tin-plate [not shown in the drawing], stopping all communication from one to another, except through the spaces between the different slips composing the conductor. A constant stream of cold water (§ 22), introduced by the compartment nearest to the middle of the conductor, and drawn off by an overflow pipe, from a

compartment next the end, was thus forced to flow all through among the different slips, and, as I found by placing thermometers in various positions in each compartment, gave a very satisfactory effect in fixing the temperature of the whole section of the conductor.

34. The experiments were made in other respects exactly as described above (§§ 28 and 29); the electric current, however, not being often again kept up for a longer time than ninety-six minutes, since the fumes, which always began to rise from the battery after the current had been flowing for about an hour, began after half an hour more to occasion great irregularities and inconvenience by causing the liquid (which sometimes became very hot), to foam and overflow in some of the iron cells. The atmosphere had been in previous experiments sometimes rendered intolerable for the observers, by the acid vapour; but this evil was done away by covering the battery with cloths kept moist with ammonia and water, and by moistening other surfaces in the neighbourhood in the same way, so that the fumes never got far without meeting vapour of ammonia and combining into white clouds, which were perfectly innocuous.

§§ 35—38. *First Experiments with Multiple Sheet Copper Conductor.*

35. The copper conductor on the new plan was first used in an experiment on the 28th of October, 1853; with the central vessel heated by steam, and a current from the eight large iron cells kept flowing for seventy-two minutes, alternately in contrary directions, six times six minutes each way. The thermometers were noted every half-minute. The observations thus recorded, when thoroughly examined, indicated a slight differential cooling effect in the part of the conductor in which the nominal current was from cold to hot, and a heating effect where it passed from hot to cold; that is to say, a convection of heat in the nominal direction of the current, or as I shall call it to avoid circumlocution, *a convection of heat by vitreous electricity.*

36. A second experiment with the same conductor was made on the 2nd of November, 1853, in which the current was kept flowing for ninety-six minutes, eight times six minutes each way, and the thermometers were noted every quarter-minute. An examination of the recorded results indicated still the same kind of effect, but to a much smaller extent. Thus the final average,

for the alteration of difference, between the temperatures at A
and B due to the flow of the current for six minutes in one
direction, after it had been flowing for six minutes in the contrary
direction, amounted to $°·039$ Cent. in the first experiment, and to
only $°·0143$ in the second experiment. A full analysis of the
progress of the differential variation of temperature during the
flow of the current is given in Tables I. and II., § 56 below, and
shows through what fluctuations the final alterations are reached.
The temperatures, at the ends of the successive times of flow in
one direction or the other, and the evaluation of the mean final
effect, are shown, for each experiment, in the following abridged
Tables.

37. The observations made during the first period (that is
the time from starting till the second reversal of the current) are
rejected from the average in every case of experiments on the new
conductors, because they were found to show so great absolute
elevations of temperature (due to the frictional generation of heat
by the current) that no alteration of difference between the thermometer observed during them could be relied on as an effect
depending on the direction of the current.

38. Conductor composed of thirteen slips of sheet copper.

Experiment I.   October 28th, 1853.

Periods.	(Current six times six minutes in each direction.) Differences of temperatures after six minutes of current entering		Augmentations of differences from middles to ends of periods.
	By end next A.	By end next B.	
	$T_A - T_B = D.$	$T_A - T_B = D'.$	$D' - D.$
I.	$\overset{\circ}{2}·18$	$\overset{\circ}{2}·09$	$-\overset{\circ}{·}09$
II.	2·07	2·10	·03
III.	2·15	2·11	$-$ ·04
IV.	2·16	2·13	$-$ ·03
V.	2·01	1·99	$-$ ·02
VI.	1·80	2·02	·22
Means for five periods...	2·038	2·070	·032

Augmentation of differ-
ence during periods } $-$ ·07
included ....)

Deduct average augmentation per half-
period .... } ·007

Effect due to reversal of current .... $0°·039$,
in favour of *Vitreous Electricity.*

Experiment II. November 2nd, 1853.

(Current eight times six minutes in each direction.) Temperatures and differences of temperature after six minutes of current entering

Periods.	By end next A.			By end next B.			Augmentations of differences from middles to ends of periods.
	T_A.	T_B.	$T_A - T_B = D$.	T_A.	T_B.	$T_A - T_B = D'$.	$D' - D$.
I.	53°·79	52°·43	1°·36	54°·00	52°·70	1°·30	-0°·06
II.	53·68	52·42	1·26	53·51	52·23	1·28	0·02
III.	53·86	52·57	1·29	53·79	52·63	1·16	-0·13
IV.	53·90	52·60	1·30	53·99	52·63	1·36	0·06
V.	54·09	52·72	1·37	54·05	52·68	1·37	0·00
VI.	55·10	54·00	1·10	54·15	52·86	1·29	0·19
VII.	53·99	52·80	1·19	54·10	52·80	1·30	0·11
VIII.	54·00	52·50	1·50	54·00	52·60	1·40	-0·10
Means for seven periods.	54·0886	52·8014	1·2871	53·9414	52·6329	1·3086	0·02143

Augmentation of difference during periods included 0·10
Deduct average augmentation per half-period .... 0·00714

Effect due to reversal of current .... 0°·01429, in favour of *Vitreous Electricity*.

§§ 39—43. *Decisive Experiments with Multiple Sheet Iron Conductor.*

39. These experiments seemed therefore on the whole to establish a probability in favour of the convection of heat by the so-called positive electricity, when a current is kept up through an unequally heated conductor. The convective effect, if of this

kind, ought (§ 18) to be less in iron than in copper, and I therefore had little expectation of finding an indication of it in the iron conductor which (§ 31) had been in the course of preparation; but as soon as it was ready for use I made the following experiments, and was much surprised by the result, which became manifest before the first of them was finished.

40. Conductor composed of thirty slips of sheet iron.

Experiment III. November 12th, 1853.

(Current six times eight minutes in each direction.) Temperatures and differences of temperature after eight minutes of current entering

Periods.	By end next A.			By end next B.			Augmentations of differences from middles to ends of periods.
	T_A.	T_B.	$T_B - T_A = D$.	T_A.	T_B.	$T_B - T_A = D'$.	$D' - D$.
I.	51°·43	53°·56	2°·13	51°·48	53°·49	2°·01	-0°·12
II.	51·62	53·30	1·68	51·41	53·21	1·80	0·12
III.	51·73	53·26	1·53	52·03	53·87	1·84	0·31
IV.	52·01	53·80	1·79	51·32	53·42	2·10	0·31
V.	51·30	53·00	1·70	51·00	52·95	1·95	0·25
VI.	51·14	52·98	1·84	50·69	52·80	2·11	0·27
Means for five periods...	51·56	53·268	1·708	51·29	53·25	1·96	0·252

Augmentation of difference during periods included 0·10
Deduct average augmentation per half-period .... 0·010

Effect due to reversal of current .... 0°·242, in favour of *Resinous Electricity*.

Experiment IV.   November 19th, 1853.

(Current seven times eight minutes in each direction.) Temperatures and differences of temperature after eight minutes of current entering

Periods.	By end next A.			By end next B.			Augmentations of differences from middles to ends of periods.
	T_A.	T_B.	$T_B - T_A = D$.	T_A.	T_B.	$T_B - T_A$.	$D' - D$.
I.	57°·50	59°·30	1°·80	58°·02	59°·84	1°·82	0°·02
II.*	48·20	51·15	2·95	46·82	49·79	2·97	0·02
III.	46·49	49·13	2·64	47·01	49·95	2·94	0·30
IV.	48·41	51·69	3·28	48·31	51·99	3·68	0·40
V.	48·36	51·74	3·38	48·18	51·80	3·62	0·24
VI.	48·00	51·20	3·20	48·00	51·49	3·49	0·29
VII.	48·31	51·51	3·20	48·06	51·60	3·54	0·34
Means for five periods...	49·3243	52·2457	3·14	49·20000	52·35143	3·454	0·314

Augmentation of difference during periods included. 0·57

Deduct average augmentation per half-period............ 0·057

Effect due to reversal of current .... 0°·257, in favour of *Resinous Electricity.*

* Rejected because of a failure in the water-supply through the coolers during the whole of Period I.

14—2

41. A full analysis of the differential variations throughout each of these experiments, derived from observations of the thermometer taken every quarter of a minute, was made in each case immediately after the conclusion of the experiment (see Tables I. and II. § 47 below), and was sufficient to convince me that the true effect in the iron conductor is of the kind indicated by the preceding summary of the effects apparent at the ends of the periods.

42. To try whether or not the very considerable effect thus discovered depended on some inequality in the conductor itself, I made an experiment on the 25th of November, 1853, exactly like the two preceding, with the exception that the middle vessel previously used as a heater was filled with cold water at the commencement. The current was sent six times eight minutes in each direction; the thermometers were noted every quarter of a minute; and the observations were reduced and compared in the usual way. The result gave no effect of the kind observed in the preceding experiments, but (probably because of a temporary failure in the water-supply for the coolers) showed, on the contrary, a deviation in the mean difference of temperature amounting to $0°\cdot029$ Cent., being about a tenth part of the amount of that effect, but in the opposite way according to the direction of the current through the conductor. Before the experiment was concluded boiling water was poured through the central vessel and left filling it, but with no lamp below. The two thermometers (A and B) being thus raised to about 27° Cent., the current was again started and was sent through the conductor for three times four minutes in each direction. The thermometers rose each nearly 2°, but fell again by nearly $5\frac{1}{2}°$ before the conclusion. The mean differential result, whether from these three periods (amounting to $0°\cdot05$ Cent.), or from the last two of them without the first ($0°\cdot025$ Cent.), was of the same kind as in the first two experiments. This experiment then conclusively demonstrated that the effect previously discovered was really owing to the heat in the central part of the conductor, and not to any inequality in the metal of the conductor itself, nor to any accidental disturbing agency.

43. It was thus established, that the *Resinous Electricity carries heat with it in an iron conductor.*

§§ 44, 45. *Experiment with Copper Conductor, repeated.*

44. The very small effect I had discovered of the opposite
kind in the copper conductor required confirmation; and indeed
the analysis of the progress of the variation (see Tables I. and II.
§ 56 below) was so unsatisfactory, that I felt it quite an open
question, whether it was the true effect, or merely an accidental
coincidence of irregularities; and I thought it improbable that
contrary effects should really exist in copper and in iron. I made
on the 26th of November another experiment on the copper conductor, with the current flowing six times eight minutes each way
(instead of eight times six minutes, as before, because the analysis
seemed to show that the effects which had chanced to appear in
the average results of the six minutes might disappear with
longer periods*); but I got still a very small result of the same
kind. The full analysis (Table III.) was equally unsatisfactory
with those of the two preceding experiments on the same conductor. The following numbers show the temperatures at the
reversals, and the final result, as in the previous abridged tables.

---

* I now believe that a true effect, amounting to from 0°·01 to 0°·02, was really
reached in three or four minutes, and that in the latter parts of the half-periods
there was no sensible augmentation of this effect, but only irregular fluctuations,
sometimes counteracting and reversing the true effect, but generally only diminishing it and increasing it alternately, and always maintaining, during the whole latter
half of the aggregate of the half-periods, an average deviation of the kind noted as
the final result. A careful consideration of the Tables I., II. and III. given below,
§ 56, for the copper conductor and of their graphical representation (see Diagram,
§ 57), is, I think, sufficient to establish this view. [April 9, 1856.]

45. Conductor composed of thirteen slips of sheet copper.

Experiment V. November 26, 1853.

(Current six times eight minutes in each direction.) Temperatures and differences of temperature after eight minutes of current entering

Periods.	By end next A.			By end next B.			Diminutions of differences from middles to ends of periods.
	T_A.	T_B.	$T_B - T_A = D$.	T_A.	T_B.	$T_B - T_A = D'$.	$D' - D$.
I.	50°·88	52°·78	1°·90	50°·88	52°·72	1°·92	− 0°·02
II.	50·64	52·49	1·85	50·53	52·32	1·79	+ 0·06
III.	50·38	52·29	1·91	50·01	52·00	1·99	− 0·08
IV.	50·14	52·08	1·94	49·90	51·83	1·93	+ 0·01
V.	49·60	51·61	2·01	49·48	51·52	2·04	− 0·03
VI.	49·11	51·22	2·11	48·80	50·92	2·12	− 0·01
Means for five periods...	49·974	51·938	1·964	49·744	51·718	1·974	− 0·01

Augmentation of difference during periods included 0·20
Add average augmentation per half-period............................... + 0·02

Effect due to reversal of current................................... 0°·01, in favour of *Vitreous Electricity.*

46. Another experiment was also made on the new iron conductor, and results as decisive as those in the first two experiments were obtained. The following abridged Table [Exper. VI.] shows sufficiently the character of the effect demonstrated; and the analysis of the progress of variation is given in the full table (Table III. § 47) below.

Experiment VI.   December 2, 1853.

Conductor composed of thirty slips of sheet iron.

Periods.	By end next A. (Current six times eight minutes in each direction.) Temperatures and differences of temperature after eight minutes of current entering			By end next B.			Augmentations of differences from middles to ends of periods.
	$T_{A'}$	$T_{B'}$	$T_B - T_A = D$	$T_{A'}$	$T_{B'}$	$T_B - T_A = D'$	$D' - D$
I.	54°·76	56°·33	1°·57	54°·80	56°·66	1°·86	0°·29
II.	54·97	56·68	1·71	54·89	56·80	1·91	0·20
III.	55·01	56·70	1·69	54·93	56·86	1·93	0·24
IV.	55·22	56·90	1·68	55·08	57·10	2·02	0·34
V.	55·31	57·08	1·77	55·07	57·12	2·05	0·28
VI.	55·12	57·00	1·88	54·84	57·03	2·19	0·31
Means for five periods...	55·126	56·872	1·746	54·962	56·982	2·020	0·274

Augmentation of difference $T_B - T_A$ during included periods .... } 0·33
Deduct average augmentation per half-period ....   0·033

Effect due to reversal of current .... 0°·241, in favour of Resinous Electricity.

47.   The following Tables [I., II., and III.] show the progress of variation of the difference between the temperatures of the two tested localities (A, B) of the iron conductor, during each of the three regular experiments [Expers. III., IV., and VI.] referred to above, as derived directly from the quarter-minute or half-minute observations actually made in the course of each experiment.

TABLE I.

Conductor composed of thirty slips of sheet iron. Middle

Streams of water at temperature

Temperatures at middle points A, B of the parts between heater and coolers:

Augmentation of difference $T_B - T_A$ (in hundredths of a degree Cent.), during Periods	1. First half-minute of current entering by end next		2. Second half-minute of current entering by end next		3. Third half-minute of current entering by end next		4. Fourth half-minute of current entering by end next		5. Fifth half-minute of current entering by end next		6. Sixth half-minute of current entering by end next	
	A.	B.	A.	B.	A.	B.	A.	B.	A.	B.	A.	B.
I.	0	− 2	− 1	1	2	0	0	0	2	0	0	1
II.	− 4	1	− 4	− 2	− 2	7	− 6	4	− 1	1	− 1	1
III.	0	0	− 1	4	− 6	2	− 2	4	− 2	3	0	− 1
IV.	2	− 2	− 1	2	− 2	3	− 3	5	− 3	4	− 1	5
V.	1	5	− 3	3	− 5	0	− 9	3	− 6	5	− 5	0
VI.	− 2	2	− 3	0	1	3	− 3	3	− 4	1	− 1	6
Augmentations during half-minutes, summed for five periods	− 3	6	− 12	7	14	15	− 23	19	− 16	14	− 8	+ 11
Differences of augmentation during corresponding half-minutes of first and second halves of a period, summed for five periods..............	9		19		29		42		30		19	
Differences of augmentation in equal intervals from beginning and middle of a period, summed for five periods	9		28		57		99		129		148	
Mean augmentation of difference in favour of thermometer next entering current	0°·009		0°·028		0°·057		0°·099		0°·129		0°·148	
......after reversal and flow for	½ min.		1 min.		1½ min.		2 min.		2½ min.		3 min.	

November 12, 1853.

of conductor in water kept hot by steam blown into it.

6° running through the coolers.

—Initial $T_A=51°·23$, $T_B=53°·30$; Final $T_A=50°·69$, $T_B=52°·30$.

7.		8.		9.		10.		11.		12.		13.		14.		15.		16.	
Seventh half-minute of current entering by end next		Eighth half-minute of current entering by end next		Ninth half-minute of current entering by end next		Tenth half-minute of current entering by end next		Eleventh half-minute of current entering by end next		Twelfth half-minute of current entering by end next		Thirteenth half-minute of current entering by end next		Fourteenth half-minute of current entering by end next		Fifteenth half-minute of current entering by end next		Sixteenth half-minute of current entering by end next	
A.	B.	A.	B.	A.	B.	A.	B.	A.	B.	A.	B.	A.	B.	A.	B.	A.	B.	A.	B.
0	3	− 3	− 1	5	− 1	5	− 1	1	− 1	1	− 3	0	− 2	− 1	− 5	− 1	0	− 4	1
− 2	5	− 1	2	− 2	− 1	1	− 5	− 10	2	− 2	− 2	− 1	0	− 5	− 2	− 1	1	8	0
− 6	3	0	2	− 3	0	− 2	4	− 3	2	− 1	4	5	0	− 3	− 1	− 3	1	0	4
− 1	4	0	5	2	− 2	1	2	0	3	− 2	− 6	− 1	2	1	− 3	2	10	1	− 1
− 2	3	− 1	0	− 3	0	− 1	1	− 3	0	1	3	− 2	− 3	0	1	− 2	− 1	0	5
0	0	− 1	4	0	− 1	− 2	0	0	4	0	4	1	0	0	1	2	0	1	0
− 11	15	− 3	13	− 6	− 4	− 3	2	− 16	11	− 4	3	2	− 1	− 7	− 4	− 2	11	10	8
26		16		2		5		27		7		− 3		3		13		− 2	
174		190		192		197		224		231		228		231		244		242	
0°·174		0°·190		0°·192		0°·197		0°·224		0°·231		0°·228		0°·231		0°·231		0°·242	
3½ min.		4 min.		4½ min.		5 min.		5½ min.		6 min.		6½ min.		7 min.		7½ min.		8 min.	

[For TABLES II. AND III. see after page 218.]

48. The gradual augmentation of the difference $T_A - T_B$ from its value at a time when the current had been flowing for eight

Mean augmentation of difference $T_A - T_B$	Time from instant of reversal in quarter-minutes
0°·000	0
0·001	1
0·006	2
0·0185	3
0·032	4
0·042	5
0·054	6
0·0695	7
0·091	8
0·0985	9
0·1167	10
0·1235	11
0·138	12
0·1435	13
0·1578	14
0·1590	15
0·175	16
0·1825	17
0·1883	18
0·196	19
0·202	20
0·212	21
0·2167	22
0·2195	23
0·229	24
0·225	25
0·2263	26
0·2305	27
0·2343	28
0·2355	29
0·2403	30
0·245	31
0·2467	32

Current entering by end next A

minutes entering by the end next B, consequent upon reversing the current and letting it flow continuously entering by the end next A, is shown by the numbers at the foot of each table, as a mean result derived from a single experiment. The mean of the

Continuation of § 47.   *To follow Table I.*, p. 217.]

Conductor composed of

Temperatures at middle points A,

Augmentations of difference $T_B - T_A$ (in hundredths of a degree Cent.), during Periods	1. First quarter-minute of current entering by end next		2. Second quarter-minute of current entering by end next		3. Third quarter-minute of current entering by end next		4. Fourth quarter-minute of current entering by end next		5. Fifth quarter-minute of current entering by end next		6. Sixth quarter-minute of current entering by end next		7. Seventh quarter-minute of current entering by end next		8. Eighth quarter-minute of current entering by end next	
	A.	B.	A.	B.	A.	B.	A.	B.	A.	B.	A.	B.	A.	B.	A.	B.
I.	-3	-2	-1	-1	-4	-2	-4	-1	4	-2	-4	4	3	1	-1	2
II*.	3	5	-1	1	0	-5	6	4	5	-2	15	1	7	0	10	-2
III.	0	1	-2	3	-3	-3	-1	1	-1	1	0	-3	-1	5	-3	2
IV.	0	-3	0	1	3	3	-3	2	5	2	-1	3	4	4	5	1
V.	1	0	-1	0	-4	3	-3	2	-2	1	0	3	-1	1	-1	1
VI.	0	2	-3	2	-5	3	-2	3	0	1	-3	1	-2	3	-6	3
VII.	0	1	1	0	-3	1	-2	0	-2	2	-2	3	-1	3	0	2
Augmentations during quarter-minutes, summed for five periods	1	1	-5	6	-12	7	-11	8	0	7	-6	7	-1	16	-5	9
Differences of augmentation during corresponding quarter-minutes of first and second halves of a period, summed for five periods ...	0		11		19		19		7		13		17		14	
Differences of augmentation in equal intervals from beginning and middle of a period, summed for five periods	0		11		30		49		56		69		86		100	
Mean augmentation of difference in favour of thermometer next entering current)	0°·000		0°·011		0°·030		0°·049		0°·056		0°·069		0°·086		0°·100	
...... after reversal and flow for	¼ min.		½ min.		¾ min.		1 min.		1¼ min.		1½ min.		1¾ min.		2 min.	

After the conclusion of this experiment the current was started when the heater and coolers were kept regular, on either thermomet

* Period II. rejected because the water-supply, wh

TABLE II. November 19th, 1853.

thirty slips of sheet iron. Middle of conductor in water kept boiling b

Streams of water at temperature 6° running through the coolers.

of the parts between heater and coolers:—Initial $T_A = 56°\cdot44$, $T_B = 58°$

	9. Ninth quarter-minute of current entering by end next		10. Tenth quarter-minute of current entering by end next		11. Eleventh quarter-minute of current entering by end next		12. Twelfth quarter-minute of current entering by end next		13. Thirteenth quarter-minute of current entering by end next		14. Fourteenth quarter-minute of current entering by end next		15. Fifteenth quarter-minute of current entering by end next		16. Sixteenth quarter-minute of current entering by end next		17. Seventeenth quarter-minute of current entering by end next		18. Eighteenth quarter-minute of current entering by end next		19. Nineteenth quarter-minute of current entering by end next		20. Twentieth quarter-minute of current entering by end next		21. Twenty-first quarter-
A.	B.	A.	B.	A.	B.	A.	B.	A.	B.	A.	B.	A.	B.	A.	B.	A.	B.	A.	B.	A.	B.	A.	B.	A.
2	-2	1	1	-1	2	1	-1	-3	1	4	3	-4	2	1	-5	2	2	-3	2	3	-1	1	1	(
-8	2	-6	9	8	-5	0	0	13	-1	6	-3	4	4	2	2	2	-1	10	-1	-2	-1	-3	2	1
4	-2	-2	3	1	2	-2	-1	0	0	0	0	0	-3	0	-3	0	2	10	0	-7	0	5	-3	0
4	0	6	1	-3	2	4	4	4	2	4	0	-3	0	2	2	-1	1	5	3	0	2	2	1	
1	1	0	1	-1	1	-2	0	-3	1	-2	2	-1	3	-2	1	-3	0	0	1	0	2	0	0	
1	0	-2	1	-1	1	-2	-1	-1	2	1	0	-2	0	0	1	0	1	-2	1	0	3	-1	2	
1	2	-1	4	-2	3	-1	2	-2	1	0	2	-1	-2	-1	0	-2	3	-1	1	-1	1	0	1	(
11	1	1	10	-6	9	-3	4	-2	6	3	4	-10	1	-4	4	-4	15	2	-1	-1	12	-2	4	-8
12		9		15		7		8		1		11		8		19		-3		13		6		
112		121		136		143		151		[152		163		171		190		187		200		206		2
0°·112		0°·121		0°·136		0°·143		0°·151		0°·152		0°·163		0°·171		0°·190		0°·187		0°·200		0°·206		0°
min.		2½ min.		2¾ min.		3 min.		3¼ min.		3½ min.		3¾ min.		4 min.		4¼ min.		4½ min.		4¾ min.		5 min.		5¼

one direction and broken, and started in the other direction and br
amounted to about 1°·4.

had failed during the whole of Period I., only commenced giving a stream through the co

Table II. November 19th, 1853.

thirty slips of sheet iron. Middle of conductor in water kept boiling

Streams of water at temperature 6° running through the coolers.

B of the parts between heater and coolers: —Initial $T_A = 56°·44$, $T_B = \ $

9.		10.		11.		12.		13.		14.		15.		16.		17.		18.		19.		20.	
Ninth quarter-minute of current entering by end next		Tenth quarter-minute of current entering by end next		Eleventh quarter-minute of current entering by end next		Twelfth quarter-minute of current entering by end next		Thirteenth quarter-minute of current entering by end next		Fourteenth quarter-minute of current entering by end next		Fifteenth quarter-minute of current entering by end next		Sixteenth quarter-minute of current entering by end next		Seventeenth quarter-minute of current entering by end next		Eighteenth quarter-minute of current entering by end next		Nineteenth quarter-minute of current entering by end next		Twentieth quarter-minute of current entering by end next	
A.	B.	A.	B.	A.	B.	A.	B.	A.	B.	A.	B.	A.	B.	A.	B.	A.	B.	A.	B.	A.	B.	A.	B.
2	−2	1	1	−1	2	1	−1	−3	1	4	3	−4	2	1	−5	2	2	−3	2	3	−1	1	1
3	2	−6	9	8	−5	0	0	13	−1	6	−3	4	2	2	−1	10	−1	−2	−1	−3	2	−1	−2
−4	−2	−2	3	1	2	−2	−1	0	0	0	0	0	3	−3	0	2	10	0	−7	0	5	−3	0
−4	0	6	1	−3	2	4	4	4	2	4	0	−3	0	2	2	−1	1	5	3	0	2	2	1
−1	1	0	1	−1	1	−2	0	−3	1	−2	2	−1	3	−2	1	−3	0	0	1	0	1	0	0
−1	0	−2	1	−1	1	−2	−1	−1	2	1	0	−5	0	0	1	0	1	−2	1	0	3	−1	2
−1	2	−1	4	−2	3	−1	2	−2	1	0	2	−1	−2	−1	0	−2	3	−1	1	−1	1	0	1
−11	1	1	10	−6	9	−3	4	−2	6	3	4	−10	1	−4	4	−4	15	2	−1	−1	12	−2	4
12		9		15		7		8		1		11		8		19		−3		13		6	
112		121		136		143		151		152		163		171		190		187		200		206	
0°·112		0°·121		0°·136		0°·143		0°·151		0°·152		0°·163		0°·171		0°·190		0°·187		0°·200		0°·206	
2¼ min.		2½ min.		2¾ min.		3 min.		3¼ min.		3½ min.		3¾ min.		4 min.		4¼ min.		4½ min.		4¾ min.		5 min.	

in one direction and broken, and started in the other direction and

·r amounted to about 1°·4.

ich had failed during the whole of Period I., only commenced giving a stream through the steam blown into it.

8; Final $T_A = 48°·06$, $T_B = 51°·59$.

…ering by end next B.	22. Twenty-second quarter-minute of current entering by end next		23. Twenty-third quarter-minute of current entering by end next		24. Twenty-fourth quarter-minute of current entering by end next		25. Twenty-fifth quarter-minute of current entering by end next		26. Twenty-sixth quarter-minute of current entering by end next		27. Twenty-seventh quarter-minute of current entering by end next		28. Twenty-eighth quarter-minute of current entering by end next		29. Twenty-ninth quarter-minute of current entering by end next		30. Thirtieth quarter-minute of current entering by end next		31. Thirty-first quarter-minute of current entering by end next		32. Thirty-second quarter-minute of current entering by end next	
B.	A.	B.	A.	B.	A.	B.	A.	B.	A.	B.	A.	B.	A.	B.	A.	B.	A.	B.	A.	B.	A.	B.
-1	-5	2	2	-2	2	0	0	1	-1	-3	-1	-1	2	-1	2	1	-1	-1	2	1	0	2
1	5	-1	1	0	-1	1	-12	0	5	-2	5	-2	-7	-1	1	0	16	2	-11	0	-4	1
1	1	-7	2	8	2	7	-1	-3	-2	-1	-3	5	0	0	2	-2	-3	0	-2	5	-3	0
1	4	1	-1	0	2	0	-1	0	1	3	2	2	-5	1	4	1	-1	1	-2	-1	4	0
-1	1	0	0	0	-1	0	0	1	0	0	1	-1	-1	2	-2	0	0	0	0	0	-1	0
-1	-2	0	-1	0	-1	1	0	1	3	0	0	1	-1	-2	-2	1	-2	-2	-1	1	-2	0
2	-2	0	-1	-1	0	0	-2	0	0	-1	-3	1	0	1	1	0	0	0	1	1	-1	1
2	2	-6	-1	7	2	8	-4	-1	2	1	-3	8	-7	2	3	0	-6	-1	-4	6	-3	1
	-8		8		6		3		-1		11		9		-3		5		10		4	
	205		213		219		222		221		232		241		238		243		253		257	
	0°·205		0°·213		0°·219		0°·222		0°·221		0°·232		0°·241		0°·238		0°·243		0·253		0°·257	
	5½ min.		5¾ min.		6 min.		6¼ min.		6½ min.		6¾ min.		7 min.		7¼ min.		7½ min.		7¾ min.		8 min.	

en, several times; and it was found that its absolute heating effect,

s at the commencement of the second Period.

Continuation of § 47. *To follow Table II.*, p. 217.]

Conductor

Temperatures at middle points A

Augmentations of difference $T_B - T_A$ (in hundredths of a degree Cent.), during Periods.	1. First quarter-minute of current entering by end next		2. Second quarter-minute of current entering by end next		3. Third quarter-minute of current entering by end next		4. Fourth quarter-minute of current entering by end next		5. Fifth quarter-minute of current entering by end next		6. Sixth quarter-minute of current entering by end next		7. Seventh quarter-minute of current entering by end next	
	A.	B.	A.	B.	A.	B.	A.	B.	A.	B.	A.	B.	A.	B.
I.	0	0	1	2	0	1	-2	2	-1	1	3	1	-1	2
II.	4	0	-1	0	-2	1	-1	3	-2	1	-1	1	-2	2
III.	1	0	0	0	0	0	-1	1	0	1	-1	0	0	2
IV.	0	1	0	1	-2	0	1	0	-2	1	-4	1	-3	3
V.	0	1	-1	-3	0	3	-1	4	-1	0	0	-1	-1	4
VI.	0	0	-1	0	0	1	-1	1	-1	0	-1	0	0	0
Augmentations during quarter-minutes, summed for five periods	5	2	-3	-2	-4	5	-3	9	-6	3	-7	1	-6	11
Differences of augmentation during corresponding quarter-minutes of first and second halves of a period, summed for five periods	-3		1		9		12		9		8		17	
Differences of augmentation in equal intervals from beginning and middle of a period, summed for five periods......	-3		-2		+7		19		28		36		53	
Mean augmentation of difference in favour of thermometer next entering current	0°·003		0°·002		0°·007		0°·019		0°·023		0°·036		0°·053	
...... after reversal and flow for	¼ min.		½ min.		¾ min.		1 min.		1¼ min.		1½ min.		1¾ min	

TABLE III. December 2nd, 1853.

...osed of thirty slips of sheet iron.   Middle of conductor at temperat...

Streams of water at temperature 7°·5 running through the coolers.

of the parts between heater and coolers:—Initial $T_A = 54°·20$, $T_B = 55°$...

	9.		10.		11.		12.		13.		14.		15.		16.		17.		18.		19.		20.	
	Ninth quarter-minute of current entering by end next		Tenth quarter-minute of current entering by end next		Eleventh quarter-minute of current entering by end next		Twelfth quarter-minute of current entering by end next		Thirteenth quarter-minute of current entering by end next		Fourteenth quarter-minute of current entering by end next		Fifteenth quarter-minute of current entering by end next		Sixteenth quarter-minute of current entering by end next		Seventeenth quarter-minute of current entering by end next		Eighteenth quarter-minute of current entering by end next		Nineteenth quarter-minute of current entering by end next		Twentieth quarter-minute of current entering by end next	
	A.	B.	A.	B.	A.	B.	A.	B.	A.	B.	A.	B.	A.	B.	A.	B.	A.	B.	A.	B.	A.	B.	A.	B
2	2	2	0	2	-2	2	3	3	-1	1	0	2	-2	0	3	0	2	-1	1	1	0	1	-1	1
	0	1	-1	0	0	0	1	0	-4	2	-1	1	0	0	-1	0	-1	1	0	1	0	0	-1	1
	0	3	0	3	-1	3	-1	2	-2	1	-3	0	-4	1	-1	1	0	0	-2	1	-1	1	-1	1
	0	3	-2	2	-2	1	-2	2	-1	0	-1	2	-1	0	0	1	-1	1	0	2	0	0	-1	2
-1	1		-2	2	-1	1	-1	2	-2	-1	-1	1	-1	0	-2	1	-3	1	-1	1	0	0	-1	1
-1	1		-1	2	0	2	0	3	-2	0	0	1	0	1	-1	1	-1	2	-1	2	-1	3	0	2
8	-2	9	-6	9	-4	7	-3	9	-11	2	-6	5	-6	2	-5	4	-6	5	-4	7	-2	4	-4	7
	11		15		11		12		13		11		8		9		11		11		6		11	
	85		100		111		123		136		147		155		164		175		186		192		203	
4	0°·085		0°·100		0°·111		0°·123		0°·136		0°·147		0°·155		0°·164		0°·175		0°·186		0°·192		0°·20	
	2¼ min.		2½ min.		2¾ min.		3 min.		3¼ min.		3½ min.		3¾ min.		4 min.		4¼ min.		4½ min.		4¾ min.		5 mi	

TABLE III. December 2nd, 1853.

mposed of thirty slips of sheet iron.   Middle of conductor at tempe

Streams of water at temperature 7°·5 running through the coolers.

B of the parts between heater and coolers:—Initial $T_A = 54°·20$, $T_B =$

8.	9.		10.		11.		12.		13.		14.		15.		16.		17.		18.		19.		2
minute of current entering by end next	Ninth quarter-minute of current entering by end next		Tenth quarter-minute of current entering by end next		Eleventh quarter-minute of current entering by end next		Twelfth quarter-minute of current entering by end next		Thirteenth quarter-minute of current entering by end next		Fourteenth quarter-minute of current entering by end next		Fifteenth quarter-minute of current entering by end next		Sixteenth quarter-minute of current entering by end next		Seventeenth quarter-minute of current entering by end next		Eighteenth quarter-minute of current entering by end next		Nineteenth quarter-minute of current entering by end next		Twentieth quarter-
B.	A.	B.	A.	B.	A.	B.	A.	B.	A.	B.	A.	B.	A.	B.	A.	B.	A.	B.	A.	B.	A.	B.	A.
2	2	2	0	2	−2	2	3	3	−1	1	0	2	−2	0	3	0	2	−1	1	1	0	1	−
1	0	1	−1	0	0	0	1	0	−4	2	−1	1	0	0	−1	0	−1	1	0	1	0	0	−
3	0	3	0	3	−1	3	−1	2	−2	1	−3	0	−4	1	−1	1	0	0	−2	1	−1	1	−
2	0	3	−2	2	−2	1	−2	2	−1	0	−1	2	−1	0	0	1	−1	1	0	2	0	0	−
5	−1	1	−2	2	−1	1	−1	2	−2	−1	−1	1	−1	0	−2	1	−3	1	−1	1	0	0	
2	−1	1	−1	2	0	2	0	3	−2	0	0	1	0	1	−1	1	−1	2	−1	2	−1	3	
13	−2	9	−6	9	−4	7	−3	9	−11	2	−6	5	−6	2	−5	4	−6	5	−4	7	−2	4	−
21	11		15		11		12		13		11		8		9		11		11		6		
74	85		100		111		123		136		147		155		164		175		186		192		
·074	0°·085		0°·100		0°·111		0°·123		0°·136		0°·147		0°·155		0°·164		0°·175		0°·186		0°·192		0
min.	2¼ min.		2½ min.		2¾ min.		3 min.		3¼ min.		3½ min.		3¾ min.		4 min.		4¼ min.		4½ min.		4¾ min.		5

); Final, $T_A = 54°\cdot84$, $T_B = 57°\cdot03$.

21.		22.		23.		24.		25.		26.		27.		28.		29.		30.		31.		32.	
Twenty-first quarter-minute of current entering by end next		Twenty-second quarter-minute of current entering by end next		Twenty-third quarter-minute of current entering by end next		Twenty-fourth quarter-minute of current entering by end next		Twenty-fifth quarter-minute of current entering by end next		Twenty-sixth quarter-minute of current entering by end next		Twenty-seventh quarter-minute of current entering by end next		Twenty-eighth quarter-minute of current entering by end next		Twenty-ninth quarter-minute of current entering by end next		Thirtieth quarter-minute of current entering by end next		Thirty-first quarter-minute of current entering by end next		Thirty-second quarter-minute of current entering by end next	
A.	B.	A.	B.	A.	B.	A.	B.	A.	B.	A.	B.	A.	B.	A.	B.	A.	B.	A.	B.	A.	B.	A.	B.
-1	1	1	1	3	1	3	-1	0	-1	0	-1	-1	0	1	2	-1	-1	-1	1	0	1	0	1
0	0	1	0	0	1	0	0	0	0	0	1	0	0	1	1	0	0	0	0	0	0	0	0
-1	1	0	1	-1	0	0	0	-1	-2	-1	-2	0	1	1	0	0	0	0	0	1	0	0	-1
0	0	0	2	-1	1	0	1	0	1	-1	0	0	1	1	0	0	0	0	1	0	0	-1	1
-2	1	-5	1	0	0	1	1	1	1	1	1	0	-1	0	0	0	0	1	1	0	-1	1	1
-2	1	-1	0	0	1	0	0	-1	0	0	1	0	-2	0	1	-1	1	0	0	0	3	0	1
-5	3	-6	4	-2	3	1	2	-1	0	-1	1	0	-1	1	3	-1	1	1	2	0	3	0	4
8		10		5		1		1		2		-1		2		2		1		3		4	
211		221		226		227		228		230		229		231		233		234		237		241	
°·211		0°·221		0°·226		0°·227		0°·228		0°·230		0°·229		0°·231		0°·233		0°·234		0°·237		0°·241	
¼ min.		5½ min.		5¾ min.		6 min.		6¼ min.		6½ min.		6¾ min.		7 min.		7¼ min.		7½ min.		7¾ min.		8 min.	

results of the three experiments is shown by the following numbers, and is exhibited by a curve in the Diagram (fig. 4) of § 57 below.

49.   That Vitreous Electricity carries heat with it in copper is indicated by each of the three experiments on the thirteen-slip conductor adduced above, but by so narrow an effect; amounting on an average to only 0°·02 Cent., which corresponds to a reading of half that amount, or $\frac{1}{100}$th of a degree, being $\frac{1}{10}$th of a division on the scale of each thermometer; with such discrepancies among the results of the different experiments (Oct. 28th, effect 0°·039, Nov. 2nd, 0°·0143, Nov. 26th, 0°·01); and with so great fluctuations in the course of each experiment (see Tables I., II., III., § 56 below), that I did not venture to draw from them so seemingly improbable a conclusion, as that the convective effects in copper and in iron should be in contrary directions.   The dynamic theory (§ 18) was fully satisfied by the demonstration which the experiments gave, that the convective effect is undoubtedly in iron a *conveying of heat in the direction of the Resinous Electricity*, and that it is *less in amount* in copper, whether in the same direction as in iron or in the contrary direction.   But it was still an object of great interest, (in fact an object of much greater interest than any verification of conclusions from the dynamic theory, which were in reality as certain before as after the experiments directly demonstrating them,) to ascertain the actual nature of the convective effect in copper, and I therefore endeavoured to make more decisive experiments for discovering it.

50.   The three experiments which had been made, were quite sufficient to prove that the convective effect, whatever its true nature might be, was nearly insensible to my thermometers without either more powerful currents or a more sensitive conductor. To work with more powerful currents would have increased immensely the labour of carrying out the experiments, and would besides have involved a large addition to the battery which had been used hitherto.   I preferred therefore to make the conductor more sensitive, which I saw could be done by diminishing the body of metal in the tested parts, and so preventing the looked-for thermal effect from being so much conducted away from the localities of the thermometers as it had been.   I accordingly had several slips of copper cut away from each side of the conductor in the parts between the heater and the coolers, leaving the parts within these vessels unchanged.

51. Several experiments were made on the conductor thus reduced successively to smaller and smaller numbers of slips; but the results did not appear much more decided than they had been in the experiments on the unreduced conductor, until it was tried with all the slips but two cut away. Thus with four slips left, the following results were obtained :—

52. Copper conductor reduced to four slips.

Experiment VII. February, 1854.

(Current six times eight minutes in each direction.) Temperatures and differences of temperature after eight minutes of current entering

Periods.	By end next A.			By end next B.			Augmentations of differences from middles to ends of periods.
	T_A.	T_B.	$T_A - T_B = D$.	T_A.	T_B.	$T_A - T_B = D'$.	$D' - D$.
I.	46°·00	45°·24	0°·76	45°·92	45°·27	0°·65	-0°·11
II.	45·82	45·19	0·63	46·01	45·43	0·58	- 0·05
III.	46·00	45·50	0·50	46·02	45·54	0·48	- 0·02
IV.	46·20	45·77	0·43	46·21	45·76	0·45	0·02
V.	46·18	45·80	0·38	46·09	45·71	0·38	0·00
VI.	46·09	45·72	0·37	46·05	45·68	0·37	0·00
Means for five periods...	46·058	45·596	0·462	46·076	45·624	0·452	- 0·010

Diminution of difference during periods included... 0·28

Add average diminution per half-period............ + 0·028

Effect due to reversal of current............ 0°·018, in favour of *Vitreous Electricity.*

The effect shown here is of the same kind as had been found in all the previous experiments, but was still too small to be very satisfactory. Some unknown cause made the difference $T_A - T_B$ to diminish so much through the whole experiment as to overpower the apparent tendency of the current from B to A to increase it, and the abridged table has on this account a very unsatisfactory appearance as regards the conclusion drawn from it after the proper correction for their diminution is applied: but the full examination of the progress of variation in the course of the experiment shown in Table IV. below is much less unsatisfactory, and shows undoubtedly the true convective effect in copper.

53. The following results [Exper. VIII.], derived from the first experiment made on the conductor reduced to two slips, show a very marked increase in the effect, and make the result quite apparent even without the full analysis given below in Table V., § 56.

Experiment VIII.   February 23rd, 1854.

Copper conductor reduced to two slips.

Periods.	(Current seven times six minutes in each direction.) Temperatures and differences of temperature after eight minutes of current entering						Diminutions of differences from middles to ends of periods.
	By end next A.			By end next B.			
	T_A.	T_B.	$T_B - T_A = D$.	$T_{A'}$.	$T_{B'}$.	$T_B - T_A = D'$.	$D' - D$.
I.	54°·81	56°·31	1°·50	55°·47	56°·88	1°·41	0°·09
II.	55·90	57·32	1·42	56·37	57·63	1·26	0·14
III.	56·70	57·99	1·29	57·16	58·29	1·13	0·16
IV.	57·49	58·70	1·21	57·80	58·91	1·11	0·10
V.	57·80	58·98	1·18	58·18	59·20	1·02	0·16
VI.	58·29	59·80	1·51	58·37	59·80	1·43	0·08
VII.	58·27	59·80	1·53	58·08	59·50	1·42	0·11
Means for six periods ...	57·4083	58·7650	1·3567	57·663	58·888	1·2283	0·1283

Augmentation of differences during periods included 0·01

Add average augmentation per half-period .... 0·0008

Effect due to reversal of current .... $\overline{0°·1291,}$

in favour of *Vitreous Electricity*.

54. The effect here shown is, as regards amount, unmistakeable, and it appeared quite to establish the conclusion I had not
ventured to draw from the previous experiments on the copper
conductor. As it was clear that diminution in the power of the
conductor had now begun to augment its sensibility for the effect
under investigation, I had it further reduced by cutting away
from its breadth, above and below, between the heaters and the
bends for the thermometer-bulbs on the two sides, and on the

other sides of those bends as far as the coolers. An experiment was then made which led to the following results:—

Experiment IX. March 7th, 1854.

Copper conductor diminished in breadth.

(Current seven times six minutes in each direction.) Temperatures and differences of temperature after six minutes of current entering

Periods.	By end next A.			By end next B.			Diminutions of differences from middles to ends of periods.
	T_A.	T_B.	$T_B - T_A = D$.	T_A.	T_B.	$T_B - T_A = D'$.	$D' - D$.
I.	56°·61	58°·00	1°·39	57°·56	58°·89	1°·42	−0°·03
II.	58·19	59·69	1·50	58·74	60·01	1·27	0·23
III.	59·12	60·70	1·58	59·61	61·08	1·47	0·11
IV.	59·91	61·53	1·62	60·18	61·75	1·57	0·05
V.	60·50	62·18	1·68	61·15	62·53	1·38	0·30
VI.	61·71	63·02	1·31	61·87	62·91	1·04	0·27
VII.	61·07	62·80	1·73	61·07	62·62	1·55	0·18
Means for six periods...	60·083	61·653	1·57	60·4367	61·8167	1·38	0·19

Augmentation of differences during periods included 0·13 = 0·01083

Add average augmentation per half-period ....

0°·20083

After this experiment I considered it quite established that *Vitreous Electricity carries heat with it in copper.*

55. The conductor was still further diminished in breadth (so as to be only an inch broad in the parts between the heater and coolers on each side), and an experiment was made before my

class on the 19th April, 1854, leading to the following results, shown as in the abridged tables of the preceding experiments.

Experiment X. April 19th, 1854.

Copper conductor of two slips, further diminished in breadth.

(Current seven times six minutes each way.) Temperatures and differences of temperatures after six minutes of current entering

Periods.	By end next A.			By end next B.			Diminutions of differences from middles to ends of periods.
	T_A.	T_B.	$T_B - T_A = D$.	T_A.	T_B.	$T_B - T_A = D'$.	$D - D'$.
I.	74°·30	76°·60	2°·30	74°·81	77°·50	2°·69	-0°·39
II.	73·80	76·48	2·68	75·32	78·10	2·78	-0·10
III.	76·25	79·42	3·17	76·17	79·18	3·01	-0·16
IV.	76·33	79·51	3·18	76·28	79·37	3·09	0·09
V.	75·60	78·69	3·09	75·20	78·07	2·87	0·22
VI.	74·80	77·70	2·90	75·00	77·75	2·75	0·15
VII.	74·10	76·84	2·74	75·42	78·20	2·78	-0·04
Means, Period I. off	75·147	78·107	2·96	75·565	78·445	2·88	0·08

Augmentation of differences during periods included .... 0·09
Add average augmentation per half-period .... 0·0075
Effect due to reversal of current .... 0·0875

| Means, Periods I. and VII. off | 75°·356 | 78°·36 | 3°·004 | 75°·594 | 78°·494 | 2°·90 | 0°·104 |

Augmentation of differences during periods included .... 0·06
Deduct average augmentation per half-period .... 0·006
Effect due to reversal of current .... 0°·110

The effect here obtained, although of quite a decisive character, does not appear to show any increased sensibility resulting from the further diminution in the breadth of the conductor.

56. The following Tables [I.—VII.] show a complete analysis of the results of the seven experiments [Expers. I., II., V., VII., VIII., IX. and X.] on the copper conductor which have been adduced.

57. The average progress towards the final effect of a reversal, as indicated by the numbers at the ends of these Tables, is exhibited graphically for the copper conductor in the different states in which it was used in the experiments, in the following diagram, along with the curve exhibiting the corresponding reverse effect in the iron conductor.

Fig. 4.

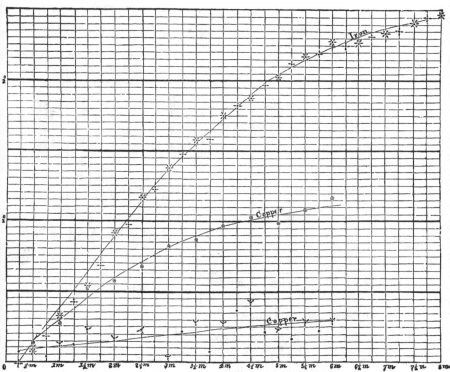

The uppermost curve represents the results of three experiments with the Iron conductor (thirty slips), the points marked ✳ representing the mean of three days' observation, and the points marked ┼ that of two days' observation. The lowest curve represents the results of three experiments with the Copper conductor (thirteen slips), the points marked ⋎ denoting three days', and the simple dots • two days' observation. The middle curve represents three experiments with the Copper conductor (two slips), the points marked ⊙ denoting the mean of three days' observation.

TABLE I.

Conductor composed of thirteen slips of sheet copper.

Streams of water at temperature

Temperatures at middle points A, B of the parts between heater and

Augmentations of difference $T_A - T_B$ (in hundredths of a degree Cent.), during Periods.	1. First half-minute of current entering by end next		2. Second half-minute of current entering by end next		3. Third half-minute of current entering by end next		4. Fourth half-minute of current entering by end next	
	A.	B.	A.	B.	A.	B.	A.	B.
I.	− 5	10	12	2	− 23	17	16	− 16
II.	− 1	− 6	− 4	7	6	2	2	− 3
III.	2	5	6	− 10	0	13	0	4
IV.	− 5	0	1	7	4	− 7	6	− 1
V.	− 1	15	7	− 7	− 8	1	8	0
VI.	− 8	10	9	8	− 5	12	1	− 3
Augmentations during half-minutes, summed for five periods	− 13	24	19	5	− 3	21	17	− 3
Differences of augmentation during corresponding half-minutes of first and second halves of a period, summed for five periods	37		− 14		24		− 20	
Differences of augmentation in equal intervals from beginning and middle of a period, summed for five periods................................	37		23		47		27	
Mean augmentation of difference in favour of thermometer remote from entering current	0°·037		0°·023		0°·047		0°·027	
...............after reversal and flow for	½ min.		1 min.		1½ min.		2 min.	

October 28th, 1853.

Middle of conductor in water kept boiling.

6° running through the coolers.

coolers:—Initial $T_A = 55°·20$, $T_B = 52°·94$; Final $T_A = 54°·70$, $T_B = 52°·68$.

5.		6.		7.		8.		9.		10.		11.		12.	
Fifth half-minute of current entering by end next		Sixth half-minute of current entering by end next		Seventh half-minute of current entering by end next		Eighth half-minute of current entering by end next		Ninth half-minute of current entering by end next		Tenth half-minute of current entering by end next		Eleventh half-minute of current entering by end next		Twelfth half-minute of current entering by end next	
A.	B.	A.	B.	A.	B.	A.	B.	A.	B.	A.	B.	A.	B.	A.	B.
−17	4	22	−5	−13	2	−4	2	8	−14	−10	1	2	−4	4	−8
−2	9	−2	−6	−2	−1	−6	7	5	−1	0	3	−5	−8	7	0
11	−4	−2	−15	−12	3	9	−3	−21	1	13	1	−5	0	4	1
−6	3	6	−2	5	−8	−5	14	1	3	0	−11	0	4	−2	−5
1	−1	−1	−11	−7	6	−3	−2	−6	−4	3	1	3	5	−8	−5
3	3	−10	−9	1	5	1	2	0	3	−7	−2	2	−6	−6	−1
7	10	−9	−43	−15	5	−4	18	−21	2	9	−8	−5	−5	−5	−10
3		−34		20		22		23		−17		0		−5	
30		−4		16		38		61		44		44		39	
0°·030		−0°·004		0°·016		0°·038		0°·061		0°·044		0°·044		0°·039	
2½ min.		3 min.		3½ min.		4 min.		4½ min.		5 min.		5½ min.		6 min.	

[For TABLES II. AND III. see after p. 236.]

15—2

TABLE IV.

Copper conductor reduced to four slips.

Streams of water at temperature

Temperatures at middle points A, B of the parts between heaters and

Diminutions of difference $T_B - T_A$ (in hundredths of a degree Cent.), during Periods.	1. First half-minute of current entering by end next		2. Second half-minute of current entering by end next		3. Third half-minute of current entering by end next		4. Fourth half-minute of current entering by end next		5. Fifth half-minute of current entering by end next		6. Sixth half-minute of current entering by end next	
	A.	B.	A.	B.	A.	B.	A.	B.	A.	B.	A.	B.
I.	−1	4	−1	5	−2	7	1	4	1	6	−2	5
II.	4	−4	−1	−3	1	2	3	1	−3	2	6	1
III.	−1	−1	3	0	−1	2	−1	1	1	1	0	0
IV.	−2	0	−1	3	3	2	−2	−2	−1	−2	2	1
V.	−1	1	2	0	1	0	1	1	−1	1	0	1
VI.	−1	0	−1	2	1	−1	0	1	−1	1	0	0
Diminutions during half-minutes summed for five periods	−1	−4	2	2	5	5	1	2	−5	3	8	3
Differences of diminution during corresponding half-minutes of first and second halves of a period, summed for five periods..............	−3		0		0		1		8		−5	
Differences of diminution in equal intervals from beginning and middle of a period, summed for five periods	−3		−3		−3		−2		6		1	
Mean augmentation of difference in favour of thermometer remote from entering current	−0°·003		−0°·003		−0°·003		−0°·002		0°·006		0°·001	
.....after reversal and flow for	½ min.		1 min.		1½ min.		2 min.		2½ min.		3 min.	

February, 1854.

Stream of water at 60° Cent. through middle of conductor.

5°·9 running through coolers.

coolers:—Initial $T_A = 44°·17$, $T_B = 45°·36$; Final $T_A = 45°·68$, $T_B = 46°·05$.

7.		8.		9.		10.		11.		12.		13.		14.		15.		16.	
Seventh half-minute of current entering by end next		Eighth half-minute of current entering by end next		Ninth half-minute of current entering by end next		Tenth half-minute of current entering by end next		Eleventh half-minute of current entering by end next		Twelfth half-minute of current entering by end next		Thirteenth half-minute of current entering by end next		Fourteenth half-minute of current entering by end next		Fifteenth half-minute of current entering by end next		Sixteenth half-minute of current entering by end next	
A.	B.	A.	B.	A.	B.	A.	B.	A.	B.	A.	B.	A.	B.	A.	B.	A.	B.	A.	B.
0	7	0	1	2	0	-1	0	2	1	0	3	2	1	4	-2	4	0	2	1
0	2	-5	0	-1	1	-1	0	1	0	1	0	-1	0	2	1	-1	0	2	-1
0	2	0	2	-2	1	0	-1	1	0	1	1	-1	0	0	1	2	-1	0	0
-1	3	0	2	0	-1	0	-1	-1	1	1	0	0	1	-1	-1	1	0	0	-1
0	2	-1	0	0	0	1	0	0	0	-1	0	1	0	-1	1	-1	0	0	0
0	-1	0	-1	0	0	-1	0	0	-1	0	1	0	0	3	0	0	0	0	0
-1	8	-6	3	-3	1	-1	-2	1	0	0	2	-1	1	3	2	1	-1	-2	-2
9		9		4		-1		-1		2		2		-1		-2		-4	
10		19		23		22		21		23		25		24		22		18	
0°·010		0°·019		0°·023		0°·022		0°·021		0°·023		0°·025		0°·024		0°·022		0°·018	
3½ min.		4 min.		4½ min.		5 min.		5½ min.		6 min.		6½ min.		7 min.		7½ min.		8 min.	

TABLE V.

Copper conductor reduced to two slips.

Streams of water at temperature

Temperatures at middle points A, B of the parts of conductor between heater

Diminutions of difference $T_B - T_A$ (in hundredths of a degree Cent.), during Periods.	1. First half-minute of current entering by end next		2. Second half-minute of current entering by end next		3. Third half-minute of current entering by end next		4. Fourth half-minute of current entering by end next	
	A.	B.	A.	B.	A.	B.	A.	B.
I.	− 5	− 7	− 14	5	− 14	4	1	6
II.	− 5	1	4	5	− 5	4	2	0
III.	− 1	3	1	5	− 3	5	2	− 1
IV.	− 3	2	0	− 1	0	5	− 5	1
V.	1	0	− 1	3	0	5	0	2
VI.	− 3	− 1	− 3	6	− 9	0	− 5	2
VII.	− 1	3	3	2	− 2	1	− 1	− 3
Diminutions during half-minutes, summed for six periods	− 12	6	4	20	− 19	20	− 7	1
Differences of diminution during corresponding half-minutes of first and second halves of a period, summed for six periods	18		16		39		8	
Differences of diminution in equal intervals from beginning and middle of a period, summed for six periods	18		34		73		81	
Mean augmentation of difference in favour of thermometer remote from entering current	0°·015		0°·0283		0°·06088		0°·0675	
.... after reversal and flow for	½ min.		1 min.		1½ min.		2 min.	

February 23rd, 1854.

Water kept at temperature 99°·5 Cent. in heater.

5°·6 through coolers.

and coolers:—Initial $T_A = 51°·23$, $T_B = 52°·48$; Final $T_A = 58°·10$, $T_B = 59°·61$.

5.		6.		7.		8.		9.		10.		11.		12.	
Fifth half-minute of current entering by end next		Sixth half-minute of current. entering by end next		Seventh half-minute of current entering by end next		Eighth half-minute of current entering by end next		Ninth half-minute of current entering by end next		Tenth half-minute of current entering by end next		Eleventh half-minute of current entering by end next		Twelfth half-minute of current entering by end next	
A.	B.	A.	B.	A.	B.	A.	B.	A.	B.	A.	B.	A.	B.	A.	B.
2	0	3	−1	−7	−1	−1	2	−4	−3	5	4	−6	−2	5	2
−4	2	4	2	3	0	3	−1	−2	3	1	−2	−2	−1	0	5
0	3	−3	−1	3	0	−1	0	−2	−1	3	3	0	1	−2	−1
2	0	1	−1	0	4	1	0	−3	−1	2	−3	−2	1	−1	3
−6	1	−3	2	−1	1	2	−1	−1	−1	1	3	0	0	1	1
−6	−2	−12	0	−1	2	−4	2	−1	1	−5	−2	1	2	−1	−2
−6	−6	0	−1	0	2	0	1	0	0	0	1	0	2	−3	9
−20	−2	−13	1	4	9	1	1	−9	1	2	0	−3	5	−6	15
18		14		5		0		10		−2		8		21	
99		113		118		118		128		126		134		155	
0°·0825		0°·09416		0°·0983		0°·0983		0°·107		0°·105		0°·11175		0°·1291	
2½ min.		3 min.		3½ min.		4 min.		4½ min.		5 min.		5½ min.		6 min.	

TABLE VI.

Copper conductor to two slips, and

Water kept at temperature 99° Cent. in heater.

Temperatures at middle points A, B of the parts of the conductor between heater

Diminutions of difference $T_B - T_A$ (in hundredths of a degree Cent.), during Periods.	1. First half-minute of current entering by end next		2. Second half-minute of current entering by end next		3. Third half-minute of current entering by end next		4. Fourth half-minute of current entering by end next	
	A.	B.	A.	B.	A.	B.	A.	B.
I.	− 1	− 5	− 4	4	− 26	2	− 13	− 2
II.	− 2	2	0	7	0	4	− 2	1
III.	− 3	2	− 3	6	− 3	1	− 2	− 2
IV.	1	10	− 4	1	0	7	− 2	− 11
V.	1	4	− 1	− 1	− 2	5	− 3	− 3
VI.	1	2	2	3	− 8	3	1	7
VII.	− 12	− 1	− 4	6	− 16	1	− 15	1
Diminutions during half-minutes, summed for six periods ....}	− 14	19	− 10	22	− 29	21	− 23	− 7
Differences of diminution during corresponding half-minutes of first and second halves of a period, summed for six periods ....}	33		32		50		16	
Differences of diminution in equal intervals from beginning and middle of a period, summed for six periods ....}	33		65		115		131	
Mean augmentation of difference in favour of thermometer remote from entering current...............}	0°·0275		0°·0542		0°·0958		0°·1092	
..............after reversal and flow for	½ min.		1 min.		1½ min.		2 min.	

March 7th, 1854.

diminished in breadth to 1½ inch.

Streams of water about temperature 6° through coolers.

and coolers:—Initial $T_A = 50°·33$, $T_B = 50°·70$; Final $T_A = 61°·07$, $T_B = 62°·62$.

5.		6.		7.		8.		9.		10.		11.		12.	
Fifth half-minute of current entering by end next		Sixth half-minute of current entering by end next		Seventh half-minute of current entering by end next		Eighth half-minute of current entering by end next		Ninth half-minute of current entering by end next		Tenth half-minute of current entering by end next		Eleventh half-minute of current entering by end next		Twelfth half-minute of current entering by end next	
A.	B.	A.	B.	A.	B.	A.	B.	A.	B.	A.	B.	A.	B.	A.	B.
− 38	2	− 9	− 1	− 7	− 4	0	3	− 7	0	− 3	− 1	0	− 1	6	0
1	1	− 1	1	− 3	1	4	2	− 2	3	1	− 4	0	2	− 4	3
− 3	5	− 4	− 2	1	− 1	− 4	− 3	− 2	1	0	0	− 7	2	− 1	2
− 2	− 2	− 4	2	3	3	− 3	− 3	0	− 1	− 2	0	0	1	− 2	− 2
6	3	− 9	9	2	1	− 5	9	2	− 2	5	− 7	− 4	8	− 3	4
7	− 2	− 3	3	5	2	− 4	4	2	0	4	0	− 1	4	1	1
− 8	5	− 4	3	− 4	0	0	4	− 9	0	− 3	0	− 3	− 1	9	0
1	10	− 25	16	4	6	− 12	13	− 9	1	5	− 11	− 15	16	0	8
9		41		2		25		10		− 16		31		8	
140		181		183		208		218		202		233		241	
0°·1167		0°·1508		0°·1525		0°·1733		0°·1817		0°·1683		0°·1933		0°·2008	
2½ min.		3 min.		3½ min.		4 min.		4½ min.		5 min.		5½ min.		6 min.	

TABLE VII.

Copper conductor two slips further diminished in breadth.

Streams of water at about

Temperature at middle points A, B of the parts of the conductor between heater

Diminutions of difference $T_B - T_A$ (in hundredths of a degree Cent.), during Periods.	1. First half-minute of current entering by end next		2. Second half-minute of current entering by end next		3. Third half-minute of current entering by end next		4. Fourth half-minute of current entering by end next	
	A.	B.	A.	B.	A.	B.	A.	B.
I.	0	− 25	− 11	− 12	− 40	− 2	− 29	8
II.	− 1	− 2	2	12	− 1	1	6	− 5
III.	− 2	− 5	− 2	1	− 8	12	− 5	6
IV.	− 2	0	− 2	5	1	2	− 2	4
V.	1	1	6	7	− 1	7	− 1	3-
VI.	0	7	5	4	0	5	− 1	1
VII.	1	3	− 5	1	4	6	0	4
Diminutions during half-minutes, summed for six periods	− 3	4	4	30	− 5	33	− 3	13
Differences of diminution during corresponding half-minutes of first and second halves of a period, summed for six periods	7		26		38		16	
Differences of diminution in equal intervals from beginning and middle of a period, summed for six periods	7		33		71		87	
Mean augmentation of difference in favour of thermometer remote from entering current	0°·0058		0°·0275		0°·0582		0°·0725	

Periods I. and VII. off.

	1.		2.		3.		4.	
Diminutions during half-minutes, summed for five periods	− 4	− 1	9	29	− 9	27	− 3	9
Differences of diminution during corresponding half-minutes of first and second halves of a period, summed for five periods	5		20		36		12	
Differences of diminution in equal intervals from beginning and middle of a period, summed for five periods	5		25		61		73	
Mean augmentation of difference in favour of thermometer remote from entering current	0°·005		0°·025		0°·061		0°·073	
............after reversal and flow for	½ min.		1 min.		1½ min.		2 min.	

Water kept at temperature 99° Cent. in heater.
temperature 6° through coolers.
and coolers:—Initial $T_A=63°·65$, $T_B=63°·81$; Final $T_A=72°·90$, $T_B=75°·05$.

5.		6.		7.		8.		9.		10.		11.		12.	
Fifth half-minute of current entering by end next		Sixth half-minute of current entering by end next		Seventh half-minute of current entering by end next		Eighth half-minute of current entering by end next		Ninth half-minute of current entering by end next		Tenth half-minute of current entering by end next		Eleventh half-minute of current entering by end next		Twelfth half-minute of current entering by end next	
A.	B.	A.	B.	A.	B.	A.	B.	A.	B.	A.	B.	A.	B.	A.	B.
− 37	− 2	− 24	− 6	− 15	11	− 39	1	− 9	0	− 11	− 3	− 9	− 7	18	− 2
− 2	2	− 7	− 3	3	− 6	1	− 5	− 2	− 2	2	− 3	1	1	− 1	0
− 4	− 1	− 4	4	7	1	− 14	0	− 2	1	1	− 1	− 5	− 1	− 1	− 1
− 2	− 3	− 6	1	− 7	− 1	0	− 3	− 1	2	0	1	3	0	1	1
0	6	1	1	− 1	− 2	− 4	4	− 1	− 2	0	4	1	− 6	− 1	− 1
− 1	0	− 2	2	1	2	− 4	− 1	− 2	− 2	− 2	− 3	− 1	3	4	− 3
3	5	− 6	− 2	4	− 5	− 4	8	53	− 16	− 5	− 5	3	− 3	− 47	0
− 6	9	− 24	3	7	− 11	− 25	3	45	− 19	− 4	− 7	2	− 6	− 45	− 4
15		27		− 18		28		− 64		− 3		− 8		41	
102		129		111		139		75		72		64		105	
0°·0850		0°·1075		0°·0925		0°·1158		0°·0625		0°·0600		0°·0533		0°·0875	

A.	B.	A.	B.	A.	B.	A.	B.	A.	B.	A.	B.	A.	B.	A.	B.
− 9	4	− 18	5	3	− 6	− 21	− 5	− 8	− 3	1	− 2	− 1	− 3	2	− 4
13		23		− 9		16		5		− 3		− 2		− 6	
86		109		100		116		121		118		116		110	
0°·086		0°·109		0°·100		0°·116		0°·121		0°·118		0°·116		0°·110	
2½ min.		3 min.		3½ min.		4 min.		4½ min.		5 min.		5½ min.		6 min.	

58. The diminution of the conducting power in the copper conductor had so markedly augmented the looked-for indication of a convective effect, that it was to be expected a corresponding augmentation might be obtained by treating the iron conductor similarly. Instead, however, of cutting up the iron conductor, which, as it stood, possessed sensibility enough to give a very decided result, I prepared a new iron conductor on a much smaller scale. It appeared that the smaller the conducting power for the same strength of current, and the same difference of temperatures between hot and cold, the greater would be the indication of convective effect; and the greatest indication would therefore be obtained by reducing the conductor so much that the current through it would generate heat enough to keep up the required difference of temperatures without any external heater.

59. The new conductor was therefore made of just two slips of sheet iron broad enough to admit the whole length of the

Fig. 5.

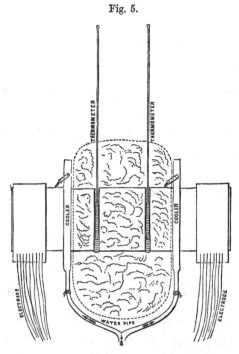

thermometer-bulbs in the same manner as in the conductor previously used; these slips were bent in the places for the thermo-

Conductor composed of thir[

Temperatures of middle points A, B

	1. First quarter-minute of current entering by end next		2. Second quarter-minute of current entering by end next		3. Third quarter-minute of current entering by end next		4. Fourth quarter-minute of current entering by end next		5. Fifth quarter-minute of current entering by end next	
Augmentations of difference $T_A - T_B$ (in hundredths of a degree Cent.), during Periods.	A.	B.	A.	B.	A.	B.	A.	B.	A.	B.
I.	−6	−7	5	1	0	−4	−3	−2	1	−5
II.	0	13	0	−1	−8	−3	4	−6	10	−3
III.	7	−1	6	0	−5	3	−6	7	−2	0
IV.	1	−1	3	4	0	−1	1	−2	−1	−1
V.	−5	9	2	−5	13	0	−5	7	−2	−1
VI.	4	10	1	−15	1	18	−2	2	−2	0
VII.	−5	9	−1	3	−5	−10	4	−1	7	5
VIII.	7	−9	−2	−1	−5	0	−1	−2	−8	4
Augmentations during quarter-minutes, summed for seven periods	9	30	9	−15	−9	7	−5	5	2	4
Differences of augmentation during corresponding quarter-minutes of first and second halves of a period, summed for seven periods	21		−24		16		10		2	
Differences of augmentation in equal intervals from beginning and middle of a period, summed for seven periods	21		−3		13		23		25	
Mean augmentation of difference in favour of thermometer remote from entering current......	0°·015		−0°·002		0°·009		0°·016		0°·018	
...........after reversal and flow for	¼ min.		½ min.		¾ min.		1 min.		1¼ min.	

TABLE II. November 2nd, 1853.

n slips of sheet copper. Middle of conductor in water kept boiling

Streams of water at temperature 10°·4 running through the coolers.

f the parts between heater and coolers:—Initial $T_A = 54°·10$, $T_B = 5$[...]

	7.		8.		9.		10.		11.		12.		13.		14.		15.	
	Seventh quarter-minute of current entering by end next		Eighth quarter-minute of current entering by end next		Ninth quarter-minute of current entering by end next		Tenth quarter-minute of current entering by end next		Eleventh quarter-minute of current entering by end next		Twelfth quarter-minute of current entering by end next		Thirteenth quarter-minute of current entering by end next		Fourteenth quarter-minute of current entering by end next		Fifteenth quarter-minute of current entering by end next	
B.	A.	B.	A.	B.	A.	B.	A.	B.	A.	B.	A.	B.	A.	B.	A.	B.	A.	B
9	0	22	-1	-5	-1	0	2	-11	5	3	-6	10	-1	9	-7	-6	18	-
4	-1	2	-4	7	2	9	-5	-10	4	-1	0	-4	4	-2	-2	5	0	-
8	7	-1	11	-9	-3	7	-7	11	3	-3	-6	-11	-2	2	6	7	14	-
2	-2	9	7	-3	10	-2	-3	-5	-7	6	6	0	-5	0	-4	1	4	-
6	5	-1	-3	8	-4	2	-7	-7	0	-5	3	-8	-2	8	0	-4	8	
0	-3	3	3	-1	10	0	-6	0	10	-3	-10	1	-46	7	-24	5	57	-
9	5	4	-8	15	-4	1	-6	-7	2	-2	4	4	5	-8	-8	0	-1	
5	12	-1	-10	-5	2	-6	-1	0	0	-5	-1	1	0	-2	3	1	3	
20	23	15	-4	12	13	11	-35	-18	12	-13	-4	-17	-46	5	-29	15	85	-
	-8		16		-2		17		-25		-13		51		44		-86	
	4		20		18		35		10		-3		48		92		6	
	0°·003		0°·014		0°·013		0°·025		0°·007		-0°·002		0°·034		0°·066		0°·004	
	1¾ min.		2 min.		2¼ min.		2½ min.		2¾ min.		3 min.		3¼ min.		3½ min.		3¾ min.	

TABLE II. November 2nd, 1853.

…en slips of sheet copper. Middle of conductor in water kept boilin…

Streams of water at temperature 10°·4 running through the coolers.

…of the parts between heater and coolers:—Initial $T_A = 54°·10$, $T_B =$

	7.		8.		9.		10.		11.		12.		13.		14.		15
minute of current entering by next	Seventh quarter-minute of current entering by end next		Eighth quarter-minute of current entering by end next		Ninth quarter-minute of current entering by end next		Tenth quarter-minute of current entering by end next		Eleventh quarter-minute of current entering by end next		Twelfth quarter-minute of current entering by end next		Thirteenth quarter-minute of current entering by end next		Fourteenth quarter-minute of current entering by end next		Fifteenth quarter-minute of current
B.	A.	B.	A.	B.	A.	B.	A.	B.	A.	B.	A.	B.	A.	B.	A.	B.	A.
− 9	0	22	− 1	− 5	− 1	0	2	−11	5	3	− 6	10	− 1	9	− 7	−6	18
− 4	− 1	2	− 4	7	2	9	− 5	−10	4	1	0	− 4	4	−2	− 2	5	0
− 8	7	− 1	11	− 9	− 3	7	− 7	11	3	− 3	− 6	−11	− 2	2	6	7	14
2	− 2	9	7	− 3	10	− 2	− 3	− 5	− 7	6	6	0	− 5	0	− 4	1	4
− 6	5	− 1	− 3	8	− 4	2	− 7	− 7	0	− 5	3	⊢ 8	− 2	8	0	−4	8
0	− 3	3	3	− 1	10	0	− 6	0	10	− 3	−10	1	−46	7	−24	5	57
− 9	5	4	− 8	15	− 4	1	− 6	− 7	2	− 2	4	4	5	−8	− 8	0	− 1
5	12	− 1	−10	− 5	2	− 6	− 1	0	0	− 5	− 1	1	0	−2	3	1	3
− 20	23	15	− 4	12	13	11	−35	−18	12	−13	− 4	−17	−46	5	−29	15	85
13	−8		16		−2		17		−25		−13		51		44		− 8
12	4		20		18		35		10		−3		48		92		6
009	0°·003		0°·014		0°·013		0°·025		0°·007		−0°·002		0°·034		0°·066		0°·0
min.	1¾ min.		2 min.		2¼ min.		2½ min.		2¾ min.		3 min.		3¼ min.		3½ min.		3¾

ot by steam blown into it.

80; Final $T_A = 54°·68$, $T_B = 53°·38$.

16.		17.		18.		19.		20.		21.		22.		23.		24.	
Sixteenth quarter-minute of current entering by end next		Seventeenth quarter-minute of current entering by end next		Eighteenth quarter-minute of current entering by end next		Nineteenth quarter-minute of current entering by end next		Twentieth quarter-minute of current entering by end next		Twenty-first quarter-minute of current entering by end next		Twenty-second quarter-minute of current entering by end next		Twenty-third quarter-minute of current entering by end next		Twenty-fourth quarter-minute of current entering by end next	
A.	B.	A.	B.	A.	B.	A.	B.	A.	B.	A.	B.	A.	B.	A.	B.	A.	B.
− 3	4	− 5	− 4	− 2	3	10	6	− 4	− 5	4	9	− 6	− 13	7	− 2	4	11
− 5	6	− 4	9	10	0	6	− 2	3	− 2	− 7	− 5	− 7	1	− 4	− 6	4	1
− 10	− 10	− 4	− 3	− 4	− 9	− 2	7	5	3	10	− 1	− 2	− 4	− 12	4	4	− 4
5	8	4	3	− 4	− 4	1	− 8	− 4	1	− 2	13	3	− 1	10	− 5	− 3	− 3
− 4	7	− 7	3	1	− 8	13	2	− 4	− 12	− 4	8	− 6	0	0	− 2	6	1
− 12	− 1	− 9	− 4	− 17	0	24	10	− 3	3	4	− 6	3	− 3	− 2	4	− 5	− 4
− 1	6	7	− 1	5	− 1	− 3	− 3	8	− 1	− 9	13	− 3	− 3	− 7	2	0	− 8
3	− 1	4	6	− 1	2	− 3	2	2	4	13	0	− 9	0	6	− 4	0	− 1
− 24	15	− 9	13	− 10	− 20	36	4	7	− 4	5	22	− 21	− 10	− 9	− 7	6	− 18
39		22		− 10		− 32		− 11		17		11		2		− 24	
45		67		57		25		14		31		42		44		20	
0°·032		0°·048		0°·041		0°·018		0°·010		0°·022		0°·030		0°·031		0°·014	
4 min.		4¼ min.		4½ min.		4¾ min.		5 min.		5¼ min.		5½ min.		5¾ min.		6 min.	

Conduct

Temperatures at the middle points A

Diminutions of difference $T_B - T_A$ (in hundredths of a degree Cent.), during Periods.	1. First quarter-minute of current entering by end next		2. Second quarter-minute of current entering by end next		3. Third quarter-minute of current entering by end next		4. Fourth quarter-minute of current entering by end next		5. Fifth quarter-minute of current entering by end next		6. Sixth quarter-minute of current entering by end next		7. Seventh quarter-minute of current entering by end next		Eighth quarter-
	A.	B.	A.	B.	A.	B.	A.	B.	A.	B.	A.	B.	A.	B.	A
I.	-1	-3	0	2	3	0	0	1	1	2	1	1	-1	1	-
II.	0	0	0	0	0	1	0	1	0	0	1	1	0	0	
III.	1	-1	2	-2	0	1	-1	0	-1	0	-1	0	1	0	
IV.	0	0	0	-1	-1	0	0	1	0	0	0	0	1	-2	
V.	1	0	0	0	-1	1	0	-1	-2	0	-2	-1	-2	0	-
VI.	0	0	0	2	-1	-1	0	-1	-2	1	-1	0	-1	-1	
Diminutions during quarter-minutes, summed for five periods..	2	-1	2	-1	-3	2	-1	0	-5	1	-3	0	-1	-3	-
Differences of diminution during corresponding quarter-minutes of first and second halves of a period, summed for five periods.	-3		-3		5		1		6		3		-2		
Differences of diminution in equal intervals from beginning and middle of a period, summed for five periods................	-3		-6		-1		0		6		9		7		
Mean augmentation of difference in favour of thermometer remote from entering current......	-0°·003		-0°·006		-0°·001		0°·000		0°·006		0°·009		0°·007		C
............after reversal and flow for	¼ min.		½ min.		¾ min.		1 min.		1¼ min.		1½ min.		1¾ min.		2

TABLE III. November 26th, 1853.

composed of thirteen slips of sheet copper.   Middle of conductor at 9…

B of the parts between heater and coolers:—Initial $T_A = 50°·96$, $T_B = $ …

	9.		10.		11.		12.		13.		14.		15.		16.		17.		18.		19.		20.	
	Ninth quarter-minute of current entering by end next		Tenth quarter-minute of current entering by end next		Eleventh quarter-minute of current entering by end next		Twelfth quarter-minute of current entering by end next		Thirteenth quarter-minute of current entering by end next		Fourteenth quarter-minute of current entering by end next		Fifteenth quarter-minute of current entering by end next		Sixteenth quarter-minute of current entering by end next		Seventeenth quarter-minute of current entering by end next		Eighteenth quarter-minute of current entering by end next		Nineteenth quarter-minute of current entering by end next		Twentieth quarter-minute of current entering by end next	
3.	A.	B.	A.	B.	A.	B.	A.	B.	A.	B.	A.	B.	A.	B.	A.	B.	A.	B.	A.	B.	A.	B.	A.	B
1	1	−2	1	0	−1	1	0	0	0	1	1	1	0	0	0	0	2	1	0	−2	−1	−1	1	−
1	0	1	0	0	−1	−1	1	1	1	0	0	1	0	1	−1	−1	1	2	1	−1	0	−3	−1	−
2	0	−2	0	0	−2	2	1	1	0	0	1	1	−4	1	−1	0	−1	1	0	−1	−1	0	1	−
1	0	1	0	1	1	2	1	1	2	0	3	1	1	0	−1	0	−3	0	−2	0	−1	−2	0	−
0	−3	−1	−2	1	1	1	2	−1	0	−2	−2	−2	0	−1	1	1	0	1	1	1	0	0	1	
1	0	−2	−3	0	0	4	1	−1	1	−2	1	−2	0	0	0	2	−4	1	−1	0	0	−2	1	
1	−3	−3	−5	2	−1	8	6	1	4	−4	3	−1	−3	1	1	2	−7	5	−1	−1	−2	−7	2	−
	0		7		9		−5		−8		−4		4		4		12		0		−5		−6	
	7		14		23		18		10		6		10		14		26		26		21		15	
7	0°·007		0°·014		0°·023		0°·018		0°·010		0°·006		0°·010		0°·014		0°·024		0°·026		0°·026		0°·01₅	
	2¼ min.		2½ min.		2¾ min.		3 min.		3¼ min.		3½ min.		3¾ min.		4 min.		4¼ min.		4½ min.		4¾ min.		5 min.	

TABLE III. November 26th, 1853.

r composed of thirteen slips of sheet copper.   Middle of conductor at

B of the parts between heater and coolers:—Initial $T_A = 50°·96$, $T_B =$

	8.	9.		10.		11.		12.		13.		14.		15.		16.		17.		18.		19.		20
	minute of current entering by end next	Ninth quarter-minute of current entering by end next		Tenth quarter-minute of current entering by end next		Eleventh quarter-minute of current entering by end next		Twelfth quarter-minute of current entering by end next		Thirteenth quarter-minute of current entering by end next		Fourteenth quarter-minute of current entering by end next		Fifteenth quarter-minute of current entering by end next		Sixteenth quarter-minute of current entering by end next		Seventeenth quarter-minute of current entering by end next		Eighteenth quarter-minute of current entering by end next		Nineteenth quarter-minute of current entering by end next		Twentieth quarter-minute of current
	B.	A.	B.	A.	B.	A.	B.	A.	B.	A.	B.	A.	B.	A.	B.	A.	B.	A.	B.	A.	B.	A.	B.	A.
	1	1	−2	1	0	−1	1	0	0	0	1	1	1	0	0	0	0	2	1	0	−2	−1	−1	1
	1	0	1	0	0	−1	−1	1	1	1	0	0	1	0	1	−1	−1	1	2	1	−1	0	−3	−1
	−2	0	−2	0	0	−2	2	1	1	0	0	1	1	−4	1	−1	0	−1	1	0	−1	−1	0	0
	1	0	1	0	1	1	2	1	1	2	0	3	1	1	0	0	−1	0	−3	0	−2	0	−1	−2	0
	0	−3	−1	−2	1	1	1	2	−1	0	−2	−2	−2	0	−1	1	1	0	1	1	1	0	0	1
	−1	0	−2	−3	0	0	4	1	−1	1	−2	1	−2	0	0	0	2	−4	1	−1	0	0	−2	1
	−1	−3	−3	−5	2	−1	8	6	1	4	−4	3	−1	−3	1	1	2	−7	5	−1	−1	−2	−7	2
	0	0		7		9		−5		−8		−4		4		4		12		0		−5		−
	7	7		14		23		18		10		6		10		14		26		26		21		1
	·007	0°·007		0°·014		0°·023		0°·018		0°·010		0°·006		0°·010		0°·014		0°·024		0°·026		0°·026		0°·
	min.	2¼ min.		2½ min.		2¾ min.		3 min.		3¼ min.		3½ min.		3¾ min.		4 min.		4¼ min.		4½ min.		4¾ min.		5 m

9° Cent.

52°·89 ; Final $T_A = 48°·80$, $T_B = 50°·92$.

	21.		22.		23.		24.		25.		26.		27.		28.		29.		30.		31.		32.	
	Twenty-first quarter-minute of current entering by end next		Twenty-second quarter-minute of current entering by end next		Twenty-third quarter-minute of current entering by end next		Twenty-fourth quarter-minute of current entering by end next		Twenty-fifth quarter-minute of current entering by end next		Twenty-sixth quarter-minute of current entering by end next		Twenty-seventh quarter-minute of current entering by end next		Twenty-eighth quarter-minute of current entering by end next		Twenty-ninth quarter-minute of current entering by end next		Thirtieth quarter-minute of current entering by end next		Thirty-first quarter-minute of current entering by end next		Thirty-second quarter-minute of current entering by end next	
	A.	B.	A.	B.	A.	B.	A.	B.	A.	B.	A.	B.	A.	B.	A.	B.	A.	B.	A.	B.	A.	B.	A.	B.
1	0	-1	0	-1	1	0	0	-1	0	0	0	0	0	-1	-1	-1	-1	2	-1	2	0	0	1	0
2	0	1	1	2	1	1	2	-2	1	-2	1	-1	0	1	0	0	-1	2	-1	0	1	1	0	1
3	0	-2	-1	-2	-1	0	-1	2	-2	-2	-1	0	0	-1	0	0	0	0	0	0	-1	1	0	0
1	1	0	0	2	-1	-1	1	1	0	0	-1	0	1	-1	1	-1	0	-1	1	-1	0	1	1	0
2	-1	0	3	0	0	-1	0	0	1	0	1	0	-1	-1	0	0	0	0	-2	0	0	-1	-1	0
0	1	0	0	1	0	-1	0	0	-1	-3	-1	0	-1	1	4	0	-1	1	0	1	0	2	1	0
4	1	-1	3	3	-1	-1	2	1	-1	-7	-1	-1	-1	-1	5	-1	-2	2	-2	0	0	4	1	1
	-2		0		0		-1		-6		0		0		-6		4		2		4		0	
	13		13		13		12		6		6		6		0		4		6		10		10	
	0°·013		0°·013		0°·013		0°·012		0°·006		0°·006		0°·006		0°·000		0°·004		0°·006		0°·010		0°·010	
	5¼ min.		5½ min.		5¾ min.		6 min.		6¼ min.		6½ min.		6¾ min.		7 min.		7¼ min.		7½ min.		7¾ min.		8 min.	

meter-bulbs, but were kept straight and bound close together elsewhere. Gutta-percha pipes were cut and cemented upon the iron slips near their ends, so as to lead streams of cold water across them. The part of the conductor between these coolers was packed round with a large mass of cotton-wool, the thermometer-bulbs being steadied in the apertures prepared for them by means of corks, as before (§ 31). The breadth of the conductor was $2\frac{1}{2}$ inches, the length between the coolers only $3\frac{1}{2}$ inches (instead of 10 inches, as in the iron conductors used previously), so that too great a time might not elapse before such a nearly permanent state of temperature as depended on the heating effect of the current would be reached.

60.   On the 25th of March, 1854, an experiment was made with this conductor in the following manner:—A constant stream of cold water was maintained through each of the coolers; a current from the full nitric acid battery of eight large iron cells was sent through the conductor for twelve times four minutes in each direction, that is for ninety-six minutes in all, and the thermometers were noted every half-minute.

The actual observations of temperature are required to show the circumstances of this experiment, and I therefore give them as follows; instead of an analytical table, such as those by which the results of the preceding experiments were exhibited:—

61.   Iron conductor heated by electric current. Ends kept cool by streams of water about 6° Cent. March 25, 1854.

The differences of temperatures here tabulated, and the half sums of the same temperatures, are graphically represented in Plate I. The result is obvious, either with or without the graphical representation, and affords a striking confirmation of the conclusion first arrived at by so different an apparatus (§ 31), that *the Resinous Electricity carries heat with it in iron.*

62.   About the same time another form of the experiment was tried on a copper tube, with a vessel of oil fitted round it in the middle, and kept hot by a lamp below it, and with gutta-percha tubes fitted to conduct streams of cold water round it. A current from the battery was sent alternately in the two directions through it, as in the previous experiments, and it was attempted

Time in minutes and half-minutes from commencement of current.	Current.	Observed temperature at A by Kew, No. 114, T_A.	Observed temperature at B by Kew, No. 91, T_B.	Difference of observed temperatures $T_A - T_B = D$.	Time in minutes and half-minutes from commencement of current.	Current.	Observed temperature at A by Kew, No. 114, T_A.	Observed temperature at B by Kew, No. 91, T_B.	Difference of observed temperatures $T_A - T_B = D$.
m					m				
0	Started.	8°·40	8°·20	0°·20	24 0	Change.	66°·61	63°·71	2°·90
1		12·70	11·00	1·70	1		66·49	63·70	2·79
2	Entering by end next A.	18·40	14·70	3·70	2	Entering by end next A.	66·50	63·50	3·00
3		22·60	18·00	4·60	3		66·50	63·32	3·18
4		28·00	22·30	5·70	4		66·58	63·11	3·47
5		32·30	26·20	6·10	5		66·56	62·92	3·64
6		35·84	29·70	6·14	6		66·60	62·80	3·80
7		39·60	33·50	6·10	7		66·69	62·73	3·96
4 0	Change.	41·30	36·00	5·30	28 0	Change.	66·53	62·59	3·94
1		44·10	38·60	5·50	1		65·72	62·31	3·41
2	Entering by end next B.	46·80	41·00	5·80	2	Entering by end next B.	65·63	62·32	3·31
3		50·00	44·00	6·00	3		65·54	62·50	3·04
4		51·30	45·40	5·90	4		65·64	62·66	2·98
5		52·40	47·20	5·20	5		65·91	62·93	2·98
6		53·80	48·90	4·90	6		66·19	63·28	2·91
7		54·00	50·50	3·50	7		66·64	63·71	2·93
8 0	Change.	55·00	52·00	3·00	32 0	Change.	66·82	64·15	2·67
1		55·60	52·40	3·20	1		67·09	64·01	3·08
2	Entering by end next A.	54·70	51·90	2·80	2	Entering by end next A.	67·75	64·13	3·62
3		53·50	51·10	2·40	3		68·18	64·30	3·88
4		53·42	50·80	2·62	4		68·43	64·39	4·04
5		53·65	50·70	2·95	5		68·57	64·41	4·16
6		53·84	50·75	3·09	6		68·71	64·43	4·28
7		54·19	50·88	3·31	7		68·83	64·48	4·35
12 0	Change.	54·50	51·10	3·40	36 0	Change.	69·03	64·48	4·55
1		54·71	51·92	2·79	1		68·28	64·32	3·96
2	Entering by end next B.	55·66	52·41	3·25	2	Entering by end next B.	68·00	64·38	3·62
3		56·41	53·22	3·19	3		67·85	64·50	3·35
4		57·50	54·23	3·27	4		67·91	64·72	3·19
5		58·65	55·30	3·35	5		68·16	65·03	3·13
6		59·63	56·23	3·40	6		68·59	65·50	3·09
7		60·31	57·19	3·12	7		69·02	65·91	3·11
16 0	Change.	60·65	57·80	2·85	40 0	Change.	69·45	66·40	3·05
1		60·40	57·70	2·70	1		69·92	66·65	3·27
2	Entering by end next A.	60·65	57·63	3·02	2	Entering by end next A.	70·92	67·08	3·84
3		60·92	57·63	3·29	3		72·00	67·70	4·30
4		61·30	57·79	3·51	4		72·59	68·00	4·59
5		61·87	58·09	3·78	5		73·01	68·20	4·81
6		62·41	58·45	3·96	6		73·45	68·50	4·95
7		62·82	58·76	4·06	7		73·90	68·80	5·10
20 0	Change.	63·39	59·18	4·21	44 0	Change.	73·91	68·95	4·96
1		63·30	59·37	3·93	1		73·14	68·72	4·42
2	Entering by end next B.	63·70	60·13	3·57	2	Entering by end next B.	72·70	68·82	3·88
3		64·19	60·75	3·44	3		72·61	69·08	3·53
4		64·84	61·64	3·20	4		72·76	69·53	3·23
5		65·18	62·08	3·10	5		73·28	70·15	3·13
6		65·71	62·70	3·01	6		73·81	70·70	3·11
7		66·14	63·14	3·00	7		74·18	71·10	3·08

Time in minutes and half-minutes from commencement of current.	Current.	Observed temperature at A by Kew, No. 114, T_A.	Observed temperature at B by Kew, No. 91, T_B.	Difference of observed temperatures $T_A - T_B = D$.	Time in minutes and half-minutes from commencement of current.	Current.	Observed temperature at A by Kew, No. 114, T_A.	Observed temperature at B by Kew, No. 91, T_B.	Difference of observed temperatures $T_A - T_B = D$.
m 48 0	Change.	74°·56	71°·58	2°·98	m 72 0	Change.	78°·70	75°·40	3°·30
1		74·64	71·56	3·08	1		79·55	75·70	3·85
2	Entering by end next A.	74·93	71·40	3·53	2	Entering by end next A.	81·31	76·41	4·90
3		75·23	71·30	3·93	3		82·82	77·21	5·61
4		75·21	71·00	4·21	4		83·90	78·00	5·90
5		75·30	70·80	4·50	5		84·81	78·58	6·23
6		75·32	70·71	4·61	6		85·47	79·09	6·38
7		75·33	70·63	4·70	7		85·73	79·42	6·31
52 0	Change.	75·12	70·40	4·72	76 0	Change.	85·60	79·58	6·02
1		73·61	70·24	3·37	1		84·50	79·45	5·05
2	Entering by end next B.	72·60	69·20	3·40	2	Entering by end next B.	84·20	79·61	4·59
3		71·88	68·90	2·98	3		84·11·	79·90	4·21
4		71·53	68·74	2·79	4		84·20	80·30	3·90
5		71·25	68·63	2·62	5		84·49	80·82	3·67
6		71·14	68·60	2·54	6		84·68	81·20	3·48
7		71·13	68·61	2·52	7		84·91	81·50	3·41
56 0	Change.	71·12	68·59	2·53	80 0	Change.	85·10	81·71	3·39
1		71·48	68·56	2·92	1		84·96	80·20	4·76
2	Entering by end next A.	72·30	68·80	3·50	2	Entering by end next A.	85·11	80·92	4·19
3		73·19	69·22	3·97	3		85·19	80·63	4·45
4		73·71	69·53	4·18	4		85·20	80·35	4·85
5		74·32	69·92	4·40	5		85·09	79·95	5·14
6		74·82	70·30	4·52	6		84·66	79·54	5·12
7		75·20	70·60	4·60	7		84·31	79·07	5·24
60 0	Change.	75·56	70·75	4·81	84 0	Change.	84·00	78·70	5·30
1		75·12	71·18	3·94	1		82·48	77·93	4·55
2	Entering by end next B.	75·16	71·61	3·55	2	Entering by end next B.	81·30	77·50	3·80
3		75·23	72·03	3·20	3		80·50	77·18	3·32
4		75·47	72·46	3·01	4		79·81	76·84	2·97
5		75·82	72·90	2·92	5		79·42	76·58	2·84
6		76·14	73·35	2·79	6		79·11	76·40	2·71
7		76·15	73·60	2·55	7		79·02	76·22	2·80
64 0	Change.	76·40	73·72	2·68	88 0	Change.	78·80	76·05	2·75
1		76·90	73·55	3·35	1		77·89	75·03	2·86
2	Entering by end next A.	77·60	73·60	4·00	2	Entering by end next A.	77·50	74·36	3·14
3		78·12	73·60	4·52	3		77·10	73·57	3·53
4		78·40	73·57	4·83	4		76·70	72·97	3·73
5		78·50	73·50	5·00	5		76·30	72·40	3·90
6		78·54	73·40	5·14	6		76·00	71·90	4·10
7		78·50	73·31	5·19	7		75·50	71·45	4·05
68 0	Change.	78·40	73·16	5·24	92 0	Change.	75·18	71·10	4·08
1		76·80	72·46	4·34	1		75·14	71·20	3·94
2	Entering by end next B.	76·61	72·70	3·91	2	Entering by end next B.	75·91	71·95	3·96
3		76·90	73·10	3·80	3		76·68	72·50	4·18
4		77·30	73·65	3·65	4		77·43	73·55	3·88
5		77·91	74·22	3·69	5		78·10	74·40	3·70
6		78·41	74·82	3·59	6		78·70	75·00	3·70
7		78·90	75·40	3·50	7		78·40	75·70	2·70
					96 0		78·70	76·20	

to observe the thermal effects by means of two open thermometer-tubes with small spherical bulbs, pushed into the copper tube from each end, and bent down at right angles outside it, with their lower ends immersed in two cups of spirits of wine. The want of any sufficient regulation of temperature to keep the liquid column of these air-thermometers within range, made it impossible to get any clear indication of a result by this experiment; but on the whole, there appeared to be an effect of the same kind as had been previously discovered in copper.

63. A few weeks ago, I began again to make direct experiments on electrical convection with a view to obtaining additional evidence in support of the conclusion which I had arrived at previously, and to investigate methods by which the nature of the quality in other metals could be discovered more readily, and the specific heat of electricity in any metal determined in absolute units. I had determined to give up the use of the nitric acid battery in consequence of the inconveniences which had been alluded to above (§ 34), and accordingly I had constructed a large Daniell's battery; consisting altogether of eight wooden cells lined with gutta-percha, and fitted with sheet copper, suitably arranged with shelves to bear crystals of sulphate of copper; sixteen porous cells, some of which had served previously in the iron battery; and sixteen zinc plates of the same dimensions as those previously used. Each wooden cell had sheet copper not only round its interior, but also a portion of the same sheet carried across it so as to divide it into two spaces, each completely surrounded by the metallic surface. A porous cell is put into each of these spaces, and a zinc plate into each porous cell; the two zincs in the porous cells contained in the same wooden cell being always united. The ordinary liquids of a Daniell's battery, acidulated solution of sulphate of copper and dilute sulphuric acid, are used. The whole battery power thus consists of eight independent cells, which, with the connexions in ordinary use, may be arranged either in one or in two elements, but which may also, should there be occasion, be readily enough set up in four or in eight elements. Any power may of course be used down to the lowest, with only a single porous cell and a single zinc plate in one of the wooden cells. The sulphate of copper solution is kept constantly in the wooden cells, which remain in a fixed

position on a shelf. Electrodes from the large commutator (§ 27), which is fixed to the wall in an adjoining apartment as near as possible to the middle of the wooden cells, are brought through the partition between the two rooms, and kept always ready to be put in communication with the two poles of the battery, however arranged.

This battery, or parts of it, has been used in nearly all the experiments described below in Parts IV. and V., and it has been found very convenient. Some of the wooden cells have contained the acidulated solution of sulphate of copper now for more than a year [*for more than two years now, Nov.* 1856], and as yet their gutta-percha linings have shown no signs of injury.

64. The first of the recent experiments on electrical convection was made with an iron conductor prepared as follows :—

The conductor, XY (Fig. 6), consists of two pieces of thin sheet iron 8½ inches long and ¾ of an inch broad, and bent so that when put together they form three tubular spaces, A, B, C, fig. 6. The iron is cut so as to make prolongations of these tubes of about

Fig. 6.

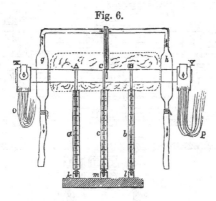

an inch beyond one side of the conductor. The slips thus put together are soldered so as to make the tubes perfectly air-tight, one end being closed, and the other left open to receive the thermometer-tubes *a, b, c,* which were cemented air-tight with wax. In soldering, great care was taken to prevent the solder from spreading between the iron slips. Copper electrodes were now soldered to the ends of the conductor, and the junctions were enclosed within pieces of gutta-percha tube, *g, h,* through which a continuous stream of cold water was made to flow. The distance between the coolers was 7½ inches, and they were placed so that the four spaces between them and A, B, C were all equal. Divided scales were attached to the thermometer-tubes, of which the lower ends were immersed in small vessels, k, l, m, containing spirits of wine. The conductor between the coolers was wrapped in a large quantity of cotton wool represented by the space within the dotted line.

To send a current through the conductor thus prepared, the whole battery, arranged, as described below, in two elements, each exposing ten square feet of zinc surface to seventeen square feet of copper, was employed:—

Description and Drawing of Battery with Connexions.

R, R and S, S (Fig. 7), two series of cells, each containing eight porous cells and eight zinc plates.

Fig. 7.

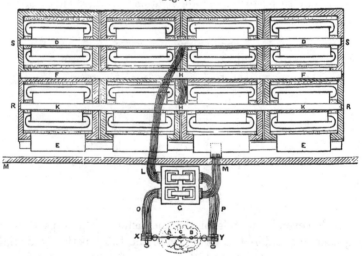

K, K and D, D, thick copper supports for the zinc plates, the zincs of R, R being firmly clamped to K, K, and those of S, S to D, D.

E, E, a thick conductor connecting the coppers of series R, R together.

F, F, a similar conductor connecting the coppers of S, S.

H, H, a bundle of wires connecting the coppers of S, S with the zincs of R, R.

M, M, a wooden partition separating the battery-room from the experimenting-room.

L and M are two bundles of wire which pass through holes in the partition and connect the commutator G respectively with the coppers of series R, R, and the zincs of series S, S. The bundles of wires O and P complete the circuit through XY, the conductor to be tested.

The dotted spaces round the porous cells represent shelves for holding crystals or sulphate of copper.

65. After about an hour and a half, the thermometer at the middle of the conductor indicated 170° Fah. (76°·7 Cent.); and one of the brass bridges of the commutator was then lifted so as to break the circuit. Immediately the liquid mounted rapidly in each of the three glass tubes of the air-thermometers, and it was prevented from rising above a certain point in the middle one by completing the circuit again. The column of liquid was kept as steady as possible at this point in the middle air-thermo-meter by a person observing it, and making and breaking the circuit by means of the brass bridge, while two other persons noted the indications of the two lateral air-thermometers. The current was reversed every three or four minutes, and the liquid in the middle air-thermometer brought back to the same point, and kept as nearly as possible to it. The imperfection of the regulating system was such as to make it very difficult to prevent great oscillations in the thermometers, but the instantaneous manner in which their indications followed the operations of the break, made it certain that the plan would be perfectly successful when a continuously acting regulator should be introduced.

66. As it was, the result afforded a most striking and im-mediate confirmation of the conclusion previously arrived at regarding the electrical convection of heat in iron. Every time the current was reversed, the liquid fell rapidly (showing a rise of temperature) in the thermometer next the end by which the current nominally entered, and rose rapidly (showing a fall of temperature) in the other.

Mr Joule assisted in this experiment, and was satisfied with

16—2

the evidence it afforded in favour of the conclusion that *the Resinous Electricity carries heat with it in iron.*

67.   Unsuccessful attempts were next made with tubular conductors of different metals; and in endeavouring to get decisive
results regarding the qualities of copper and brass, I again had
recourse to the form of conductor used in the preceding experiment.   The new conductors were, however, made of much thinner
sheet metal than those of the iron, to admit of a less powerful
battery being used; and consequently, in each case, a frame-
work had to be arranged to hold the conductor steady.   Great
difficulties were met with in continually repeated failures of the
air-thermometers.   It was therefore found necessary to have
metallic tubes continued downwards several inches from the bulbs,
so as to prevent the wax by which the glass was cemented from
being melted by the heat.   The battery, however, had also to be
reduced to a single zinc plate in one of the wooden cells, as with
more of the battery than this, the heating action had been found
to be so sudden in the thin copper and brass conductors, as almost
immediately to melt the solder about some of the bulbs, and so
make one or more of the thermometers fail before the regulating
action of the break was applied.   Notwithstanding all precautions,
the central thermometer failed in each case, and the action of the
lateral thermometers was very unsatisfactory both in the copper
and in the brass conductors.   The central thermometer could,
however, be well dispensed with, by regulating by the break one
or other of the lateral thermometers; and thus, after many unsuccessful attempts, experiments were made on copper and brass
conductors, which, although still unsatisfactory, showed decidedly
the looked-for convective effect.   In each case, the thermometer
which was not kept to one point by the regulator, always showed
an increase of temperature, both in the copper and in the brass
conductor, when the current was reversed so as to enter by the
end remote from it, and showed a diminution of temperature
when the current was again reversed so as to enter by the end
next it.   Hence it appeared that *the Vitreous Electricity carried
heat with it in both copper and brass.*

68.   The lateral metallic tubes branching down from the conductor to carry the glass tubes of the air-thermometer, constituted
a great defect in the plan of apparatus used in the experiments

just described; and the only way of avoiding it appearing to be to make the glass tubes pass through the body of the conductor itself, so as to admit of their being cemented air-tight at its cool ends, I again had recourse to the tubular form of conductor which had been tried unsuccessfully before.

69. A tube made of very thin sheet platinum, soldered with gold, was arranged in the following manner :—A glass rod, $2\frac{1}{8}$ inches long, wrapped closely round with thin cotton-thread, was pushed into the central part of the tube, in which it fitted closely, and was carefully luted with red lead. After keeping it for several days heated by a stove, gutta-percha coolers, A, A (Fig. 8) were fitted on it, leaving a length of six inches of tube between them. Wooden troughs, B, B, were then fitted on outside the coolers, and fastened to the ends of a piece of wood, C, C; straps of thick copper, about an inch broad, were bent to form conducting linings for the troughs, their ends turned round, firmly fastened to C, C and brought together at D, D, thus forming connexions with the

Fig. 8.

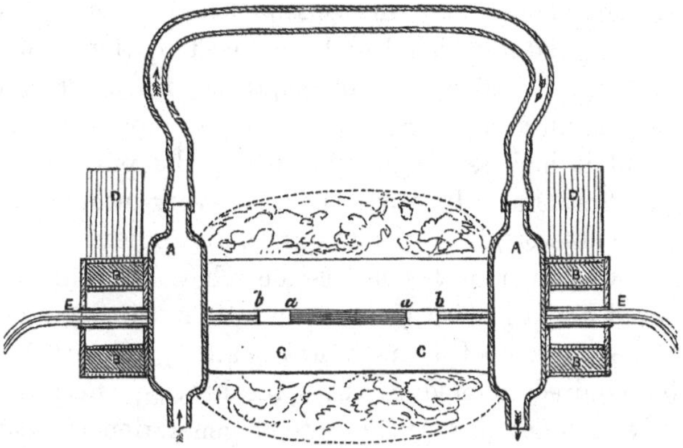

electrodes of the commutator (for this part of the arrangement see also fig. 9). Two pieces of thermometer-tube, bent to right angles, had their short arms rolled with thread, and were pushed into the tube from its ends, as far as b, b, leaving spaces ab, ab, each two-thirds of an inch, between them and the stopper aa in the middle of the tube, and made air-tight by cement applied at E, E. The dotted line represents the space round the tube and its wooden stand C, C, filled with cotton-wool. A conducting communication was established between the platinum tube and D, D, by pouring mercury into the troughs B, B (see figs. 8 and 9).

Fig. 9.

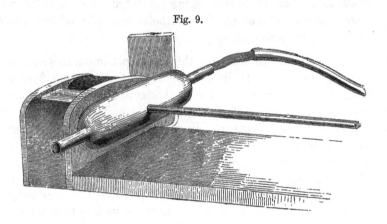

70.　The system of regulating the temperature in one part of the conductor by breaking and making the circuit, had been adopted only as a temporary expedient in the experiment on the iron conductor (§ 65), in consequence of the failure of a continuous regulator which had been fitted up for that experiment. It had the advantage of requiring no other apparatus than the commutator, in regular use in all applications of the battery, and it had been found to answer the purpose tolerably well in the first trial. It proved, however, very inconvenient with the finer conductors, from the too great abruptness of its action. Besides, it was open to this very serious objection, that it kept up the required heating effect by an intermittent current, and therefore by the passage of a much less quantity of electricity than would be required to produce the same heating effect if flowing in a nearly constant current (the rate of generation of heat being proportional to the square of the strength of the current at each instant, while the looked-for convective effect is proportional simply to the strength of the current at each instant, and is therefore, on the whole, proportional to the whole quantity of electricity that passes).

In order, therefore, that the current might be kept as nearly as possible constant at the particular strength required to maintain the heating effect used, I had the following regulator constructed.

71. Two iron tubes, AB, CD (Fig. 10), 20 inches long and ¾ths of an inch in diameter, open at the top but closed at the bottom, are bound firmly together with insulating blocks of wood, AC and BD, so as to be parallel to one another. Pieces of thin sheet copper are bent into cylinders; to their tops pieces of thick copper, E, E, are soldered, and the copper cylinders are put into the iron tubes.

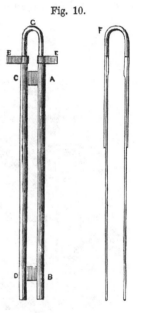

Fig. 10.

To each end of a piece of thick copper wire, bent as shown at F, two pieces of No. 18 iron wire are fixed, one of the same length as the iron tubes, and the other less than half that length, and the two branches are parallel, and at such a distance that when their ends are introduced into the two tubes, they move along their axes. To use the regulator, the tubes are filled with mercury, the apparatus is put into the circuit by connecting with EE, and the requisite amount of resistance is introduced by raising G, which is kept in any position by having one end of a cord fixed to its upper part, carried over a pulley, and stretched by a counterpoise hung at its other end. [Great improvement has been since made in the regulator, by using, instead of No 18 iron wire, thick copper wire tapering to points at the lower ends [the improved form being shown separately in the diagram (Fig. 10)]; and by attaching cups of gutta-percha to the tops of the iron tubes allowed to communicate with the interior by small holes, to serve as overflow cisterns for the mercury. By this arrangement the tubes were kept always full of mercury, and irregular contacts between the connecting conductor and the interior of empty parts of the tubes were prevented.—*Nov.* 1856.]

72. The apparatus was set up as shown in the accompanying view (Fig. 11). The battery connexions were completed with the regulating break partly up, so as to check the current somewhat, and prevent injury from sudden overheating in any part of the conductor.

After a few minutes the break was raised further so as to reduce the current very much, and the liquid began to rise in

the stem of each of the glass tubes, showing that both air-thermometers at first acted perfectly. One of the thermometers was
then steadied with great ease to a small fraction of a scale-division
by using the regulator. The liquid in the other thermometer was
observed, and its position occasionally noted. The direction of
the current was reversed every few minutes, as before, by means
of the ordinary commutator.

Fig. 11.

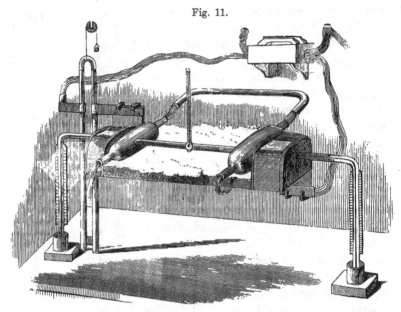

73. Slight differences were observed in the free thermometer
after the reversals, but as yet no very decisive indications of the
looked-for effect appeared. The mercurial thermometer beside the
central conductor indicated less than 80° Fah. (27° Cent.), its column
of mercury having not yet become visible, after the experiment
had been continued in this way for several reversals.

74. The regulating break was then pushed down until a
somewhat further elevation in the temperature of the platinum
was indicated by a considerable escape of air in bubbles from the
open ends of the thermometer-tube. The break was again drawn
up until the liquids again mounted in the stems. One of the
thermometers was again steadied by the regulator, and, the other
being observed, the experiment was continued as before. A
decided effect now appeared almost immediately after each re-

versal. The free thermometer regularly indicated a higher temperature when the current nominally entered by the end next it, and a lower temperature when the current nominally entered by the remote end. After four reversals this part of the experiment had lasted about twenty minutes, and the mercury thermometer beside its middle showed 104° Fah. (40° Cent.).

75. The regulating break was again pushed down for some time, and again raised till the liquid rose in each thermometer-tube, and the experiment was continued as before for four reversals, the central mercury thermometer rising to about 150° Fah. (66° Cent.). The free thermometer rose and fell alternately through several scale-divisions almost immediately after each reversal, and showed the same convective effect as had previously been observed by smaller indications.

76. The regulating break was again pushed down and air again escaped copiously from the thermometers, but very soon beads of liquid began to appear following one another rapidly down the capillary tubes from the interior of the conductor.

As the spirits of wine had not once been allowed to run up into the bulb of either thermometer, these beads of liquid could be nothing but products of the distillation of the oil which had been used in the luting of the central plug; and on taking away the cups of spirits of wine from below the tubes, the smell and taste of the small quantities of liquid which continued to descend gave unmistakeable evidence of their origin. After this it was scarcely possible to get any satisfactory indication from either of the air-thermometers; but the experiment was continued, and one or other of them, when by any means the beads of disturbing liquid could be sufficiently got rid of for a time, was steadied to a constant temperature; the other thermometer being observed when possible, and the reversals repeated as before.

The same result was still obtained; and on the whole, notwithstanding the defect which caused so much inconvenience, it was very decidedly established by the experiment that *the Resinous Electricity carries heat with it in platinum.*

77. [*Added Dec.* 1856.—After many unsuccessful trials on short brass tubes, first with air-thermometers of the metal itself and capillary glass tubes arranged as in the platinum tube (§ 69), and latterly with glass air-thermometers (§ 62) having very small

cylindrical bulbs, the following conclusive experiment was made a few days ago. Four of the large double cells, connected to form a single Daniell's element, exposing 10 square feet of zinc to 17 square feet of copper, were used to send a current through a piece of brass telescope tube six inches long, $\frac{1}{4}$ of an inch diameter, and ground as thin as it could be, without breaking it up, by emery-paper, over the length of $3\frac{1}{2}$ inches which was left between the near sides of gutta-percha coolers, fitted to it in the manner represented above (see fig. 11, § 72). Streams of water being, as in the other experiments, kept running through the coolers, and the regulating break (§ 71) being used to keep the liquids within range in the tubes of the air-thermometers, a small mercurial thermometer pressed against the middle of the brass tube, with its stem and scale projecting out through the cotton-wool, indicated from 190° to 195° Fah. (90°·6 Cent.).

The regulator was not used so much as it might have been with advantage; but, notwithstanding great unsteadiness in the indications of the two air-thermometers, the observations showed decidedly, after each reversal of the current, a cooling effect on the thermometer next the entering stream, in every case in which the irregularities were not so great as to make a comparison impossible. This effect is manifest from the following four cases, selected merely as being those in which one of the thermometers was most nearly steady during a few minutes of flow of the current, first in one direction and then in the other.

Current entering by end next		Readings, in arbitrary scale-divisions, of Thermometer A.	Thermometer B.
I.	A	43	$57\frac{1}{2}$
	B	42	$44\frac{1}{2}$
II.	A	41	41
	A	41	$34\frac{1}{2}$
III.	A	$31\frac{1}{2}$	$26\frac{1}{2}$
	B	$29\frac{1}{4}$	$16\frac{1}{2}$
IV.	A	$27\frac{1}{2}$	22
	B	$20\frac{1}{2}$	$12\frac{1}{4}$

Hence the conclusion (see below, §§ 102 and 103), that *the Vitreous Electricity carries heat with it in brass*, which I anticipated three years ago from the mechanical theory*, is now established by a direct experimental demonstration.]

* See "Dynamical Theory of Heat," § 132 [Art. XLVIII. Vol. I. above].

PART II. ON THERMO ELECTRIC INVERSIONS.

78. Cumming's discovery of thermo-electric inversions having afforded the special foundation of that part of the theory by which I ascertained the general fact of electric convection in metals, and every observation of a thermo-electric inversion being a perfect test as to the relative positions of the two metals between which it is observed in the Table of Convections (see below, § 103), I was induced to make experiments with a view to finding new instances of inversion, and to determine in each case, with some degree of precision, the temperature at which the two metals are thermo-electrically neutral to one another.

79. In the experiments on thermo-electric inversion described by Cumming, and by Becquerel, the only other experimenter, so far as I am aware, who has published observations on the subject, one junction between the two metals is generally kept cool, while the other is raised until the current indicated by the galvanometer, instead of going on increasing, begins to diminish, comes to a stop, and then sets in the reverse direction *.

80. In this way Cumming found that " if gold, silver, copper, brass, or zinc wires be heated in connexion with iron, the deviation [indicating the current], which is at first positive, becomes negative at a red heat †." Many other experimenters have professed themselves unable to verify these extraordinary results, and have attempted to explain them away by attributing them to coatings of oxide formed on the metals, or to other causes supposed with equally little reason to exercise sensibly disturbing influences; but the descriptions, given by the original observers, of their experiments, leave no room for such doubts. It is certainly not easy to get the inversion between copper and iron (with such specimens as I have tried) by the heat of a spirit-lamp, applied as described by Becquerel to one junction while the other is left cool; but I readily obtained it by raising the other junction somewhat in temperature, with the first still kept at a red heat. Probably if the atmospheric temperature had been higher, or if a somewhat more intense red heat had been obtained from the

---

* Cumming's *Electro-Dynamics*, section 104, p. 193. Cambridge, 1827.
† *Cambridge Philosophical Transactions*, 1823, addition to p. 61.

spirit-lamp, I should at once have obtained the result simply in the manner described by the previous observers.

81.   The easiest way to verify the thermo-electric inversion of iron and copper is to take a piece of iron wire a foot or two long, and twist firmly round its ends two copper wires connected with the electrodes of any ordinary astatic-needle galvanometer. Then first heat one of the junctions with the hand, or by holding it at some height over a flame and note the deflection, which will be found such as to indicate a current from copper to iron through the hot junction. Again, heat both junctions in flame, or in sand at any temperature above 300° Cent., and withdraw one a little from the hottest place, so that while both junctions are at temperatures above 300° Cent., that which was heated in the first experiment may be still decidedly hotter than the other. The deflection will now be found to be the reverse of what it was before, and will be such as to indicate a current from iron to copper through hot. The reversal of the current may be very strikingly exhibited by allowing the two junctions gradually to cool, while ensuring that the same one remains always somewhat above the other in temperature. When the mean of the temperatures of the two junctions falls below 280° Cent. or thereabouts, the primitive deflection will be again observed. All these phenomena are observed indifferently whether the copper wires be simply twisted on the ends of the iron wire, or brazed to them, or tied to them by thin platinum or iron wire.

82. Similar phenomena may be observed without the necessity of going to so high temperatures, by soldering galvanometer electrodes of copper to the ends of a double platinum and iron wire, and treating this compound circuit in the manner just described, only with a more moderate application of heat (see Fig. 12). If the platinum wire be very thin in comparison with the iron one connected along with it, the circumstances will be but little altered from those observed when iron simply is used. By taking a thicker platinum wire, or several thin ones together, in connexion with the same iron wire, or by using a thinner iron wire and the same platinum, the neutralization and reversal may be

Fig. 12.

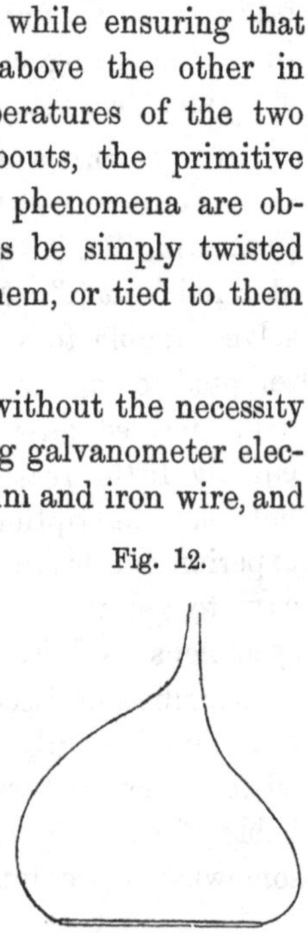

shown with temperatures below the boiling-point. Most specimens of platinum wire thus applied reduce the neutral point of copper and the compound platinum-and-iron wire much below the temperature of melting ice, when the proportion of platinum to iron in the bundle is sufficiently increased (the limit, of course, being the neutral point of copper to the platinum itself. See below, §§ 83, 84).

83. A certain specimen of platinum wire in my possession, when tested by such elevations of temperature as could be produced by the hand, was found to lie in the thermo-electric series, on the other side of copper from the position in which platinum is placed in all statements of the thermo-electric qualities of metals previously published. That is to say, when connected by copper electrodes with the circuit of a galvanometer, and when heated at one junction up to ten or twenty degrees above the atmospheric temperature, a current set from copper to platinum through hot. On raising the temperature of the hot junction towards the boiling-point of water, the strength of the current began to diminish; came to a stop when a temperature, I suppose little above that of boiling water, was reached; and set in the reverse direction with increasing strength when the temperature of the hot junction was further raised, the other junction being kept all the time at the atmospheric temperature. I afterwards found that this specimen of platinum wire (referred to under the designation of P_1 in what follows) became neutral to ordinary copper wire at the temperature 64° Cent.

84. Of two other specimens of platinum wire which I tried with copper, one (marked P_2) gave indications of a neutral point about the zero of Fahrenheit's scale, but the other (P_3) remained, for the lowest temperatures I reached, always on the same side of copper as that on which platinum appears, at ordinary and high temperatures, generally to lie. When these three platinum wires were tried with one another thermo-electrically, they gave, as was to be expected, the mutual thermo-electric indications of different metals lying in the order Bismuth, P_3, P_2, P_1, Iron, Antimony. They retained all the same qualities after being heated to redness; and in a great many experiments performed upon them, in which I have found them extremely convenient as thermo-electric standards, have exhibited perfect constancy in their thermo-electric bearings.

I have not yet discovered on what their differences depend, but in all probability it is on the different degrees to which they are alloyed with other metals.

85.   The fact of copper changing in the thermo-electric series from below the position of the platinum specimen P_2 to above that of iron, when the temperature is raised from $-30°$ or $-20°$ Cent. to $300°$ C., proves that every metal which lies between P_2 and iron for any intermediate temperature, must become neutral to either P_2, or copper, or iron, at some temperature between these limits.   Now nearly all the common metals, for instance, lead, tin, brass, zinc, silver, cadmium, gold, lie between platinum and iron in the thermo-electric series at ordinary temperatures, and no doubt many of the rarer metals (I have found aluminium to lie between P_3 and P_2 *) are to be ranked within the same limits. Hence at temperatures easily reached and tested, neutral points may be looked for with the certainty of finding them, between each of those metals and one or other, if not several, of the metals and metallic specimens (P_2, P_1, Copper, Iron) referred to above. Taking then the platinum specimens P_1, P_2, P_3 as standards, and using besides ordinary copper and iron wires, I commenced investigating their thermo-electric relations to as many other metals as I could obtain.

86.   In experiments to determine temperatures of neutrality, the first apparatus which I employed for regulating the temperatures of the two junctions, consisted of copper vessels placed side by side, in which oil could be raised by gas-burners as high in temperature as the mercurial thermometer can be used, that is to $340°$ or $350°$ Cent., or somewhat above the boiling-point of mercury.   To do away with irregularities from the flame, and cold air playing unequally on the sides of these vessels, smaller ones were placed on wire stands within them, and were completely filled and surrounded with oil.   In each experiment a wire or slip, about 18 inches long, of one of the metals to be tested, had somewhat longer wires or slips of the other soldered to its ends.   The compound conductor thus constituted was bent into such a shape that the two junctions of the metals could be placed near the centres of the oil-baths; it was supported in this position, carefully insulated from touching the copper vessels and from all

* See Art. LXXXIX. above, also Art. XLVIII. Part VI. Vol. I. above.

other metallic contacts; and thermometers were put with their bulbs in the oil as close to the junctions as possible. The gas-furnaces were applied below and round the sides of the large copper vessels, so that they could be regulated to any desired temperatures.

Fig. 13.

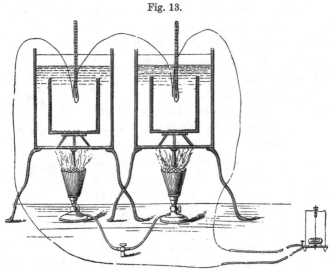

87. After this apparatus had been used in several experiments, and neutral points between copper and iron, copper and P_1, lead wires and P_1, and brass and P_1 had been determined, I saw reason to alter the arrangements in various respects, and had another apparatus constructed, according to the following description.

88. Two small oil-baths (Fig. 14) were made, each of an outside, partly cylindrical and partly plane, sheet of copper, and a concentric copper tube five inches long and $\frac{6}{10}$ths of an inch diameter brazed to it by ends of sheet copper, shaped as shown in the diagram. The space round the inner tube and within the outer sheet and ends was filled with oil completely covering the inner tube, and, when heated, rising into the space

Fig. 14.

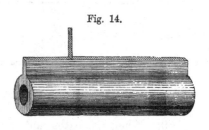

between the upper parallel plane parts of the outer sheet. A narrow ring of sheet metal with a long slip projecting from one side for holding it by, was put in the inner tube before the other parts were brazed on, and during an experiment was kept as constantly moving from one end of the bath to the other and back as was required to keep the whole mass of oil at one temperature. The second drawing (Fig. 15) represents, on the actual scale, a section of either bath through the position occupied by its stirrer. This diagram also shows a section of an external case of sheet metal which supports the bath, and serves as a flue to carry the flame and products of combustion round its sides. The rows of gas-burners for the two baths were fixed in a line, and each burner was regulated by a separate stop-cock. The outer cases are screwed to the same stand, and the copper vessels holding the oil are pushed into them and rest with their axes in a line over the burners. The ends of the baths and of the outer cases are kept about one-fourth of an inch apart, and their supports are also made quite separate, which was found to be necessary to allow one of the baths to be kept cool, while the other was raised to a high temperature. (In the third drawing (Fig. 16), the stems by which the stirrers are held are accidentally omitted.)

Fig. 15.

Fig. 16.

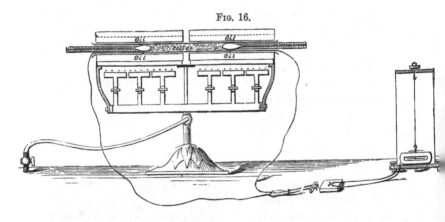

89. When the baths and their furnaces are all fixed in their proper positions, a tube of thin glass, about $10\frac{1}{2}$ inches long, just

small enough to enter easily, is pushed into the inner tubes of the baths, and is left resting there, with its ends projecting a little outside their remote ends. In recent experiments I have substituted a simple roll of paper for the glass tube, and have found it answer quite as well.

90. A compound conductor, to be tested thermo-electrically in this apparatus, consists of a wire or a thin bar of one metal, or a bundle of wires of two different metals, 5½ inches long, with wires of from 18 to 30 inches long of another metal soldered to its ends. To avoid circumlocution I shall call the former the *mean conductor*, and the wires soldered to it the *electrodes*, of the thermo-electric arrangement. The connexions between the mean conductor and the electrodes are generally made by brazing, or by hard-silver solder, when temperatures much above the boiling-point of water are to be used.

91. A conductor thus prepared of the two metals to be tested, is drawn through the glass tube till the *mean conductor* occupies a position, lying on the glass or paper tube, with its centre under the centre of the tube, and consequently with its ends about the middle of the hollow spaces surrounded by the oil-baths.

92. The electrodes are carried from the ends of the insulating tube to the connexions required for completing the circuit through the coil of a galvanometer. These must essentially be maintained at the same temperature, unless the electrodes of the thermo-electric arrangement be copper, the same as those of the galvanometer. After trying several obvious, more or less troublesome plans to secure the fulfilment of this condition, I found this to be a perfectly effective way; simply to tie the connexions firmly together as close to one another as possible, only separated from contact by a fold or two of paper wrapped round each, and to tie a quantity of paper, or to make up a bundle of cotton wool, or some other bad conductor, round the two, for two or three inches on each side of the junctions. The junctions themselves, except when they are between homogeneous metals, are not made by binding-screws, but either by soldering, or by cleaning the surfaces and then tying the metals firmly together by fine twine. To avoid mistakes and prevent the necessity of disturbing the bundle round the junctions, in tracing the courses of the conductor on the two sides of it, a thread or mark of some kind is attached to one galvanometer

electrode, and a corresponding mark on the electrode of the thermo-electric apparatus to which it is joined. This system of electric insulation and thermal connexion between junctions of dissimilar metals, I have found very convenient in a great variety of thermo-electric and other electro-dynamic experiments, and when it was used I have never observed the slightest trace of a current attributable to any difference of temperatures in the parts of the circuit to which it is applied.

93. The conductor being thus arranged, two thermometers are pushed into the glass or paper tube from its ends and placed with the centres of their bulbs as close as possible to the metallic junctions, and with their graduated tubes extending nearly horizontally outside the apparatus, but inclined upwards as much as the inner diameter of the insulating tube and their dimensions permit, so as to check as much as possible the tendency (in some of the thermometers found very inconvenient) of the column of mercury to divide when sinking rapidly. All the space inside the glass or paper tube left vacant by the thermometers and the conductor is filled with cotton wool, well pressed in to prevent currents of air.

94. This apparatus has many advantages over that first used and described in (§ 86) above: the temperatures of the baths can be changed with great rapidity, in consequence of the smallness of the quantities of oil which they contain; and by watching the thermometers and adjusting the gas-burners, can be regulated as desired with great ease. I have found it not a small practical advantage to be freed from the necessity of bending the mean part of the conductor to be tested, and of making the arrangements to prevent irregular contacts and to keep the junctions and the thermometers in their proper positions immersed in the oil. When a rare metal is to be tested, or one, such as sodium or potassium, which cannot be kept in air, it will be of great consequence to be able to apply the tests to a little straight bar or slip only a few inches long, or to a small column filling a glass tube.

95. For experimenting at low temperatures a modified apparatus was made, consisting of a double wooden box, each compartment, nearly a cube of 4 inches side, fixed to a common base with a space of about $\frac{1}{4}$ inch between their sides, and a glass

tube running through them and cemented at the apertures in the sides so as to hold water-tight and resist the action of acids which might be employed in freezing mixtures. The conductor to be tested and the thermometers are arranged in this glass tube, as in that of the other apparatus (§§ 89 to 93); and while a freezing mixture is kept in one compartment, the other is either allowed to take the atmospheric temperature, or is heated by hot water or steam.

Fig. 17.

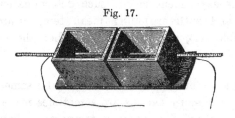

96. The way of experimenting which I followed, was to raise the temperature of one bath until a deflection of the galvano-meter-needle became sensible; then to go on raising it, and letting that of the other follow, so that the two thermometers may indicate as nearly as may be a constant difference of temperatures; and to watch the needle until a reversal is observed, or until the limit of temperature which the arrangements admit of is reached.

As soon as a reversal is obtained, the two thermometers are allowed to sink until the needle begins to return from its reverse deflection. When it approaches zero the thermometers are kept from any rapid changes, but allowed to sink very slowly, with always the same difference, or at least with a quite decided difference of the same kind as that raised between them at the beginning. The last readings of the sinking thermometers which give a sensible deflection before the original deflection is recovered, several readings when the needle appears perfectly at zero, and the first readings when the needle is discovered to deviate again in the original direction, are carefully noted. The arithmetical mean of the temperatures of the two thermometers for each of these simultaneous or nearly simultaneous readings is taken; and it is generally found that the means derived from the readings taken when no deflection can be discerned, lie within a fraction of a degree of the mean of the last sinking mean temperature of the

17—2

junctions which show one deviation, and the first which shows the deviation in the other direction. The mean of either the readings which give no deviation, or of the last and first which give the contrary deviations, or of all these readings together, according to the nature of the memoranda made by the observer, is taken as a determination of the neutral point of the two metals, that is, the temperature at which they are thermo-electrically like one metal, or thermo-electrically neutral to one another. In the course of one experiment several such determinations, both with descending and with ascending mean temperature, are made, and if possible also, with first one and then the other junction higher.

97. Either in one experiment, or with the same apparatus on successive days, determinations are sometimes made with as considerable a variety of differences of temperature between the two junctions as is attainable. Sometimes the difference of temperatures used is so small as to give very slight indications of electromotive force, even when the mean of the temperatures of the junctions differs widely from the neutral point, in which cases, of course, the test is deficient in sensibility. The best determinations are generally those derived from observations showing the galvanometer at zero, with the widest difference between the temperatures of the junctions, to which the thermometers are applicable with trustworthy indications; as, for instance, 100° or 150° Cent., which are attainable in the most favourable cases, being those in which the neutral point is at about midway between the temperatures of freezing and boiling water. The differences between these determinations sometimes amount to a degree or two, and even to several degrees when zinc was one of the metals; but generally the final mean for the neutral point does not differ by more than a degree from any single determination considered as satisfactory at the time it was made.

98. The mutual interchanges of thermo-electric order observed in various specimens of zinc, gold and silver, occasioned considerable perplexity [which has only been cleared up by observations made subsequently to the communication of this paper]. The following determinations were made at different times and by different observers, as noted :—

Observer.	Date.	Metals.	Neutral Points.
Mr C. A. Smith	Sept. 27, 1854...	P_1 mean .... Gold electrodes	− 3·06
Mr C. A. Smith	Aug. 18, 1854...	P_1 mean .... Silver electrodes...................	− 1·5
Mr C. A. Smith	Sept. 8, 1854....	P_1 mean .... Zinc electrodes	+ 8·2
Mr C. A. Smith	Sept. 20, 1854...	Silver mean .... Zinc specimen (1) electrodes..	43·9
Mr G. Chapman and Mr J. Cranston ...	Jan. 29, 1856, and Feb. 5	Silver electrodes .... Zinc specimen (2) mean......	51·5
Mr G. Chapman and Mr J. Cranston ...	Feb. 1856........	Silver mean .... Zinc (1) electrodes..............	46·55
Mr G. Chapman and Mr J. Cranston ...	Feb. 1856........	Silver mean .... Zinc (2) electrodes..............	58·18
Mr J. Murray.........	Aug. 1856	Silver mean .... Zinc electrodes	56·95
Mr G. Chapman and Mr J. Cranston ...	Feb. 1856........	Gold mean.......................... Zinc electrodes	71
Mr G. Chapman and Mr J. Cranston ...	Feb. 1856........	Gold mean.......................... Zinc electrodes	69·76
Mr G. Chapman and Mr J. Cranston ...	Feb. 27, 1856 ...	Silver electrodes .... Gold mean	70·8
Mr J. Murray.........	Aug. 21, 1856...	Silver mean...................... Gold electrodes	− 5·7

Of the two results for the neutral point between silver and gold, only the last can be reconciled with the indications derived from the previous results as to the relative positions of these and the other metals tried along with them; and accordingly $-5°·7$ has been taken as the neutral point of gold and silver in the thermo-electric diagram given below (§ 101). The first result, $70°·8$, was found as the mean of several determinations, from none of which it differed by more than $0°·7$, and the discrepance can scarcely be attributed to errors of observation, but is probably due to slight differences in the specimens of gold and silver used in the different experiments. That very slight chemical differences in specimens of gold and silver wire may make great alterations in the temperature at which they become thermo-electrically neutral to one another, is readily understood by glancing at the diagram given below (§ 101), and observing how close together the lines for gold and silver lie.

99. The question, *Does the difference between the specific heat of electricity in two metals vary with the temperature*? may be answered by experiments showing the law according to which the means of widely different temperatures of the junctions giving no electromotive force deviate from the true neutral point, which is

* See " Dynamical Theory of Heat " [Art. XLVIII. Vol. I. above], § 115, equations (15) and (17).

the mean of any infinitely small difference of temperature giving no electromotive force.

I have not yet obtained indications of such a deviation in any case, having been prevented from prosecuting the inquiry by delays in the construction of a suitable air-thermometer. The examination I have been able to give the subject is only sufficient to show that the arithmetical mean of the temperatures of the two junctions giving no current, is probably in general within a degree of the true neutral point, when the difference between those temperatures does not exceed 100° Cent.

The following summary of a series of experiments made on two consecutive days may serve as an example of the degree of consistence of the results obtained by the method which has been explained, in a case in which the two metals deviate rapidly from one another above and below their neutral point.

Sheet-lead electrodes; P_1 mean.

Determinations by Mr C. A. Smith, May 17 & 18, 1854.

Difference of temperatures.	Half sum of mercurial thermometer temperatures giving no current.	
°	°	
77½	121	
71	121¼	
71½	121¾	Mean
71½	121¾	121°·5
70	122	
185½	120¾	
158½	121¾	
143	121½	Mean
133	121½	121°·4
125½	121½	
68½	122	Mean
50	123	122°·5
	Mean of temperatures by mercurial thermometer giving no current.	
Differences from 50° to 77°......	122°·15	
Differences from 125° to 185°...	121°·4	

These results seem on the whole to show that the mean of apparent temperatures giving no current is rather less for the wide than for the narrow ranges, in the case of the two metals concerned; that is, that the mean of the apparent temperatures giving no current is somewhat below the true neutral point. I need scarcely remark, however, that even if this indication could be relied on, it would be necessary to compare the actual mercurial thermometers which were used, with an air-thermometer, before any conclusions of value could be drawn from it regarding the constancy of the difference of specific heats of electricity in lead and platinum.

100. The following Table shows the results of observations leading to actual determinations of neutral points between various pairs of metals.

	-14°C	-12°·2	-5°·7	-3°·06	-1°·5	8°·2	33°	36°	38°	44°	44°	47°...71°
	P_3	P_1	Silver.	P_1	P_1	P_1	Tin.	P_2	P_2	P_2	Lead.	Different specimens of Silver.
	Brass.	Cadmium.	Gold.	Gold.	Silver.	Zinc.	Brass.	Lead.	Brass.	Tin.	Brass.	Different specimens of Zinc.

	53°	57°*	64°	71°	99°	121°	130°	162°·5	Some temperature between 223° & 253°·5	237°	280°
	P_2	Hard steel.	P_1	Gold.	P_1	P_1	P_1	Iron.	Iron.	Iron.	Iron.
	Double wire of Palladium, 11·31 grs., and Copper, 19·41 grs.	Cadmium.	Copper.	Zinc.	Brass.	Lead.	Tin.	Cadmium.	Gold.	Gold. Silver.	Silver. Copper.

* This determination has been added in consequence of information given by Mr Joule (December, 1856), that hardened steel at ordinary temperatures differs thermo-electrically from copper by about one-tenth of the thermo-electric difference of iron from copper.

The number at the head of each column expresses the temperature Centigrade by mercurial thermometers, at which the two metals written below it are thermo-electrically neutral to one another; and the lower metal in each column is that which passes the other from *bismuth towards antimony as the temperature rises.*

It was also found that Aluminium must be neutral to either P_3 or Brass, or P_2, at some temperature between $-14°$ C. and $38°$ C ; that Brass becomes neutral to Copper at some high temperature, probably between $800°$ and $1400°$; Copper to Silver, a little below the melting-point of silver; Nickel to Palladium, at some high temperature, perhaps about a low red heat; and P_3 to impure mercury (that had been used for amalgamating zinc plates), at a temperature between $-10°$ and $0°$. P_3 appears to become neutral to pure mercury at some temperature below $-25°$ Cent.

101.   The following diagram exhibits graphically the relative thermo-electric bearings of the different metals, and may in fact be regarded as a series of tables of the thermo-electric order of metals at different temperatures from $-30°$ to $300°$ Cent.

The object to be aimed at in perfecting a thermo-electric diagram, is to make the ordinates of the lines (which will in general be curves) corresponding to the different metallic specimens be exactly proportional to their *thermo-electric differences**
from a standard metal (P_3 in the actual diagram).

§§ 102, 103.   *Theoretical inferences regarding Electrical Convections of Heat, from facts of Thermo-Electric Inversions.*

102.   The thermo-dynamic reasoning adduced above (§§ 10 to 15) leads to the conclusion (§§ 14, 15), that the *convective power of the vitreous electricity is greater,* or (which is the same thing) *the convective power of the resinous electricity is less, in each metal for which the line in the diagram cuts the line for another metal from below it on the left to above it on the right, than in this other metal.*   Now it was established in Part I., that the vitreous elec-

---

* See " Dynamical Theory of Heat " [Art. XLVIII. Vol. I. above], § 140.

Explanation of Thermo-electric Diagram.

The orders of the metals in the thermo-electric series, at different temperatures, are shown by the points in which the vertical lines marked with the temperatures Centigrade, are cut by the horizontal and inclined lines named for the different metallic specimens.

Fig. 18.

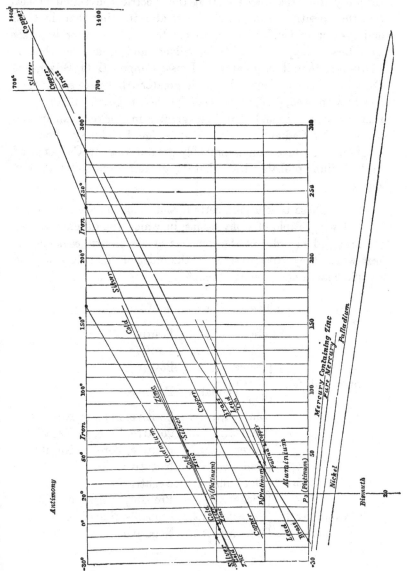

tricity carries heat with it in copper (§ 54), or, as it may be expressed, the *electric convection of heat is positive in copper.* From the diagram we infer that it is greater, and consequently positive, in Brass. That it is positive in brass has been proved also by direct experiment (§§ 67 and 77). We infer also with certainty from the diagram, that the electric convection of heat (whether positive or negative) is greater in Zinc than in Gold, and greater in Gold than in Silver; that it is greater in Brass, Tin, Lead, Copper, Zinc, Gold, Silver, and Cadmium than in Platinum; that it is greater in Brass, Copper, Gold, Silver, and Cadmium than in Iron; that it is greater (that is to say, since it has been proved, § 76, to be negative, less negative) in Platinum than in Mercury; and that it is greater in Nickel than in Palladium. In Cadmium, as we may judge by the eye from the diagram, the convection is probably greater than in Copper; and in Palladium probably less (that is, greater negatively) than in Platinum.

103. These conclusions, certain and probable, are collected in the following Table of Convections, in which the different metals are arranged in order of the *amounts of the electric convection of heat* which they experience, or in the order of the values of "the specific heat of electricity in them."

Electrical Convection of Heat

	In Cadmium.........Positive.	
	Brass .... Positive.	
Order doubtful.	Copper............Positive.	
	Lead ⎫ equal......Positive.	
	Tin ⎭	
	Zinc..............Positive, Zero, or Negative.	
	Gold ....Positive, Zero, or Negative.	
	Silver............ Positive, Zero, or Negative.	
Order doubtful.	Iron ....Negative.	
	Platinum.........Negative.	
	Nickel............Probably Negative.	
Probably nearly equal.	Palladium ...... Probably Negative.	
	Mercury ....Negative.	

PART III. EFFECTS OF MECHANICAL STRAIN AND OF MAGNETIZATION ON THE THERMO-ELECTRIC QUALITIES OF METALS.

104. Physical agencies having directional attributes and depending (as all physical agencies we know of except gravitation appear to do) on particular qualities of the substance occupying the space across or in which they are exerted, are transmitted or permitted with different degrees of facility in different directions if the substance is crystalline. The phenomenon of crystallization, exhibiting different chemical affinities on different bounding planes, between a growing crystal and the fluid from which it is being formed, and the cleavage properties (different specific capacities for resisting stress in different directions) afford the primary illustrations of this statement. It is probable that the proposition asserted is a universal proposition in this sense, that there is no kind of physical agency falling under the category referred to, which does not meet with different capacities for receiving it in different directions in some crystals. There certainly may be, and probably are, crystals which transmit certain physical agencies equally in all directions. Crystals of the cubical system, for instance (unless possessing the conceivable dipolar rotatory property*, from which some, if not all, are certainly exempt), conduct heat and electricity equally in all directions, and have equal magnetic inductive capacities and equal thermo-electric powers. But thermal and electric conductivity, magnetic inductive capacity, and "thermo-electric power†" are undoubtedly different in different directions in many, if not in all, crystals not of the cubical system. Many crystals have not shown any marked difference in their absorption of light according to the direction of its propagation through them; but some undoubtedly do show a difference of this kind, to such a degree as to give sensibly different colour to

---

* See "Dynamical Theory of Heat," § 168; also §§ 163, 166, 167, 169 to 171, *Transactions of the Royal Society of Edinburgh*, May, 1854 [Art. XLVIII. Vol. I. above]. See also Professor Stokes "On the Conduction of Heat in Crystals," *Cambridge and Dublin Mathematical Journal*, Nov. 1851.

† Or thermo-electric difference from a standard metal. See "Dynamical Theory of Heat," § 140 [Art. XLVIII. Vol. I. above].

light passing short distances through them in different directions*. Faraday had good reason, after making the discovery of the induction of electro-polarization in non-conducting substances†, to try the specific directional qualities of crystals used as dielectrics; and although he found no sensible differences in the inductive capacities of the crystals (rock crystals and Iceland spar) which he tried for this kind of action, in different directions, it appears highly probable that induced electro-polarization will sooner or later be ascertained to be no exception to the general rule.

105. Another very general principle is, that any directional agency applied to a substance·may give it different capacities in different directions for all others. Whether or not this is true as a universal proposition, events have proved that the probability of its being true in any particular case is quite sufficient to warrant an experimental inquiry. Brewster discovered that mechanical stress induces in glass directional properties with reference to polarized light, which are lost as soon as the stress originating them is removed. These properties were shown by Fresnel to be of the same kind as the property of double refraction possessed by a natural crystal. Experiments made by Sir David Brewster and Mr Clerk Maxwell prove that isinglass and other gelatinous substances dried under stress, thin sheet gutta-percha permanently strained by traction, and probably all non-brittle (or plastic) transparent solids when permanently strained otherwise than by uniform condensation or dilatation in all directions, possess double refraction as a property of the molecular alteration which they acquire under the stress and retain after the stress is removed. Again, magnetization, as Joule discovered‡, causes an elongation of iron in one direction (that of the magnetization) and a contraction in all directions perpendicular to it, with no sensible change of volume. Faraday discovered the wonderful dipolar optical property of transparent bodies in a magnetic field (the first and only case known of any dipolar qualities, other than those of magnetic

---

* Most crystals not of the cubic system, even when nearly colourless, exhibit difference. See Haidinger's *Researches*.

† *Experimental Researches in Electricity*, Series XIV. §§ 1688, 1689, 1692 to 1698. June, 1838.

‡ "On the Effects of Magnetization upon the Dimensions of Iron and Steel Bars," *Phil. Mag.*, Series 3, Vol. xxx.; also Joule's *Scientific Papers*, Vol. I. p. 245.

and electric reactive forces, called into existence by induction): Maggi discovered that magnetized iron conducts heat with a greater facility across than along the lines of magnetization *.

106. In applying the dynamical theory of heat to thermo-electric currents in conducting crystals, [see Art. XLVIII. Vol. I., above, §§ 147—155] I was led to consider the probable effects of mechanical strain, and of magnetization on the thermo-electric properties of non-crystalline metals, and in consequence entered on the investigation, of which the results, so far as I have yet advanced in it, are now laid before the Royal Society.

107. To find the effect of longitudinal tension on the thermo-electric quality of a metal, I first took eight thin copper wires

Fig. 19.

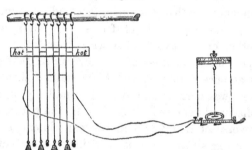

each capable of bearing about 10 lbs., and, attaching their upper ends to a horizontal wooden arm at distances of about ¼ of an inch from one another, allowed them to hang down, each kept stretched by a weight of about ¼ lb. They were connected with one another in order, and the first and last with the electrodes of a galvanometer, by nine wires soldered to them, as shown in the diagram; the junctions between the successive wires being alternately in the upper and lower of two horizontal lines four inches apart. Every alternate wire was then stretched with a weight of about 3 lbs., and a slip of hot plate glass was applied,

---

* Doubts have been thrown on this result, I believe, by other experimenters, who have not succeeded in verifying it by their own observation, but its close correspondence with a result I have recently discovered by experiments on the electric conductivity of magnetized iron, have diminished the impression such doubts produced on my own mind; and I look with much interest to a repetition of Maggi's experiment.

sometimes to the upper and sometimes to the lower row of junctions. A deflection of the galvanometer needle was observed in one direction or the other, according as the glass heater was applied to one set of junctions or the other. The deflection was also reversed when the weights were changed to the alternate set of wires, and the heater kept applied to the same set of junctions. In every case the deflection was such as to indicate a current from stretched to unstretched through hot junctions. The uniform and consistent nature of the indications was such as could leave no doubt as to the result; and I concluded that copper wire stretched by a longitudinal force, bears to copper wire of the same substance unstretched, the same thermo-electric relation as that of bismuth to antimony.

108. I next made a similar experiment on iron wire, varying the arrangement so that the weights could be rapidly shifted; and again so that equal sets of forces could be applied to one or to the other of the two sets of wires, merely by pressing with the foot upon one or another of two levers. A perfectly decided result was at once obtained; and I ascertained that the thermo-electric effect was induced and lost quite suddenly on the pressure being applied and removed. In this case the nature of the effect was the reverse of that found in the experiment on copper, the deflections being always such as to indicate a current in the iron wires from unstretched to stretched through the hot junctions.

109. The thermo-electric effect which these experiments demonstrated to accompany temporary strain produced by a longitudinal force, was, in each of the metals, the reverse of that which Magnus* had previously discovered in the same metal hardened by the process of wire-drawing, and which I ascertained for myself to be produced in each case when the metal is hardened by simple longitudinal stress without any of the lateral action inseparable from the use of the draw plate. I thus arrived at the remarkable conclusion, that when a permanent elongation is left after the withdrawal of a longitudinal force which has been applied to an iron or copper wire, the residual thermo-electric effect is the reverse of the thermo-electric effect which is induced by the force, and which subsists as long as the force acts.

* Poggendorf's *Annalen*, Aug. 1851.

110. I have made a single experiment demonstrating this conclusion for iron by means of a multiple tension apparatus,

Fig. 20.

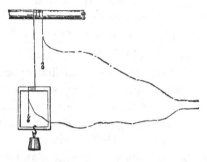

similar in principle to that described above (§ 105). But with a somewhat more sensitive galvanometer than the one I used, the result may be shown in a perfectly decided manner (for iron at least) without any multiplication of the thermo-electric elements; and a very striking experiment may be made on the following plan:—A thin iron wire is wrapped three or four times round a wooden peg held firmly in a horizontal position, and again two or three times round another parallel peg, about four inches lower. A frame is rigidly connected to this second peg, so that it may remain stably in a horizontal position, hanging from the wire and pulled down by the frame with either a light or a heavy weight attached to its lowest point. To keep the wire from slipping, the parts of it running from the pegs towards the ends are kept stretched by light weights tied to them; and the slack parts below these weights are carried away to the galvanometer electrodes, with which they are connected in the manner described above (§ 92). Any convenient source of heat is applied to the part of the wire bent round either peg, so as to keep it at some temperature, perhaps about as high as that of boiling water. If the wire be well annealed at the commencement of the experiment, and if weights be gradually added to the lower side of the frame, the galvanometer needle gradually moves to one side, indicating a current from the unstretched to the stretched round the hot peg; and the deflection goes on increasing as long as weights are added, up to the breaking of the wire. If, however, before the wire breaks, the weights are gradually removed, the needle comes back towards its zero-point, reaches zero, and remains there when a certain part of the weight is kept suspended. If this is removed the needle immediately goes to the other side of zero, and remains, indicating a current from the strained part into the unstrained part of the iron wire round the part wrapped on the hot peg; that is, from strained to unstrained through hot, or as Magnus found, "from hard to soft through hot."

111. If weights be added again, as at first, this deflection is done away with, and the deflection that first appeared is regained, when the weight which previously allowed the needle to return to zero is exceeded. We thus conclude that iron wire hardened by longitudinal tension, may, by the application of a certain longitudinal force, have its thermo-electric quality reduced to that of unstrained soft iron, and by a greater force may be made to deviate in the other direction; or *that hardened iron under a heavy stress, of the kind by which it has been hardened, and hardened iron left free from stress, are on different sides of unstrained soft iron in the thermo-electric series.* There can be no doubt but that the same property holds for copper wire, being in fact demonstrated by the experimental results described above in §§ 107 and 109.

112. I have not yet investigated the thermo-electric effects of stress (that is, the effects accompanying temporary strain) in other metals than iron and copper; but it appears probable that the same law of relation to the thermo-electric effects of permanent strain without stress will be found to hold in each case, since it has been established for two metals in which the absolute thermo-electric effects are of contrary kinds. I hope, however, before long to be able to adduce experimental evidence which will supersede conjectures on the subject. [Since this paper was read I have verified the same law for platinum wire.]

113. The object which was proposed in entering on the investigation, being to test the thermo-electric properties of a strained metal, in different directions with reference to the direction of the strain, was not attained by comparing the thermo-electric properties of a longitudinally strained metal with those of the same metal in its natural state; but it would certainly be promoted by discovering the effect of lateral pressure on a wire in modifying its longitudinal thermo-electric action. I therefore made the following experiments on the thermo-electric effects

experienced during the application of a *moderate* lateral pressure, and of permanent strain after the cessation of *excessive lateral pressure*, in various wires.

114. Experiment to discover the temporary effect of lateral pressure on the thermo-electric quality of iron wire:—A rect-

Fig. 21 a.   Fig. 21 b.

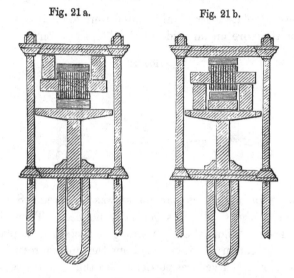

angular bar of iron (1⅔ inches square), with pieces of thin hard wood placed on two opposite sides, had fine iron wire laid in a coil of about twenty turns round it. The wood perfectly insulated the wire from the iron bar, and the different turns of the wire were kept from touching one another, by little notches cut in the edges of the pieces of wood. The whole coil was made firm, and its extreme turns tied down to the wood to prevent slipping. The ends of the wire, extending a foot or two on each side of the coil, were connected in the usual way (§ 92) with a galvanometer. The bar bearing the coil was laid with its two wooden faces horizontal, and one of them supported on a thin piece of hard wood lying on the stage of a Bramah's press. Another thin piece of hard wood was laid upon the top of the coil, to prevent the upper part of it (when, in the course of the experiment, it is forced upwards,) from touching the roof of the press. Blocks of iron were placed on the ends of the bar, so that when the stage is pushed up they may be resisted by the roof, cause a heavy stress to act on the bar, and press the lower horizontal parts of the wire coil between the two pieces of hard wood touching them above and below (see Fig. 21 a). The same blocks are afterwards shifted to rest on the stage and bear the ends of the bar upon them, so that, when the stage is forced up, the upper parts of the wire coil may be pressed against the piece of hard wood above them, which will then be resisted by the roof of the press (see Fig. 21 b).

Pieces of plate glass highly heated were applied to the vertical parts of the wire on one side of the bar, those on the other side

Fig. 22.

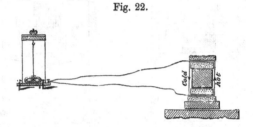

being left cool, and the galvanometer was observed. Some slight deviation of the needle was generally noticed. Then the press was worked, and immediately a strong deflection took place, indicating a current in the iron coil, from the uncompressed portions through the heated vertical portions, into the compressed portions. The pressure was relieved, and the galvanometer needle returned nearly to zero. It was reapplied, and the same powerful deflection was observed. The glass heaters were shifted to the other side, and, the pressure being continued, the deflection of the needle became reversed. The pressure was removed, and by shifting the iron blocks, and working the press again, was applied on the other horizontal side of the coil. The heating being kept unchanged, a reverse deflection was observed, powerful as at first. The current indicated was in every case from *free iron wire* to *pressed iron wire through hot*, as is illustrated in the diagram (Fig. 22), for a case in which the upper parts of the wire are compressed.

115. From this, in conjunction with the result regarding the effect of longitudinal stress previously obtained, we may nearly conclude that a longitudinal strain in iron developes reverse thermo-electric qualities in the axial direction and in directions perpendicular to it; for there can be little doubt but that a lateral

traction would produce the reverse effect of a lateral pressure, or that a portion of a linear conductor of iron pulled out on two opposite sides in a direction at right angles to its length, would acquire such a thermo-electric quality as to give rise to currents from *stretched to free through hot*. But in the former experiment (§ 108) it was demonstrated, that when part of an iron conductor is pulled out longitudinally, the thermo-electric effect gives currents from *free to stretched through hot*. The crystalline characteristic is therefore established for the thermo-electric effect of mechanical stress applied to iron, if it be true that traction produces the reverse temporary effect to that of pressure in the same direction. There seems so strong a probability in favour of this supposition, that it may almost be accepted without experimental proof; but I intend, notwithstanding, to make experiments, for the purpose of explicitly testing it, as soon as some preparations at present in progress enable me to do so. In the mean time I have made the following decisive experiment on the difference of thermo-electric quality in different directions in iron subjected to stress.

116. A piece of sheet-iron 36 inches long and 16 inches broad, was rolled round two thick iron wires ($\frac{1}{4}$-inch diam.), along its breadth at its two ends, and soldered to them. It was cut into narrow slips, each about $\frac{1}{4}$ of an inch broad and of different lengths, as shown in the diagram, so as to prevent electric conduction, except along a band about half an inch broad running across the sheet at an angle of 45° through its centre. The ends of the slips on each side of this band were clamped (as shown in the annexed sketch Fig. 23) between two flat iron bars, but insulated from them by thin pieces of hard wood*, and from one another, where necessary, by pieces of cotton cloth. These bars were each $\frac{1}{2}$ an inch thick, 3 inches broad, and 30 inches long; and the two at each side clamped together upon the pieces of hard wood, with the iron slips between

Fig. 23.

---

* The thinner the better, I believe, as a partial failure was experienced from these pieces of wood breaking at one side and allowing the ends of the iron slips to get drawn in between the iron bars.

them, formed a firm beam, [see Figure 24, which
is a section] by means of which a considerable stress would be brought to bear on the
sheet iron to stretch it in the direction of the
slips. The upper of these beams was laid resting
with its two ends on the tops of stout wooden
pillars, supported below on a very strong wooden
bar laid on the stage of a Bramah's press. The
lower double iron beam hanging down and straight-
ening the sheet iron by its weight, had strong iron
links put over its ends, and an iron bar of about
1⅜-inch square section slipped through them below, so as to hang
down a small distance below the roof of the press. Thus, when the
press is worked, the upper double iron beam is forced up, and the
sheet iron is stretched between it and the lower double iron beam,
which is held down by the links and the bar under the roof of the
press. [See Figure 25.] Before working the press, the rectangular

Fig. 24.

Fig. 25.

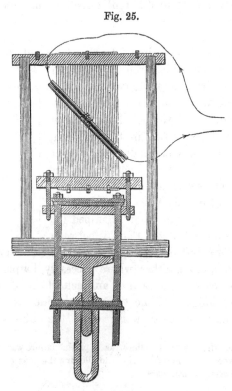

wooden frame with its iron cross-head is steadied by cords from hooks in the ceiling, and the following arrangements are made:—Two slips of sheet iron, each about 18 inches long, are soldered to the upper and lower ends of the oblique conducting channel, and their other ends are soldered to copper wires and put into the circuit of a galvanometer, with the usual precautions (§ 92) to ensure equality of temperature and electrical insulation between the two junctions of the dissimilar metals. Four tin-plate tubes, of semicircular section, each about ⅜-inch diameter, and coated with a single fold of paper pasted round it, are pressed with their flat sides on the two sides of the sheet iron against the upper and lower edges of the oblique conducting band; and are connected by india-rubber junctions, so that steam may be blown through two of them to heat one edge of the conducting band, and cold water sent through the other pair to keep the other edge of the band cold. The arrangements being thus made, a small boiler, heated by a common wire-gauze gas-lamp, is used to send steam through one pair of the tubes, and the town-supply water-pipes give a continued stream of cold water through the other pair. When the galvanometer was observed, there was at first no sensible indication of a current. The press was then worked, and the galvanometer immediately exhibited a slight deflection. The press was released, and a careful observation gave again little or no evidence of a current. Then, by an arrangement of double-branched stop-cocks, the steam and cold water were quickly reversed, so that the edge of the conducting band which was hot became cooled, and the other one became heated. Still the galvanometer showed no sign of current until the press was worked, when a reverse deflection to the former was manifested. While the press was kept up the steam and cold water were again sent along the same edges as at first. After a short time the deflection of the needle was reversed, and the same current as at first was indicated. The deflections were very slight in each case, but were unmistakeably demonstrated by the use of the reversing break (commutator) connected with the galvanometer. Had it not been for the accident noted above, a much more powerful stress would have been applied to the iron, and I have no doubt but that conspicuous deflections of the needle would have been produced.

117. The current in every case was *down the inclined channel of sheet iron when the upper edge was heated, and up the incline*

when the lower edge was heated. That is, if we imagine a rectangular zigzag, from side to side of the bar, instead of the true
rectilinear course of the current, the current would be from *transversely stretched to longitudinally stretched through hot.* Hence it
is established by this experiment, that iron, under a simple longitudinal stress, has *different thermo-electric qualities in different
directions.*

Knowing, as we do, from the first experiment on copper, described above (§ 107), that iron is not the only metal thermo-
electrically affected by stress, we may conclude with much probability that, in general, metals subjected to stresses not equal
in all directions will acquire the crystalline characteristic of having
different qualities, as regards thermo-electricity, in different directions.

118. The qualitative investigation of the thermo-electric effects
of stress, unaccompanied by permanent strain, that is, the elastic
thermo-electric effects of stress, would be complete for iron if the
thermo-electric effect of a uniform dilatation or condensation in
all directions had been ascertained. I hope before long to be able
to carry into effect various plans I have formed with this object in
view; but in the mean time it would be the merest guessing to
speculate as to the result.

119. The establishment of the crystalline characteristic for
the thermo-electric effects of stress not equal in all directions,
would make it probable that any thermo-electric effects which a
metal permanently strained by such a stress can retain after the
stress is removed, must also possess the crystalline characteristic.
That this is really the case I had in fact proved, before performing
the decisive experiment, just described, regarding the nature of
the elastic effect, which was only made a few weeks since. The
following experiments on the thermo-electric effects of permanent
strains in metals were all made more than a year ago.

120. Well-annealed iron wire was rolled in a coil of about
twenty turns on a flat bar of iron $\frac{1}{4}$-inch thick and two inches broad.
The bar was laid on an anvil, with little pieces of thicker wire laid
upon it to support the iron core and prevent the lower parts of
the coil from being pressed. The upper parts of the coil lying
on the upper flat side of the core were hammered till they were
all very much flattened. The coil was then a little loosened and

drawn off the bar of iron, and a similar wooden core was pushed into it. The ends of the iron wire were arranged, with the usual precautions (§ 92), in connexion with the electrodes of a galvanometer. A piece of hot glass (not above the boiling-point of water) was laid along one edge of the coil, so as to heat the iron wire at one set of the points separating hammered from unhammered portions. The galvanometer showed by a great deflection of its needle a current through the iron coil *from hammered* to *unhammered through hot*. When the heater was applied at the other edge of the flat coil, the deflection soon became reversed; still, and always in subsequent repetitions, *indicating a current from the strained to the soft metal through the hot junctions*.

121. The coil was next replaced on its iron core, heated to redness in the fire, and cooled slowly. It was then insulated by slipping in paper between it and the iron bar, or by putting it once more on its wooden core; and it was tested in the galvanometer circuit with the application of glass heaters as before. Not the slightest trace of a current was now found; a result verifying the conclusion arrived at by Magnus, that it is not peculiarities of form in different parts of a circuit of one uncrystallized metal, but variations in its quality as to mechanical strain, that can ever give it continuous thermo-electric action.

122. It has thus been proved that a circuit of iron permanently strained by pressure across the lines of conduction acquires the same kind of thermo-electric quality as that which Magnus first discovered to be produced by the lateral pressure compounded with longitudinal traction, which the process of wire-drawing calls into play, or as that which I had myself found to result from a simple traction, leaving a permanent elongation after the force is removed. In all these cases the iron is found to be harder than it was before acquiring the strain, or than it becomes again after being annealed. Hence the nature of the thermo-electric effect in each of the three cases falls under the designation "*current from hard to soft through hot*," by which Magnus stated his result as regards iron. This is just as is to be expected from the crystalline theory; since longitudinal extension has a common characteristic with lateral condensation in the theory of strains, and only differs from condensation uniform in all transverse directions, by a certain degree of absolute dilatation which accom-

panies it, instead of the slight absolute condensation accompanying the lateral condensation as an effect of pressure all round the sides. In fact the agreement between the characters of the thermo-electric effects due to longitudinal traction and lateral pressure, and again between the reverse characters of the effects of permanent longitudinal extension and those of permanent lateral compression established by the experiments which have been described, proves that these effects are due to distorting stress, and to permanent distortion, in the main, and leaves it quite an open question, only to be decided by further experimental investigation, what may be the effects of uniform pressure and of permanent uniform condensations or dilatations.

123. The crystalline theory is really unavoidable when it is thus established that the effect discovered is due to distortion; but

Fig. 26.

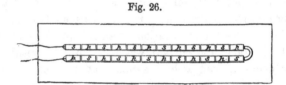

still, as the one designation "current from hard to soft through hot" applies to all the cases of permanent strain in iron as yet experimented on, I thought it necessary, for removing the possibility of objections, that an iron conductor giving a current from soft to hard through hot, should be constructed (Fig. 26). I therefore took twenty-four small soft iron bars turned in a lathe to a cylindrical form ¼th of an inch diameter, and each an inch long, with flat ends; and compressed twelve of them longitudinally in a Bramah's press, so as to permanently shorten each by about ⅛th of an inch. They were then set in a wooden board cut to hold them firmly lengthwise in two rows, those hardened by compression and those left soft, being placed alternately with their ends in contact. The end pieces towards one side were connected with one another by a little slip of iron touching each, and the other ends of the rows were connected with the electrodes of a galvanometer by slips of iron touching them. Each row was firmly wedged up between its terminal iron slips to ensure metallic contact; but after several attempts, and with all care in cleaning the surfaces meant to touch, no sufficient completeness of contact throughout the circuit could be obtained until mercury was introduced as a liquid solder to connect the pieces of iron. This was done simply by pressing them together as at first, pasting paper round the junctions, and pushing little drops of liquid mercury or small quantities of soft mercurial amalgam into apertures in the tops of these paper coverings. Twelve hollows were cut in the board under and round the junction of the iron bars, each except the last including a pair of ends of the bars in contact in each row, and the last including the ends of the extreme bars on that side and the slip of iron by which they are connected. These hollows were filled alternately with hot sand and cold sand, which was everywhere piled over the junctions; and the galvanometer gave slight indications of a current, the direction of which through the iron appeared to be generally from uncompressed to compressed through hot.

124. The result, however, was not satisfactory; and it was obvious that the plan which had been adopted for heating and cooling was quite insufficient to sustain the required differences of temperature through so considerable masses of iron; I therefore had an apparatus constructed for the purpose, consisting of two main pipes of tin-plate, each carrying six smaller pipes and leading to small cells, also of tin-plate, with cylindrical passages through them to admit the iron bars, and with short discharge pipes attached to them on the other side from that by which the former enters. These cells (Fig. 27) were fitted into the hollows cut for the sand in the board formerly used, the main pipes occupying parallel positions above them on each side several inches from one another. The iron bars, each coated with paper and united as before one to another with mercury solder, were pushed through the hollows of the cells, and were fixed in two rows, with a junction in the centre of each of these hollows, and with the terminals adjusted as before. Cold water from the town supply-pipes was then run into one of the main pipes, so as to flow through the branch pipes and cells connected with it; and steam from a boiler heated by an ordinary wire-gauze gas-burner was sent through the other system, so as to cool and heat alternately in their order of position the twelve cells with the junctions which they surround

Fig. 27.

(Fig. 28). A deflection of the galvanometer needle, amounting
to about 4°, was now observed; and when the cold water and steam

Fig. 28.

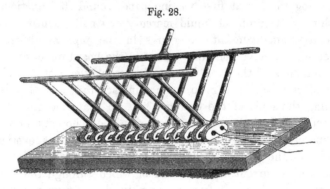

supplies were interchanged in the two sets of tubes, an equal
reverse deflection almost immediately took place. The current
indicated was always in many trials from *uncompressed to compressed through hot* in the iron of the circuit.

125. Here then we have a case of thermo-electric action in
iron giving a current *from soft to hard through hot;* not as found
before, "from hard to soft through hot." Hence it is not pieces
of hardened iron in general, but *the direction of extension* or
directions perpendicular to the direction of compression, in iron
hardened by extension or by compression, that have the thermo-
electric quality of deviating from soft iron towards bismuth; and
a line of compression, or (as we may now safely conclude) *lines
perpendicular to a line of extension,* have the reverse deviation,
that is deviate from soft iron towards antimony, in the thermo-
electric series. [*Addition, Dec.* 1856.—Subsequently to the reading of the paper, I have, in verification of this conclusion, found,
by a direct experiment, that a conductor of sheet iron, hardened
by lateral extension and softened in parts, has the thermo-electric
property of giving a current *from soft to hard through hot.*] The
crystalline theory being thus fully established for the thermo-
electric effects of mechanical strain in iron, whether temporarily
induced during the application of stress, or remaining with molecular displacement after the stress is removed, we may readily
suppose it will be found to hold equally for all thermo-electric
effects any metal can experience from mechanical action, except
the hitherto undiscovered effects of condensations or dilatations

equal in all directions. The experiments I have already made on other metals than iron, do not go further in verifying the crystalline theory than to show for copper and tin wires what I had previously shown for iron, that the same thermo-electric effect in a linear conductor is produced by permanent longitudinal extension and permanent lateral compression.

126. The process of raising to a high temperature and then cooling very suddenly, produces a marked effect on the mechanical qualities of most metals, especially on their hardness; and generally all that is necessary to do away with this effect and restore the metal to its primitive condition, is to keep it for some time at a high temperature and let it cool slowly. This process being called annealing, I shall for brevity designate as *unannealed*, any substance which has been subjected to the former process (sudden cooling) and which had not been subsequently annealed. It is not easy to judge exactly of the relation of the strains in the different parts of an unannealed piece of metal, to simple mechanical strains; but *some thermo-electric effect*, whatever its exact nature and explanation may be, is to be anticipated, with so great a change of other qualities as many metals experience in the process of sudden cooling; and it may be readily supposed that different thermo-electric qualities will be found in unannealed pieces of different shapes. I have therefore made experiments on the thermo-electric differences between unannealed and annealed linear conductors consisting of round wires, of wires flattened by hammering, and of flat slips, of one metallic substance.

127. Twisting a wire beyond its limits of elasticity hardens it perhaps as much as traction or hammering, and certainly in every case, when continued far enough, makes the metal very brittle. The nature of the mechanical strain here operative is easily expressed and explained in the theory of elasticity, in terms of simple strains different in magnitude and direction in different parts of the wire; but it is not very easy to judge by theory, from the effects of simple strains supposed known, what kind of thermo-electric effect, if any, is to be expected in a metallic wire, with strain thus heterogeneously distributed through it. I have therefore made experiments to determine this effect in various metals.

128. For experimenting on the thermo-electric differences between annealed and unannealed metallic conductors, a wire, round

or flattened, or a slip of the metal was wrapped in a coil of from ten to thirty turns on a wooden core, about two inches broad and $\frac{1}{4}$ of an inch thick, or sometimes only an inch broad, with a flat slip of thin sheet-iron laid on one side of it. The wooden core was then drawn away, and the coil, held in form by the thin iron core, was heated to redness in the fire, or to some temperature short of its melting-point, in hot oil, and was then suddenly plunged in cold water. After that, one side of the iron core was held over a flame, so as to heat the parts of the coil next it, while the parts of the coil on the other side were carefully kept cool, by the constant application of cold water with a sponge. The wooden core was then slipped in and the sheet-iron removed; and the coil was ready for testing by the galvanometer.

129. The preparations for an experiment on the thermo-electric effect of permanent torsion, were commenced by bending a short portion at each end of a length of two or three yards of the wire to be examined, holding these end portions so as to keep the wire between them firmly stretched, and twisting it till it became brittle. It was then wound on a flat iron core (unless it was too brittle, as often proved to be the case, and then another wire was similarly prepared but not twisted quite so much); the parts of the coil on one side were carefully annealed by flame or hot oil, while those on the other side were kept cool by sponging with cold water. The iron core was then drawn out and the wooden core slipped into its place; and the coil was ready for testing by the galvanometer.

130. In making the thermo-electric experiments on the coils prepared in these various ways, glass heaters were first used, but I afterwards substituted two tubes of horse-shoe section made of tin-plate and coated with paper, which were applied with their concave parts touching the coil round its two edges (Fig. 29). Steam from the small boiler was sent through one of these, and cold water from the town supply-pipes through the other.

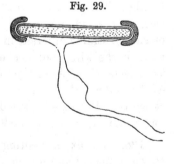

Fig. 29.

131. The wires used, with the exception of the iron, steel and brass, were all supplied by Messrs Matthey and Johnson, as chemically pure. The results of the experiments (made as described in §§ 120 and 121) on the effects of lateral hammering were, in every other kind of wire tried, the reverse of those found for iron. Thus in steel, copper, tin, brass, lead, cadmium, platinum, zinc, the current was always found to be from the unhammered to the hammered portions through hot. All the wires except zinc were carefully annealed by myself, before they were coiled and hammered (§ 120); but the process of annealing by heating in oil and cooling slowly made the zinc very brittle and crystalline, instead of softening it as in the other cases, and it was therefore taken as supplied by the manufacturers, and coiled on the core and hammered in the manner described.

132. The experiments on the coils differently tempered in their different parts (§ 126), in the cases of tin and cadmium, gave only doubtful galvanometer indications; zinc wire proved so brittle *in the annealed parts* as to defeat some attempts to test the thermo-electric effects of temper. I have little doubt but that results may be obtained in all these cases by a careful repetition of the experiments, with perhaps some modification to meet the peculiarity of zinc. Slips of sheet-iron and of sheet copper were tried without any thermo-electric indication being noticed. [*Addition, Dec.* 1856.—I have recently found in slips of sheet iron the same thermo-electric effect of temper as in round and flattened iron wires.] All the other conductors tried gave very decided results. In the cases of round iron wires of very different diameters, of iron wire flattened through its whole length by hammering, of round steel wire, and of steel wire flattened through its whole length by hammering, and of steel watch-spring, the thermo-electric effect of annealing portions of the coil after the whole had been suddenly cooled, was *a current from unannealed to annealed through hot.* In round wires of copper and brass, the thermo-electric effect of the same process was *a current from annealed to unannealed through hot.*

133. The effects of permanent torsion were decisively tested only for iron and copper wires; and they proved to be in each case the same as the effects of hardening by longitudinal extension, by lateral compression, or by rapid cooling, being quite

decidedly *from brittle to soft through hot in the iron*, and *from soft to brittle through hot in the copper.*

134. The views explained above (§ 105), by which I was led to look for the thermo-electric qualities of a crystal in a non-

Fig. 30.

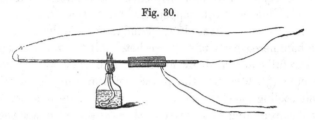

crystalline metal subjected to mechanical strain, show the probability of finding such properties also developed along with magnetism, by external magnetic force, especially in the few metals, iron, nickel and cobalt, which have high capacities for magnetic induction. Towards verifying this idea I tried first the following simple experiment, analogous to the first experiment (§ 107) which I had made on the thermo-electric effects of tension. A little helix about three inches long, consisting of 220 turns of thin covered copper wire laid on in three strands on a cylindrical core of pasteboard, about ¼ of an inch internal diameter, was slipped upon a piece of thick straight iron wire about two feet long, which was supported in a horizontal position by its ends, and through them put in the circuit of a galvanometer (Fig. 30). A spirit-lamp was held under the middle of the wire so as to raise it to a high temperature, and then a current from a few of the iron cells was sent through the helix, which was kept a little on one side of the middle of the wire. Immediately the galvanometer needle, which was not at first disturbed by the application of the spirit-lamp, experienced a deflection. The little helix was slipped rapidly through the flame of the spirit-lamp to the other side of the hot part of the wire, and a reverse deflection was immediately produced. It was easy, by moving the helix alternately to the two sides of the hot middle of the wire, to make the needle of the galvanometer to swing through an arc of 10° or more. When the needle was brought to rest there was always a most sensible permanent deflection, on one side or the other, according as the helix was left on one side or other of the heated parts. When the circuit of the galvanometer was broken, none of these effects followed from the motions of the helix. They were therefore not due to the direct force of the magnetism in the helix and iron wire, but to that of a current through the galvanometer coil. This always took place in such directions as to indicate a current *from unmagnetized* to *magnetized through hot.*

135. The decided character of the result of this experiment established it beyond doubt, that the thermo-electric quality of iron is altered by magnetization. Immediately the question arose (from the general considerations referred to above, § 104 and 105), *are the thermo-electric qualities equally or even similarly affected in all directions?* and the crystalline hypothesis suggested the answer:—no; probably even the reverse thermo-electric effect may be found across their lines of magnetization. As theory could give no more than a conjectural answer, I tried to find the truth by experiment; and, after various fruitless operations, obtained a very decided result, in the following way.

136. A piece of thin sheet iron was cut into the shape shown in the diagram (Fig. 31), the breadth everywhere being about $\frac{1}{4}$ of an inch, the length of the longer branch 45 inches, and that of the shorter six inches. The longer branch was rolled into a plane spiral, on a cylindrical core $\frac{1}{2}$ an inch diameter, the different successive turns being prevented from touching one another by a piece of narrow tape wound on along with the iron slip. The shorter branch, which stood out

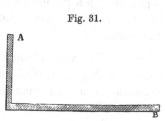

Fig. 31.

from the inner end of the coil at right angles to the plane of the spiral, was bent round into this plane, and carried out along one side of the spiral several inches beyond its circumference. Along with it, a portion of the slip next the other end which was left uncoiled, was carried out from the outer part of the spiral, and cut to such a length as to let the two ends be brought close together. Copper wires, to lead to the galvanometer electrodes, were soldered to these ends, and the junctions of dissimilar metals thus formed were arranged with the usual precautions (§ 92) to ensure equality of temperature and electrical insulation. Contrary poles of two steel bar-magnets, each about three feet long and of rectangular section four inches by $\frac{1}{2}$ inch, were placed pressing on each side of the spiral, as shown by the dark shading in the diagram, but insulated from it of course. Four rectangular pieces of thick plate glass,

Fig. 32.

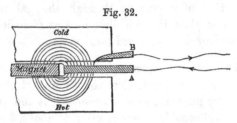

two of them very hot (perhaps about 300° C.) and two cold, were applied, touching the coil on each side, and symmetrically arranged on the two sides of the steel magnets.

The galvanometer showed a current in the direction indicated by the arrow-heads. The pieces of hot and cold plate glass were interchanged, and the current became reversed. The magnets were removed, and their effects became scarcely perceptible, or altogether ceased. On repeated trials a current was found always in the direction, from parts of the coil between the magnets towards parts touched by the hot glasses. The experiment was repeated with a powerful electro-magnet, and gave the same result, but not with the same ease, because of difficulties in applying the heaters, &c.

137. The very strong tendency iron has to assume longitudinal rather than transverse magnetization, when of any form extended in one direction more than in others, was partially done away with by the mutual influence of the different turns of the spiral used in the experiment which has been described; and the symmetrical arrangement of the heaters was such as to nearly exclude all thermo-electric action, except what is due to the thermo-electric difference between that part of the coil touched on each side by the steel magnets, and the part diametrically opposite. Any thermo-electric effect there may have been from longitudinal magnetization in the parts of the iron ribbon on each side of the steel magnets, must, so far as I could judge, have been contrary to the effect observed. The result obtained, therefore, demonstrates an electro-motive force urging a current from *transversely magnetized parts of the iron conductor, through hot parts, to comparatively unmagnetized* parts. Hence a transversely *magnetized* iron conductor deviates from unmagnetized iron to-

wards bismuth, or in the reverse direction to that of the deviation discovered in wire longitudinally magnetized, in the first experiment on the thermo-electric effects of magnetism. It may be concluded, *à fortiori*, that in uniformly magnetized iron, *directions transverse to the lines of magnetization* differ thermo-electrically from *directions along the lines of magnetization;* and differ in such a way, that if we could get an iron conductor of the shape indicated in the diagram (Fig. 33), magnetized with perfect uniformity everywhere, in the direction shown by the lines of shading, and if, when the two ends kept at the same temperature are put into the circuit of a galvanometer, the corner is heated, a current would be found to set in the direction shown by the arrow-heads, that is, *from transversely magnetized* to longitudinally magnetized *through hot.*

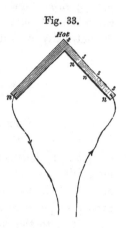

Fig. 33.

138. To test and illustrate this conclusion, I took a piece of sheet iron, cut to the shape shown in the diagram (Fig. 34), and wound it spirally on a wooden cylinder, prepared with spiral grooves and pipes for steam and cold water, as described below. The oblique edge of the iron, shown on the left boundary in the diagram, being cut at angles of 45° and 135° to the long edges coterminous with it, was bent in a plane perpendicular to the axis of the cylinder, and thus the long edges of the iron, and the cut separating it into two branches, formed spirals, each at an

Fig. 34.

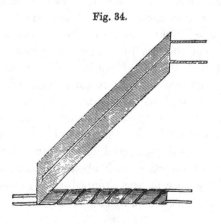

angle of 45° to the axis of the cylinder. The two long edges themselves came very nearly to coincide, the circumference of the cylinder being a little greater in length than the oblique edge of the iron which thus nearly met round it. These two edges, as well as the two edges on each side of the cut between the branches, were prevented from touching one another by being, one at least in each of the contiguous pairs, bound with cotton tape. The projecting slips (shown on the right in the diagram (Figs. 34 and 35)) came to positions parallel to the axis of the cylinder, through two diametrically opposite parts of its circumference. Their ends had

Fig. 35.

copper wires soldered to them, and were arranged with the usual precautions (§ 92) to ensure electrical insulation and equality of temperature between them. The wooden cylinder had two diametrically opposite spiral grooves, each at the same inclination of 45° to the axis, and spiral sheet copper tubes, prepared of the proper shape, were slipped into these grooves, and nearly filled up the spaces to the surface of the cylinder. The outsides of these tubes were coated with paper, so as to maintain electric insulation between them, and the sheet iron wound on outside.

The wooden cylinder bearing the spiral tubes, and the sheet iron arranged in the manner described, was slipped into the hollow of an electro-dynamic helix, steam was sent through one of the spiral tubes and water through the other, and the copper wires soldered to the ends of the iron slips were connected with the electrodes of a galvanometer. No current was at first indicated. The galvanometer circuit was broken by its own commutator, and a current was sent through the magnetizing helix. The galvanometer circuit was completed again, and immediately a strong indication of a current through it was manifested. The galvanometer circuit was broken, the magnetizing current reversed, and the galvanometer circuit again completed; again the same current as before was observed. The steam and cold water were interchanged in the spiral pipes, and the galvanometer current soon set in the reverse direction, with about the same force as before. The magnetizing current was stopped (the galvanometer circuit being broken for the time and closed again), and only slight traces of the current that had been so powerfully indicated could now be observed.

139. In this experiment the action of the electro-dynamic helix caused the double slip of iron to receive magnetization in lines nearly parallel to the axis of the cylinder (only a little disturbed in consequence of the gaps between the adjacent edges), that is to say, magnetization as nearly as may be in directions at an angle of 45° to its length. The sources of heat and cold applied along the two spirals, gave either heat along each of the outer edges of the double slip, and cold along the inner edges between the two branches, or cold along the outer edges and heat along the inner edges. When the ends were connected with the electrodes of the galvanometer, in the case illustrated in the diagram, the current was in the direction indicated by the arrow-heads; and it was

Fig. 36.

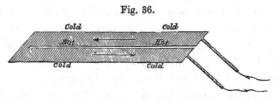

always in such a direction, that if a zigzag line be traced through the two slips from side to side of each, on the whole in the same direction as the current, the changes of direction at the sides of the slips are from *transversely to longitudinally magnetized through hot, and from longitudinally to transversely magnetized* through cold ; which is the conclusion that was anticipated.

140. I also experimented on the thermo-electric effects of retained magnetism in steel, after the magnetizing force is removed, and obtained very decided results, showing that at least in the case of magnetization along the lines of current, the effect is of the same quality as in soft iron or in the steel itself while under a magnetic force which induces such a state of magnetization.

141. In one of these experiments, thirty-nine pieces of steel wire, each about $\frac{1}{18}$th of an inch diameter and two inches long, soft tempered, were connected by thirty-eight pieces of copper wire, each an inch long, placed between each two of the pieces of steel, and hard soldered to their ends. Pieces of copper wire of

the same length were soldered to the outer ends of the first and last pieces of steel, and several feet of steel wire to the ends of each of these. A little electro-dynamic helix was made, two inches long and wide enough internally to slide freely over this compound steel and copper conductor; and by means of it every second piece of the two inch steel wires, commencing with the first and ending with the thirty-ninth, were magnetized alternately with their poles in dissimilar directions, while the other short wires, and the longer steel terminals, were left as free from magnetism as possible. The magnetizing helix was then removed, and the compound conductor was made into a flat coil on a wooden core (two inches broad and ¼-inch thick), by bending the short copper wires, and arranging the two inch steel wires alternately on the two sides of the wood. The terminals were joined, with the usual precautions (§ 92), to the galvanometer electrodes, and one edge of the coil was immersed nearly an inch below the surface of a vessel of oil at the temperature of about 100° C. Immediately a strong deflection of the needle showed a current, of which the direction in the coil was *from unmagnetized to magnetized through hot*. When the other edge of the coil was similarly heated, a contrary deflection of the needle has decidedly showed the same thermo-electric difference of quality between the magnetized and the unmagnetized steel wires.

142. The object of the peculiar arrangement just described, was to prevent the magnetism from spreading to those of the steel portions of the circuit which were to be kept as free from magnetism as possible in order to be compared with those which were magnetized. The introduction of the connecting pieces of a different metal from steel into the circuit, cannot give rise to any thermo-electric disturbances *, provided the two ends of each are at the same temperature, a condition which was nearly enough fulfilled in the way the experiment was made, and which was very much favoured by the shortness and the high thermal conductivity of the little copper arcs.

The same result was demonstrated in an experiment made with a homogeneous coil of steel wire, of which parts had been magnetized, by ordinary steel magnets, before it was bent on the core.

---

* "Dynamical Theory of Heat," [Art. XLVIII. Vol. I., above] § 138, Cor. 1.

[§ 143. Received, May 10, 1856.]

§ 143. *Experiment.—On the Effect of Magnetization on the Thermo-electric Quality of Nickel.*

Through the kindness of Dr George Wilson, I have been able to experiment on a bar of nickel, about half an inch in diameter and about eight inches long, in the form of a horse-shoe magnet, belonging to the Industrial Museum of Edinburgh. The accompanying sketch (Fig. 37) and description show the plan of the experiment.

DESCRIPTION OF SKETCH.

N, nickel horse-shoe.

B B, double tubes of sheet copper, electrically connected with one another by a copper band, and insulated from the nickel by silk paper, laid on with shell-lac varnish; serving to drain all electrical leakage from the magnetizing coil, without causing the slightest sensible current through the nickel, and serving also to convey a stream of cold water to maintain the lower parts of the two branches of the horse-shoe at as nearly as possible equal temperatures.

A A A, india-rubber pipes to lead a stream of cold water through the coolers.

C, magnetizing coil, wrapped on one of the copper coolers.

E E, electrodes of magnetizing battery of twenty iron cells, charged with nitric acid, &c.

F, commutator for interrupting and reversing the connexion between the magnetizing battery and coil, or reversing the current.

M M, mercury cups, in which the extremities of the nickel were immersed (mercury being both very convenient for the purpose, and the metal least thermo-electrically removed from nickel of all that have been tried by any experimenter).

m m, mercury electrodes joining copper galvanometer electrodes D D, at G G.

K, commutator for interrupting and reversing the connexions of the galvanometer electrodes.

Fig. 37.

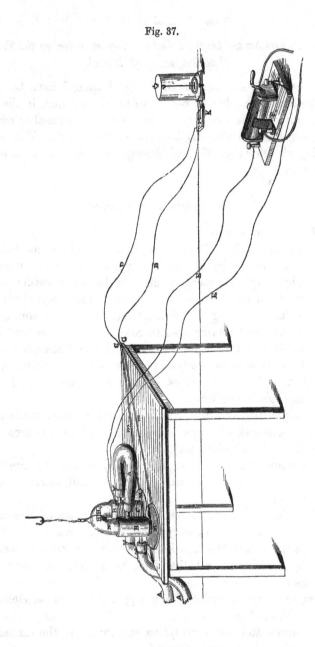

Heat was applied at H H by means of a gas-lamp and blow-pipe. A current from magnetized to unmagnetized through hot, was indicated by a considerable galvanometer effect, which, by management of the galvanometer break, K, was readily directed to give oscillations of the needle through three or four degrees.

The same conclusion had been indicated in several previous attempts, with various defects of arrangement remedied in the experiment just described. In this last experiment the result was made most manifest; and, being completely separated from all effects of induced currents (which were quite insensible), of electrical leakage, and of unequal heating of the junctions of mercury and nickel, and of the junctions of mercury and copper, was set beyond all doubt. I therefore conclude, *that longitudinally magnetized nickel in a thermo-electric circuit deviates from nickel not under magnetizing force, in the same direction as bismuth.* This is the reverse of the deviation which I formerly found to be produced in iron by longitudinal magnetization.

144. The results of the various experiments which have been described in Part III. are collected in the following Tables.

TABLE I.—*Effects of Stresses and Strains on the Thermo-electric Qualities of Metals.*

Description of Conductor.	Thermo-electric Order reckoned from Bismuth towards Antimony.		
Iron	. . . .	Free . . .	Under longitudinal traction.
Iron	. . . .	Free . . .	Under transverse compression.
Iron	Under transverse traction. . . .	. . . .	Under longitudinal traction.
Iron	Permanently strained by longitudinal traction, and left free from stress.	Soft . . .	Permanently strained by longitudinal compression, or by lateral extension, and left free from stress.
Iron	Hardened by transverse hammering .	Soft . . .	Hardened by longitudinal hammering.
Round iron wires of different diameters.	Made brittle by twisting	Annealed after being made brittle by twisting.	
Round and flattened iron wires.	Suddenly cooled.	Annealed.	
Steel wire	Some specimens flattened by transverse hammering.	Soft . . .	Other specimens flattened by transverse hammering.
Round and flattened steel wires.	Hardened by sudden cooling . . .	Annealed.	
Steel watch-spring	Hardened by sudden cooling . . .	Annealed.	
Copper	Under longitudinal traction . . .	Free.	
Copper	. . . .	Soft . . .	Permanently elongated by longitudinal traction, and left free from stress.
Copper	. . . .	Soft . . .	Hammered transversely.

		Annealed after being made brittle by twisting.	Made brittle by twisting.
Round copper wire	. . .	. . .	Made brittle by twisting.
Round copper wire	. . .	Annealed. . . .	Suddenly cooled.
Platinum	Under longitudinal traction . . .	Free.	
Platinum	. . .	Soft . . .	Hammered transversely.
Tin	. . .	Soft . . .	Permanently elongated by longitudinal traction, and left free from stress.
Tin	. . .	Soft . . .	Hammered transversely.
Brass	. . .	Soft . . .	Hammered transversely.
Round brass wire	. . .	Annealed. . . .	Suddenly cooled.
Cadmium	. . .	Soft . . .	Hammered transversely.
Lead	. . .	Soft . . .	Hammered transversely.
Zinc	. . .	Soft . . .	Hammered transversely.

TABLE II.—*Effects of Magnetism on the Thermo-electric Qualities of Iron and Nickel.*

Description of Conductor.	Thermo-electric Order reckoned from Bismuth towards Antimony.		
Iron . . .	Under transverse magnetizing force .	Free	Under longitudinal magnetizing force.
Steel . . .	. . . .	Unmagnetized . .	Retaining longitudinal magnetization.
Nickel . . .	Under longitudinal magnetizing force.	Free.	

PART IV.   METHODS FOR COMPARING AND DETERMIN-
ING GALVANIC RESISTANCES, ILLUSTRATED BY
PRELIMINARY EXPERIMENTS ON THE EFFECTS OF
TENSION AND OF MAGNETIZATION ON THE ELECTRIC
CONDUCTIVITY OF METALS.

145.   In endeavouring to discover the effects of magnetization
and of mechanical strain on the electric conductivity of iron and
other metals, I was led, from trying various more or less obvious
methods for testing resistances, to use a differential galvanometer
of a very simple kind, which I constructed for the purpose.   I
shall give no description of this instrument, as I now (Nov. 1856)
find it in one important quality inferior to the differential galvanometer first constructed and used by M. Becquerel*, and I do not
know that its peculiarity has compensating advantages.   I mention it only because it was with it that I made nearly the first
of my trials to find the effects of magnetism on the electric
conductivity of iron, and the very first by which I obtained a
decided result.

146.   In these experiments I used two covered iron wires,
each several yards long, coiled into circles about four inches diameter, as the two resistance branches in the divided channel
through the two conductors of the galvanometer.   Magnetizing
one of them tangentially by means of a coil of covered copper
wire wound on a copper sheath soldered round it as an electric
drain, I ascertained, on the 23rd of April, 1855, that the electric
conductivity of iron wire is diminished by longitudinal magnetization.   The arrangement however proved, as I anticipated, to
be of a very unsatisfactory kind; and the needle kept moving
across the field in one direction almost steadily, during the whole
time the current was sustained through the tested conductors,
which was for several hours.   Continually more and more resistance had to be added to the conducting channel containing the
iron wire round which there was no magnetizing coil, to keep the
needle within range.   After the magnetizing current had passed

* *Annales de Chimie et de Physique*, tome xvii. 1846.

for some time, this variation of the needle went on more rapidly, and called for more frequent adjustment by the additions to the other branch. All this was just as must be expected; and my reason for not introducing currents of cold water round the two iron coils, to maintain them in precisely similar thermal circumstances, was that the tubular systems required for the purpose could not be easily made, and that I thought I might find out the nature of the result in the first instance, notwithstanding the imperfection of the arrangements. In this hope I was not disappointed. The glass needle (carried by the little suspended magnet, which was only about ½ an inch long), while moving steadily across its field, would receive an impulse forward and make two or three very rapidly diminishing oscillations, when the current was started through the magnetizing coil: when the current was suddenly reversed, the needle would show little or no indication of any effect: when the current was broken, it would make a start backwards, and after two or three oscillations would continue advancing as before, perhaps rather more rapidly. Traces of induced currents in the iron coil under the influence of the magnetizing helix were exhibited by scarcely perceptible differences in the bearing of the needle, according as the current was made in one direction or the other, and by slight impulses it received when the magnetizing current was suddenly reversed. After the current had been kept up for some hours through the iron wires, and when, partly by the heat developed by the magnetizing current during the periods of its flow, and partly by heat conducted from the iron wire within, the outside of the magnetizing coil had become very sensibly hot to the touch, the variation of the needle in the galvanometer became much less rapid than at first; and tolerably satisfactory indications, amounting to a fraction of a degree of permanent deflection, showed with perfect consistence an increase of resistance in the iron wire under magnetic force when the magnetic current was sustained in either direction, and a diminution of resistance in the same iron wire following immediately a cessation of the magnetizing current.

147. I followed the same method in a first attempt to find the effect of transverse magnetization on the electric conductivity of iron; two spirals made on the plan described above (§ 136) being used as the resistance branches in the two channels conveying the divided current, and one of them placed between convex poles

of a Ruhmkorff electro-magnet. The induced currents in making, reversing, and breaking the magnetizing current were of course most conspicuously indicated by the galvanometer needle, but the needle came to rest after a few oscillations; and then it did not exhibit any deviations of a sufficiently marked character, when the direct effect of the electro-magnet (which, by a very trouble-some process of shifting the position of the magnet, was reduced as much as possible in preliminary arrangements,) was eliminated by reversals, to allow me to draw any decided conclusion as to the effect of the magnetic force on the conductivity of the iron spiral across which it acted.

148. Before carrying into execution various obvious improve-ments in the experimental arrangements just described, or ap-plying the system with the differential galvanometer to other investigations, I began to think of Maggi's experiment* on the relative thermal conductivities of a magnetized iron disc in direc-tions across and along the lines of magnetization. As the electrical analogue, the method which Matteucci, and I believe Kirchhoff and others, have used in tracing equipotential lines on the surface of a conductor traversed by an electric current, occurred to me. Six months later, I thought of the multiplying branch (first used in the experiment described in § 161 below) to render available the sensibility which a powerful current through the body to be tested, with the use of a moderately sensitive galvanometer, must obviously give to that method, when appplied to the investigation of differential effects on the electric con-ductivity of a body in different directions; and I succeeded with great ease in making very satisfactory experiments (§§ 161 to 165 below) by means of it, which first decided the question as to whether or not the effects of magnetization give different electric conductivity in different directions to a mass of iron. At first, however, I did not see this or any other way to render the method practicable with galvanometer electrodes, either moveable upon the sheet of metal to be tested (in which case a motion of $\frac{1}{100}$th of an inch would drive the needle from an extreme deflection on one side to an extreme reverse deflection), or by electrodes soldered to points on an equipotential line (in which case a slight alteration in temperature in different parts of the plate might drive the

* De la Rive, *Electricity*, Vol. I. part 3, chap. iii. (p. 316, English edition, 1853).

needle irrecoverably to an extreme deflection on one side or the other); but the experiments which I knew as having been made by Matteucci suggested to me the following very simple plan, which I immediately commenced trying, and which I have since found applicable with the greatest ease to a variety (I believe now to every variety) of experiments on electric conductivities*.

149. Let AB (Fig. 38) be the conductor to be tested, and let CD be another of nearly equal resistance, either a piece of the same wire continuous with the other through an arc BC, or connected with it by a thicker arc of copper, or of another metal, as may appear convenient for the particular case treated. Sometimes the experiment is arranged to test differential effects experienced alternately or simultaneously by AB and CD. But when one of them, AB, alone is acted upon, with a view to varying its resistance, it alone may be regarded as the conductor which is tested; and the other, CD, will then be called the *reference conductor*. Let a wire, AP'OPD, which will be called the *testing conductor*, be soldered by its ends to the ends A and D of the conductor to be tested and of the reference conductor, or to strong pieces of metal to which those ends are firmly attached. Let one electrode of a galvanometer be soldered to the connecting arc BC, at its middle, or at any other point of it, Q; and let the other galvanometer electrode be ready to be applied by the hand to any position on the

---

* [*Note added Nov.* 1856. An hour before the meeting of the Royal Society at which this paper was read, I learned that a method of testing resistances had been given by Mr Wheatstone which would probably be found to be the same in principle as that to which I had been led in the manner described in the text. I have since ascertained that Mr Wheatstone's "Differential Resistance Measurer" (described in § 15 of the Bakerian Lecture for 1843, see *Transactions*, June 15, 1843) is an instrument founded on precisely the same principle as all the various arrangements by which, with great and necessary alterations of detail, I have continued the investigation of effects of magnetism and of other influences, on the electric conductivity of metals, to the present time, and of which some are fully described in Parts IV. and V. of the text. Mr Wheatstone refers to "Experimental Determinations of the Laws of Magneto-electric Induction," printed in the *Philosophical Transactions* for 1833, "as containing the description of a differential arrangement of which the principle is the same as that on which" his own instrument has been devised, and adds, "To Mr Christie must therefore be attributed the first idea of this useful and accurate method of measuring resistances."

It is worth remarking, that the experiments of Matteucci and Kirchhoff, alluded to in the text, are stated to have been first suggested from Wheatstone's idea of applying the two electrodes of a galvanometer to points in separate channels through which two parts of the whole current from one battery are conducted.]

testing conductor. A current is then sent from one or more cells of
Daniell's battery through electrodes
connected with A and D. This current flows through the divided channel ABCD and AP′OPD, in quantities inversely proportional to the
resistances of the two parts. The
moveable galvanometer electrode is
then applied, first to one point and
then to another of the testing conductor (care being taken not to re-

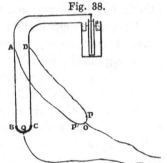

Fig. 38.

verse, nor even to diminish, the magnetism of the lower needle
in the astatic system of the galvanometer*), until by trial the
point O, that may be touched without producing any deflection in
the needle, is found. The influence to be tested, whether it be
magnetization, or tension, or elevation of temperature, is then
applied to AB, or the influences to be tested against one another
are applied to AB and CD, and the moveable galvanometer electrode is (if it has been removed) again applied at O. If the
needle remains undisturbed, no effect is indicated; that is, no
alteration in the resistance of ABQ, or only an alteration in the
same proportion as an alteration experienced by QCD, has been
indicated. If, however, a deflection is observed, in such a direction
that the moveable electrode must be moved to some point P in
the part OD, it is inferred that the ratio of the resistance of
ABQ to that of QCD has been increased; or on the other hand,
if such a deflection as requires a motion of the moveable electrode
to a point P′ in OA, the resistance of AB has been diminished
relatively to that of CD.

* In the galvanometers which I have used, the two needles of the astatic combination are of similar material (pieces of the same steel wire, tempered brittle),
and the lower one is a little longer (perhaps by about $\frac{1}{10}$) than the upper. Both
are magnetized to saturation, and consequently the lower preponderates and gives
its direction to the system. The strongest current through the coil only confirms
the required state of magnetization, provided when it is started the index is either
at zero, or on the side of zero towards which the deflection is to be. If by accident
a powerful current is admitted through the coil when the index is on the wrong
side of zero, the lower needle has its magnetism instantaneously reversed; but it
may be as instantaneously put right again by suddenly reversing the current. If at
any time, from the lower needle having either lost magnetic moment, or acquired
a reverse magnetization, the astatic system is found reversed, it may be put in
order with ease either by simply sending a powerful current through its coil, or by
doing so and then suddenly reversing the current.

150. As an example, I shall describe an experiment on the

Fig. 39.

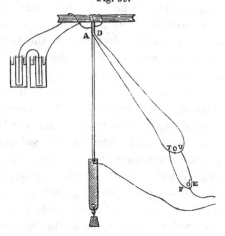

relative effects of tension on electric conductivity in copper and iron wires. Two pieces of stout copper wire, A, D (Fig. 39), were each twisted into a loop which was made fast by solder ; a couple of inches towards one end of each wire being left free from the twisted part. These loops were put upon a strong hard wood peg about $\frac{3}{4}$ of an inch diameter, at a distance of about $\frac{1}{4}$ of an inch from one another ; and to their lower ends were firmly soldered fine iron and copper wires (strong enough to bear weights of about 8 lbs. and 5 lbs. respectively) These wires were cut to the same length of $4\frac{1}{2}$ feet, and their lower ends were put into slits about $\frac{1}{4}$ of an inch deep, cut in the top of a piece of stout copper slip of the form and dimensions shown in the diagram (Fig. 40), and the copper pressed upon them, to hold them fast, by a pair of pincers. Solder was then applied to make a complete and compact metallic connexion between the wires and the copper piece. A testing conductor ATOUD (Fig. 39), consisting of seven yards of No. 18 copper wire, was soldered by its ends to the upper copper pieces A, D; and a current from six small cells of Daniell's was sent through the double channel by electrodes soldered a little higher up to the same copper pieces, A, D. One galvanometer electrode was soldered to

Fig. 40.

the lower copper slip, and the other was applied to the testing conductor till the point O, equipotential with the point of attachment of the former, was found. As from previous experiments I knew that an accidental variation of $\frac{1}{100}$th of an inch in the position of the moveable electrode on the testing conductor might lose or overbalance the effect looked for, I added a *multiplying branch*, TFO'EU, consisting of a yard of No. 18 copper wire, with its ends soldered about half an inch on each side of O. This, of course, when touched by the moveable electrode, gave about thirty-six times the motion that would be required to produce or to correct any effect on the galvanometer if the simple testing conductor were used. The point O', on the multiplying branch, that could be touched without giving any deflection, was then found; and weights were hung from the lower end of the lower copper slip, so as to stretch the copper and iron wire equally. Immediately a deflection of the needle in the galvanometer showed a current. This was corrected by sliding the moveable electrode on the multiplying branch towards U, that is, towards the parts conterminous with the copper wire. When the weights were removed, immediately a reverse deflection was observed. The conclusion is, that iron and copper wire equally extended have their resistances altered differently when under the stress; that of the iron wire being more increased, should the absolute effect in each wire be an augmentation of resistance, as other experiments I have made give me reason to suppose it is, or less diminished should it turn out that the absolute effect in each wire is a diminution of resistance.

151. Again, a heavier weight was applied so as permanently to elongate the wires. The deflection, which was in the same direction as at first, was noted, but not corrected by any motion of the moveable electrode, and the weight was again removed. The needle returned towards zero, but remained deviating in the same direction as it had done to a greater degree with the weight on. By applying the hand instead of weights and gradually pulling down the lower copper piece, at first slowly, and afterwards rather faster, the needle could be made to deviate to 7° and kept steadily there. After the wires had been stretched by rather more than an inch, the hand was removed with a gradual diminution of stress, which could easily be regulated to let the needle down without oscillation to whatever position it

would rest in, with the stress entirely off. This, in several repetitions of the experiment on the same wires, was found to be somewhere about 3° or 4° in the same direction as the deviation which was kept at 7° for a few seconds during the stress. Hence it was further concluded, that, as regards electric conductivity of the substance, the effect of permanent elongation, remaining after the stress is removed, differed between iron and copper in the same way as the effect of longitudinal stress during its action; that is, that the galvanic resistance of iron is more increased by permanent elongation than that of copper. Irregular variations to a considerable extent, obviously due to thermo-electric effects from the copper and iron in the compound conducting circuit, made me not attempt to measure with much care, the distance the moveable electrode had to be shifted to counteract the effects of tension; but I intend repeating the experiment and making it for other pairs of metals, with this source of irregularity removed by a modification of the testing conductor.

152. In the kind of experiment which has been described, the channels through the two metals experienced exactly the same elongation, and, it may be said without committing any sensible error, the same narrowing, by-the longitudinal extension. The effect observed, therefore, depends truly on variations in the conductivities of their substance. I had previously made various experiments on copper wire alone, and on iron wire alone, in which I attempted to eliminate the effects of elongation and narrowing, and had very nearly established, for the case of iron wire at least, that the augmented resistance due to tension, either temporary or permanent, is a very little more than can be accounted for by the change of form. As, however, I have other experiments in progress, by which I hope to be able to show for a single metal the absolute effect on its specific conductivity, separated perfectly from any influence on the resistance of the conductor occasioned by a change of its form, I defer in the meantime giving more details of investigation on this subject.

153. The method which has now been described has many great advantages over that by the differential galvanometer, or any other that I know of for testing or measuring galvanic resistances. In the first place, the irregularities, dependent on the electrodes, connexions, and circular conductors, of the differential

galvanometer, are entirely done away with, and only the tested and the testing conductors, all connected by compact solderings, can influence the indication from which the results are to be drawn. In the second place, the galvanometer circuit may be broken and completed, and reversed, as often as is desired, by its own commutator, without affecting to the slightest sensible degree, the strength of the current through the tested and testing branches; while in the former mode of experimenting the indicating needle was always under the action of the divided current, unless the current in one or the other of the branches was broken, which introduced irregularities, lasting for a considerable time, by the consequent changes of temperature through the conductors. This was an immense convenience in every experiment, and allowed small deflections, amounting to the tenth of a degree, to be tested with ease by using the commutator of the galvanometer, and getting oscillations. But it was of especial advantage in the experiments on the effects of transverse magnetization, since the galvanometer circuit had only to be kept broken for a few seconds during the making, breaking, or reversing of the magnetizing current, to get entirely rid of all disturbances of the needle due to induced currents; and in all experiments in which the Ruhmkorff magnet was used, since by breaking the galvanometer circuit and using a little steel magnet in the hand, the galvanometer needle could be let down in a few seconds into its position as affected by the direct action of the large magnet, before proceeding to test the current due to the change of resistance under investigation. In the third place, it is possessed of almost unlimited capacity for increase of sensibility. In some of the experiments on the influence of tension on electric conductivity, I have tested with the greatest ease effects amounting to only $\frac{1}{15000}$th of the whole resistance of the wire under examination, and I see no difficulty in testing effects amounting to only the tenth part of that, or even hundreds of times smaller effects, by using more powerful currents, and applying artificial means to keep the wires cool.

PART V. ON THE EFFECTS OF MAGNETIZATION ON THE ELECTRIC CONDUCTIVITY OF METALS.

154. The remarkable effects which I found produced in the thermo-electric quality of a metal by magnetization and by mechanical strain, appeared to render it highly probable that the same agencies would also influence their electric conductivities. To demonstrate this if I could, and to discover the nature of the anticipated effects, I commenced an experimental investigation of the subject, and, after various nugatory operations, arrived at a variety of positive results by the following processes.

155. Experiment 1. *On the longitudinal electric conductivity of longitudinally magnetized iron wire.*—A length of seventy-two yards of silk-covered copper wire was rolled in six strands, or altogether in about 860 turns on a core made up of two concentric brass tubes, connected at their ends by a ring of sheet brass, and arranged to have water sent through the space between them by suitable entrance and exit pipes soldered to apertures in the outer tube; the external diameter of the outer brass tube was about $\frac{5}{12}$ inch, and the internal diameter of the inner one about $\frac{1}{4}$ inch; the metal of both outer and inner tubes being as thin and as well smoothed as it could be got. The piece of iron wire to be tested was soldered at one end to a piece of thick copper wire, and then insulated by a thin coating of writing-paper, wrapped twice round it, and pushed into the inner brass tube, which was just large enough to admit it easily. A second iron wire of equal dimensions was similarly prepared and inserted in a second core, in all respects like the other, except that in this experiment it had no copper wire wrapped round it. The two cores being laid side by side, the free ends of the iron wires were connected as shown in the diagram (Fig. 41), by an arc of thick copper wire, C, soldered to them. A current from a single large cell of Daniell's was admitted and carried off by the electrodes A and B. Cold water was kept constantly flowing through the spaces between the concentric brass tubes round the iron wires. The testing conductor (§ 149) used in this experiment consisted principally of the following parts:—(1) Two pieces of No. 18 copper wire,

each sixteen yards long, prevented from touching one another
by a piece of twine between them, rolled together on a thin
copper cylinder, 12 inches long and three inches diameter, from
which they were insulated by a coating of two folds of silk cloth
sewed round it. (2) Soldered to two of their contiguous ends,
a connecting arc of thick copper wire, which was at first intended

Fig. 41.

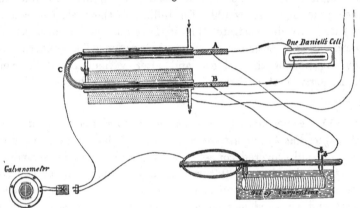

to be graduated, and will be called *the scale of the testing conductor*. (3) Separate short thick wires soldered to the other ends
of the wires coiled on the copper cylinder, to bear binding screws
for making connexions with the electrodes A and B of the conductor to be tested. One electrode of the galvanometer was
soldered to the middle of the connecting arc between the two iron
wires, and the other was held in the hand, and applied about the
middle of the scale of the testing conductor. A rather troublesome
process was then required to bring the galvanometer to zero by
adding resistance on one side or the other between the ends of
the testing conductor and A, or B. When this was done, it was
found that great deviations of the galvanometer needle were
produced by sliding its moveable electrode a few inches in either
way on the scale, and a perfectly sensible deflection by sliding it
as much as ⅛th of an inch. The point of the testing scale to
which the moveable electrode had to be brought, to give no
deflection of the galvanometer, was determined: the circuit of
the galvanometer was broken, and a current from six of the small
iron cells was sent through the magnetizing coil. Immediately on
completing the galvanometer circuit again, with its electrode held

on the same point of the testing scale as before, a very considerable deflection was observed. On breaking the galvanometer circuit, reversing the magnetizing current, and completing the galvanometer circuit again, the same deflection was observed; and when the magnetizing current was stopped the galvanometer again gave zero, or nearly so. On repeating the process as regards the magnetizing current, without breaking the galvanometer circuit, the same deflection was always observed, in whichever direction the current was sent through the magnetizing coil; and little or no either instantaneous or permanent effect was produced on suddenly reversing this current. It was found that the deflection occasioned by the magnetization was diminished by sliding the moveable electrode along the scale from its end communicating with B, towards its end communicating with A, and was corrected by such a motion through a space of about $\frac{3}{4}$ths of an inch; equivalent to $\frac{1}{10}$th of an inch of the No. 18 wire, constituting the chief part of the testing conductor. It was concluded that the iron wire had its electric resistance increased by magnetization, and that this augmentation amounted, in the particular experiment, to about $\frac{1}{3000}$ of the whole resistance of the magnetized piece.

156. Experiment 2. *On the effect of permanent magnetization on the electric conductivity of steel wire.*—The same apparatus as in Experiment 1, was used, and was in all respects similarly arranged, except that hardened steel wires, as free from magnetism as possible, were substituted in place of the soft iron cores in the brass tubes. On bringing the galvanometer to zero and sending a current through the magnetizing coil, the same deviation as before was observed, and a much smaller deviation in the same direction remained after the magnetizing current ceased. This experiment was repeated several times on fresh unmagnetic steel cores, and always with the same result. I concluded that steel when subjected to magnetic influence has, like iron, its electric conductivity diminished in the direction of the lines of force; and that it retains some of the same effect with the permanent magnetism subsisting after the magnetizing force is removed. At the same time I was not quite satisfied with the experiment, as the galvanometer needle was never very steady, and, to keep it about zero, the moveable electrode had to be shifted largely along the scale, sometimes quite to one end,

when, to get it on the scale again, additional adjustment wires had to be added to the other branch of the testing conductor. This prevented me from using more powerful currents through the wires to be tested and so getting larger indications of the results; but I determined if possible to repeat the experiment afterwards with arrangements better adapted to do away with all variations in the conductivity of the circuit except those under investigation. I still keep it in view to do so, and I have no doubt now of being able to get rid of all the unsteadiness which I had found so troublesome.

157. Experiment 3. *Attempt to discover the effect of transverse magnetization on the longitudinal conductivity of a slip of sheet iron.*—Two brass cores like those described above, and of the same length (10 inches), but of larger inner and outer diameters, were prepared, and a quantity of covered copper wire rolled on one of them in four strands, or in all 570 turns. Two slips of sheet iron, each seven feet long and $\frac{1}{8}$ inch broad, were wound upon single brass tubes coated with paper, and the successive spires of each were kept from contact by a piece of twine wound on between them. A length of nine inches of each brass tube had 84 inches of the slip iron laid upon it, and therefore the inclination of the helix to a plane perpendicular to its axis was about 6°, being the angle whose sine is $\frac{9}{84}$. Each of these iron spirals was protected outside with a coating of paper, and pushed into the interior of one of the brass cores. A copper arc, C, (Fig. 42) was soldered to each

Fig. 42.

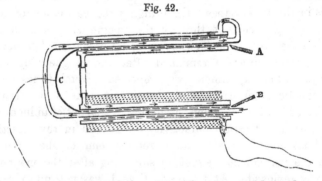

of them so as to connect their extremities on one side, and powerful copper electrodes, A and B, were soldered to their other extremities. Then, a stream of water being kept constantly flowing through each of the inner tubes and through the spaces between the concentric brass tubes outside, a current from a large cell of Daniell's (§ 63) (exposing 2˙5 square feet of zinc to 4˙4 square feet of copper) was sent through the iron spirals, and a testing conductor (the same one as before) was put in communication with their electrodes, A and B. One electrode of the galvanometer being, as before, soldered to the middle of the copper arc connecting the iron spirals, the other was applied to the scale of the testing conductor. The galvanometer being brought to zero by the insertion of adjustment wires at one end or other of the testing conductor, it was found to be rather steadier than in the former experiments, probably because of the diminution of thermal effects by the stream of water through the cores, and the greater surface of iron exposed outside and inside to refrigeration. When a current was sent and maintained through the magnetizing helix, a very decided permanent deflection was occasioned in the galvanometer; and this the same with each direction of the magnetizing current. If the galvanometer circuit was kept complete, its needle experienced a powerful impulse, sending it through a great many degrees in one direction or the other at the instant of starting, or of reversing, or of stopping, the magnetizing current, but quickly in each case showed the nature of the permanent deflection by oscillating about one position, when the current was steadily maintained, in either direction. These impulsive deflections were of course due to induced currents, and were entirely prevented by keeping the galvanometer circuit broken during the starting, the reversal, or the stoppage, of the current through the coil of the electro-magnet.

158. The deflection due to the effect of magnetic force on the substance of the iron, was corrected in each case by sliding the moveable electrode towards the part of the testing scale remote from the end connected with the iron spiral which experienced that effect, and it therefore indicated a diminution of conductivity in the iron.

159. If the lines of magnetization had been exactly perpendicular to the lines of electric current through the iron, we should now conclude that transverse magnetization diminishes the conductivity of an iron conductor; that is, that it produces the same kind of effect on the conductivity as longitudinal magnetization. But the lines of current formed spirals inclined at an angle of 84°

to the lines of the magnetizing force; and the mutual influence
of the consecutive parts of the magnetized iron spiral would have
an effect (not wholly compensated by the mutual influences between
the successive spires because of the thickness of the twine between
them,) contributing to longitudinal magnetization; and therefore
the lines of magnetization must have been inclined, not at 90°,
but at some angle less than 84°, to the direction of the lines of
current. Hence all we can conclude is, that not only longitudinal
magnetization but oblique magnetization up to some angle of
obliquity less than 84° from the lines of current, diminishes the
electric conductivity of iron.

160. It remains to be determined by experiment what is the
effect of magnetization right across the lines of current: if a
diminution of conductivity, whether a greater or a less diminution
than is caused by an equal longitudinal magnetization? or if it
is an increase of conductivity, what is the angle of obliquity of
the magnetization which gives neither increase nor diminution of
conductivity?

161. Experiment 4. *To discover the differential effect of magnetization on the conductivity of iron in different directions.*—A
square of 1½ inch each side was cut from thin sheet metal, and
powerful electrodes were soldered to two corners, A and B (Fig. 43).
A reference-electrode (§ 149) of No. 18 copper wire was soldered to
C, one of the other corners, and the two extremities of a yard of
the same kind of wire, to be used as a multiplying branch, were
soldered to points D, E, about 1/15th of an inch from one another
on each side of the remaining corner. A current being conducted
through the square by the principal electrodes A and B, the reference-electrode was used to connect C permanently with the
commutator belonging to the testing galvanometer. Another wire
used as a testing electrode, was applied to connect any point of
the plate, or of the multiplying branch, with the other galvanometer electrode. In the first place, it was found that a powerful
current was raised in the galvanometer coil if the testing electrode
was applied to any point of the multiplying branch; and it was
necessary therefore, as was anticipated, to adjust the distribution
of resistance through the square by filing, so that there might be
some point on the testing branch which would give no current
when touched by the testing electrode. (See below, § 176, where
a less troublesome way of managing this part of the arrangement,

in an analogous experiment, is described.) For this purpose, in the first place the testing electrode was applied at different places along the edges BD, EA of the square till a point was found

Fig. 43.

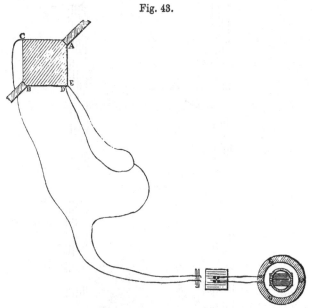

which gave no deflection of the galvanometer. If this was in BD, the plate had to be thinned in its middle parts parallel to CA and BD, or else to be thinned along the edges CB, AE, so as to increase the resistance to conduction parallel to the last-mentioned edges. Or if the neutral point was in EA, the plate had to be thinned in its middle parts parallel to CB and AD, or along its edges BD, CA. By using the file according to these directions, after a few trials the neutral point was brought upon the testing branch; that is to say, the resistance was so adjusted in the square that the line from C cutting right across the lines of conduction, or which is the same thing, the

Fig. 44.

equipotential line through C, passed between D and E. A piece of sheet copper as broad as the iron square, but rather longer, was bent as shown in the diagram (Fig. 44), so as to give a depressed space in which the iron, insulated from the copper simply by a piece of writing-paper, could rest steadily. This copper cradle was placed resting on the flat poles of a Ruhmkorff electro-magnet (Figs. 45 and 46), which were pushed together so as to hold it firmly.

Fig. 45.

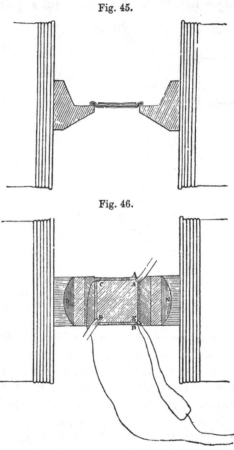

Fig. 46.

Any leakage of electric currents from the coils of the electro-magnet was thus effectually drained by the copper, so that a simple sheet of paper was quite enough to do away with all sensible indications of currents in the iron acquired otherwise than through the electrodes A and B. [This electrical drainage would be made more nearly perfect by using paper or some other non-conductor to separate the cradle from the poles of the magnet.]

162. A large single element of Daniell's (§ 63), consisting of seven zinc plates in seven porous cells, contained in four large wooden cells, and exposing in all 8·75 square feet of zinc surface to 15·3 square feet of copper, was then used to send a current

through the iron square, insulated between the poles of the electro-magnet, in the manner described.

163. The neutral point on the testing branch being got by trial, it was found to remain tolerably steady, although no doubt during the first minutes of the flow of the current it may have varied much, as the iron got heated, which it soon did to a degree very sensible to the touch. Moving the electrode along the testing branch through a quarter of an inch on either side of the neutral point, gave a very marked deflection of the galvanometer. The galvanometer circuit was then broken, and a current from six of the small iron cells was started through the coils of the electro-magnet. When the galvanometer circuit was again, after a few seconds, closed, with its electrode on the same point of the multiplying branch as before, a very considerable deflection was observed in the needle. To correct this deflection and bring the needle to zero, the testing electrode had to be moved to a position, two or three inches nearer D, on the testing branch.

164. The new neutral point was unchanged when the electro-magnet was reversed, and when the magnetizing current was broken there was a permanent deflection in the galvanometer, the reverse of that observed when the current was started in either direction. If the galvanometer circuit was completed within a second or two of any of the changes in the magnetizing current, the needle experienced, obviously from induced currents, powerful impulses in one direction or the other, according to the direction of the current made or unmade through the coils of the electro-magnet. But in every case, although from various disturbing causes the neutral points gradually shifted largely along the testing branch, the permanent effects of making and of unmaking the electro-magnet were most marked, and were uniformly as stated above.

165. Thus it appears that magnetization shifts the equipotential line through C from its position running across to the opposite corner, to a position (dotted in the diagram Fig. 47) a little nearer CB ; so much so that its end is shifted about $\frac{3}{36} \times \frac{1}{20}$, or $\frac{1}{240}$ of an inch from E towards D. This shows that the passage of electricity in the directions AE, CB has become less resisted than it was, relatively to

Fig. 47.

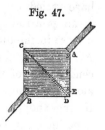

the passage in the directions AC, DB; and it therefore follows that the electric conductivity of magnetized iron is greater across than along the lines of magnetization.

166.   Still, as the preceding experiment (Exp. 3) had appeared (§ 159) to show that the absolute conductivity is diminished in all directions by magnetization, it seemed possible that the effect now observed might be caused by inequalities in the distribution of magnetism in the plate.   Thus if, from the character of the distribution of the magnetizing force, or because of non-uniformity in the plate, the parts between C and B and between A and E were less intensely magnetized, and those between C and A and between B and D more intensely magnetized, than the average, the observed effect could be accounted for without any difference in the electric conductivity of the substance in the different directions.   To test this conceivable explanation, pieces of soft iron (cubes and little square bars nearly double cubes) were laid over the square plate, being kept insulated from electric communication with it by paper, so that while the conducting mass remained unchanged, the distribution of the magnetization of its substance might be altered.   Before the magnetic force was applied, a great effect on the neutral point of the multiplying branch was observed, taking place gradually during several minutes, and obviously due, in a great measure, to variation of the distribution of temperature in the conducting square.   (See below, § 177, for an illustration of this effect.)   When a new neutral point was found, the magnet was made, reversed, unmade, &c., and always with the same effects as before.   Different arrangements of the little masses of soft iron produced different absolute effects on the neutral point, causing it to shift sometimes as much as fifteen inches on the multiplying branch, but the effects of magnetism were invariably found to be consistent with the first-mentioned result. As the distribution of the magnetism in the square plate must have varied very much under these different circumstances, and in all probability must have been in some of the cases more intense in the quarters towards AE and CB than in those towards AC and DB, the conclusion could scarcely be avoided, that the conductivity of the magnetized substance was greater across than along the lines of magnetization.   For the purpose of further testing and illustrating this conclusion I planned the following experiment, to compare directly the resistances of two equal and

similar squares of sheet iron, equally and similarly magnetized, arranged in the same circuit to conduct electricity across the lines of magnetization of one, and along those of the other.

167. *Experiment 5. To compare the conductivities of magnetized iron along and across the lines of magnetization.*—A piece of sheet copper, BCHK (fig. 48), three inches long, two inches broad and $\frac{1}{12}$ inch thick, was bent round the line FH into the form shown in fig. 49. A square of thin sheet iron, two inches wide (weighing

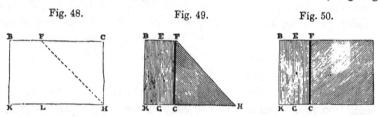

Fig. 48.   Fig. 49.   Fig. 50.

103 grains), was soldered by one side to the edge CH of the copper in the position shown in fig. 50. The projecting part, FBKL, of the copper slip was bent round its middle line EG, so as to bring its edge, BK, close over the edge of the iron square lying over FC; and to this edge, BK, in its new position (fig. 51), a second iron square, of the same dimensions and weight as the other, was soldered by one side, with its area lying in a position close to that of the former. The relative position of the two squares and

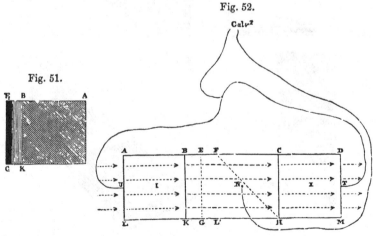

Fig. 52.

Fig. 51.

the connecting piece of copper will be understood by looking at fig. 52, which represents the iron squares as if soldered to the piece of copper before it was bent, and the iron square CDMH turned round its side CH, from the position close to the plane of the copper adjoining it, into a position in this plane continued across CH. If, now, we suppose the iron square CDMH to be turned down so as to lie below a square, FL'HC, of the copper; this square of the copper to be bent sharp round its diagonal, FH, till the part FL'H lies over HCF; and, lastly, the part FL'KB projecting beyond FC to be bent downwards round EG with a less sharp bend; the iron square, ABKL, will be brought close under the other one, CDMH, with the edges of the two which are connected to the edges of the copper perpendicular to one another, and the whole compound conductor will have exactly the position shown below in fig. 54.

168. A convenient electrode was soldered along the edge of each iron square parallel to the edge of the same square soldered to the connecting piece of copper; so that a powerful electric current entering by one of those electrodes and carried away by the other would pass through the second-mentioned square of iron in lines exactly parallel to the side AB, through the connecting piece of copper in lines which were parallel to its length, BC, before it was bent; and through the first-mentioned square

Fig. 53.

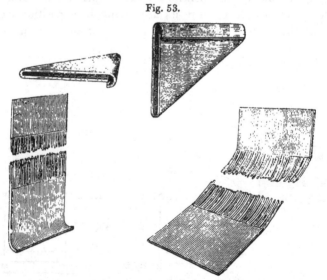

in lines exactly parallel to its side CD, and therefore perpendicular to the lines along which it traverses the second square. The course of the current will be understood by looking at figure 52, where the two squares and the copper connecting them are supposed to be opened out so as to throw the course of the current into a straight line The order followed in constructing the compound conductor was not exactly the order of the description given above; but the connecting piece of copper was first cut and bent, and other pieces, to serve as electrodes (shown in the accompanying views, fig. 53), were prepared, and the iron squares, put in their proper places, were then soldered by their edges to the edges of the connecting piece, and the electrodes were soldered to their opposite edges.   A view of the whole thus put together, with the reference and testing wires described below, is given in fig. 55.   A testing conductor of two yards of No. 18 copper wire

Fig. 54.

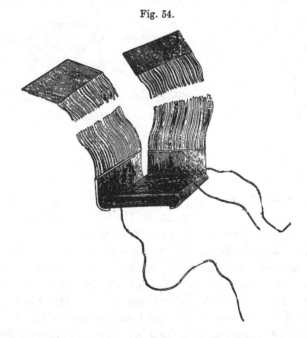

was soldered with its two extremities to the copper electrodes, close to the middle points of the edges AL, DM of the iron squares; and a fixed galvanometer electrode was soldered to the middle point, N, of the copper connecting piece.

169.   The squares, their electrodes, the connecting piece, and the testing conductor being then guarded against irregular con-

tacts by a little square of pasteboard pressed between the iron
squares, a half-square of pasteboard between the first-mentioned
iron square and the portion FCH of the connecting copper (see
fig. 49), and fragments of paper and pasteboard elsewhere, the
whole was placed, with the second-mentioned square lowest, in
a copper cradle lined with paper, and resting between the horizontal edges of the flat poles of the Ruhmkorff electro-magnet
used in the preceding experiment (Exp. 4).

170.   The positions of the magnetic poles of the squares, of
the bent connecting piece of copper, of the testing conductor,
and of the galvanometer electrodes are indicated in fig. 55, but,
to avoid confusion, the principal electrodes are not shown.   A

Fig. 55.

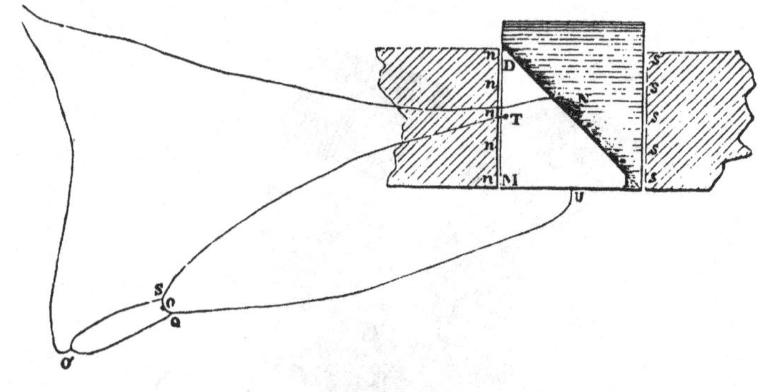

current from the four large double cells, connected so as to constitute in all a single element of Daniell's, exposing 10 square
feet of zinc surface to $17\frac{1}{2}$ square feet of copper, was then introduced by the principal electrode soldered to the edge MD of the
upper square, and drawn off by the other principal electrode,
namely, that soldered to the edge of the lower square lying
exactly below the edge MH of the upper.   The course of the
current into the principal channel between these electrodes would
be across the upper square from MD to HC, and across the lower
square from the edge below CD to that below HM; also, in the
secondary channel between the same electrodes, from T soldered
to the first through the testing conductor, to its other end U
soldered to the second.

171.   A fixed galvanometer electrode being (§ 168) soldered
to the middle point, N, of the connecting-copper, the other elec-

trode of the galvanometer was moved along the testing conductor till a point, O, was found at which it might be applied without giving any deflection. By moving it $\frac{1}{50}$th of an inch on either side of O very sensible deflections were obtained, and therefore a yard of copper wire was soldered by its ends to points S and Q a quarter of an inch on each side of O, and was used instead of the "scale" of the testing conductor described as used in the first three experiments. The neutral point, O', on this multiplying branch having been found, the galvanometer circuit was broken, and the electro-magnet was excited by six of the small iron cells. On closing the galvanometer circuit again immediately, a considerable deflection was observed, to correct which the moveable electrode had to be moved through about two or three inches from O' towards Q. On unmaking the electro-magnet a reverse deflection in the galvanometer was observed, and was corrected by bringing back the electrode to O'. The same result was obtained when the magnet was·made in the reverse way, and never failed to appear, to an unmistakeable extent and with perfect consistency, after the operation had been repeated many times and varied in every possible way.

172. It showed that the effect of the magnetization was to increase the resistance relatively in the upper square of iron, and to diminish it relatively in the lower square. I concluded with confidence that the electric conductivity of magnetized iron is greater across than along the lines of magnetization.

173. Experiment 6. *A double experiment, to test the absolute nature of the two effects of which the difference was shown in the preceding experiment.*—A divided current from the battery was made to pass through the two squares by electrodes, of which one was soldered to the middle of the copper band connecting them, and the other clamped to the now united extremities of the bundles of copper wire which had served before to lead in and out the whole current in the preceding experiment. As testing conductor was used the same piece of copper wire which had served as the fixed galvanometer electrode in the preceding experiment, with its end which had been connected with the galvanometer now soldered to the junction of the two copper branches of the divided channel (the resistance of each of which was found to be nearly equal to that of the iron square with which it was connected). The testing wire used in the preceding experiment

was cut in two, one part to serve as fixed galvanometer electrode
in one, and the other in the other, of the two experiments which
it was intended now to make. I first attempted to test the
effect on the conductivity of the upper of the two squares, produced by the magnetization which in it is along the lines of
current. I found, however, on fixing the copper wire proceeding
from one side of that square to one electrode of the galvanometer,
and applying the other to the testing conductor in the usual way,
that the circumstances were constantly varying, and that the
point to be touched to give no deflection shifted rapidly along
the testing conductor. Hence I gave up this part of the experiment, of which the result might be anticipated with certainty
from the experiment on the effect of magnetization along the line
of current described above (Exp. 1, § 155), and I gave the whole
time during which the experiment could be continued, to an
examination of the influence of the electro-magnet on the current in the branch leading through the lower square across its
lines of magnetization. Accordingly, the galvanometer electrode,
which had been united to the part of the old testing conductor
terminating in an edge of the upper square, was transferred to
the other part of the old testing conductor, that is, to the part
terminating in a side of the lower iron square. The same new
testing conductor was still used; and as soon as a point could be
found on it which gave no current when touched by the moveable
galvanometer electrode, points about ¼ of an inch on each side
of it were taken, and a multiplying branch of one yard of No. 18
copper wire was soldered by its ends to them. Before, however,
the effect of the magnetism could be decidedly tested, the zero-
point had moved off the multiplying branch, which had accordingly
to be shifted along the testing conductor to get into range again.
The same process had to be gone through a great many times,
and at last, after the current had been flowing continuously
through the two squares and the divided copper channel for
about five hours, the zero-point became sufficiently steady to
remain on the multiplying branch when fixed at the right place
on the testing conductor, and to allow a decisive experiment to
be made. The result was a very slight effect, proving *a diminution
of resistance in the iron square.*

174. The cause of the long-continued variation in the conditions of electric equilibrium between the testing conductor and

the fixed point on the edge of the lower square, was clearly the gradual warming of the long copper wires extending up from this point, due to the conduction of heat generated in the iron squares by the electric current; and it would obviously be much diminished by using a simple form of conductor with only one iron square at a time, and with the reference conductor kept near it, so as to acquire quickly whatever temperature it would rest with during the flow of the current. I accordingly made the following experiment (7. § 175), choosing first the effect of transverse magnetization, as the experiment just described had not been of a satisfactory kind, although apparently conclusive, while the first experiment of the series (Exp. 1. § 155) had been less unsatisfactory in point of steadiness, and had led decisively to a conclusion regarding the effect of longitudinal magnetization on the resistance of a conductor.

175. Experiment 7. *To test the effect on the conductivity of iron of magnetization across the line of current.*—A square of sheet iron like those used in the last experiment (four square inches, weighing 103 grains, and consequently about $\frac{1}{78}$th of an inch thick,) was soldered along one edge to a slip of lead of the same

Fig. 56.

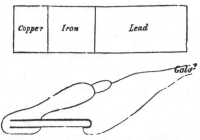

width, about twice as thick and about one-half longer. To the opposite edge of the iron square was soldered a stout copper slip an inch broad and equal in length to the side. The piece of lead was bent round, so as to give a straight part lying about $\frac{1}{4}$ of an inch from the plane of the iron, and to extend about as far as the copper slip soldered to the other edge of the square. A current from an arrangement of the cells (§§ 63 and 64) constituting a powerful single element of Daniell's was sent through the iron square and the lead band, by electrodes clamped to one end of the lead and to the copper slip fixed to the other edge of the iron. A point in the lead slip having been found, such that the galvanic resistance between it and the edge next the iron was nearly equal to the resistance in the iron square itself, a testing conductor (two yards of No. 18 copper wire) was soldered by one end to that point in the lead, and by its other end to the middle of the edge of the iron square to which the copper slip is attached. A copper wire, to serve as fixed galvanometer electrode, was soldered to the lead band, at a point in the middle of its breadth close to its edge of attachment to the iron. A copper cradle was put between the flat poles of the electro-magnet, as before (see above, § 161), and covered with a piece of paper. The iron square was supported upon it in a position with the line joining the poles perpendicular to the line of the current through it. Then, the current being kept steadily flowing through the iron and lead band, a zero-point was found on the testing conductor, and a multiplying branch (one yard of No. 18 copper wire) was soldered with its ends $\frac{1}{4}$ of an inch on each side of this point, in the usual way. The zero-point on this multiplying branch was almost immediately found, and continued on the whole very steady from the first. The galvanometer circuit being broken, a magnetizing current from six small iron cells was sent through the coils of the electro-magnet, and the needle of the galvanometer was let settle (or it could be, in a few seconds, by the aid of a little magnet held in the hand) into its position of equilibrium as affected by the direct force of the magnet. On completing the galvanometer circuit again, with its moveable electrode on the same point of the multiplying branch as before, a current was made sensible by an excessively slight deflection. The galvanometer circuit being broken, and the electro-magnet reversed, a similar deflection was found in the galvanometer on again completing its circuit. It ceased, as nearly as could be discovered, when the electro-magnet was unmade, and was uniformly observed when the magnet was made again either way, in a great many repetitions. The current indicated by the galvanometer when the magnet was made was always such as to be corrected by carrying the moveable electrode from its previous zero-point, along the multiplying branch, towards the part of the testing conductor terminating at the iron square, and therefore indicated an *increase of conductivity in the iron*. The effect was so very slight, that I could scarcely determine how much the moveable conductor had to be shifted to correct it. I intend to repeat the experiment with similar arrangements, but with two or three times as powerful a current through the electro-magnet, which ought to give about four or nine times the amount of effect. In the mean time, however, I am quite convinced that I have observed the true result, and I conclude that *the electric conductivity of iron is increased by magnetic force across the lines of current.*

176. Experiment 8. *To show the variation of a line of electric equilibrium in a circular disc of iron conducting electricity between two opposite points of its circumference, when subjected to magnetic force in a direction at an angle of 45° to the line joining these points.*—A circle 2·3 inches diameter (see Fig. 57) was cut from a piece of sheet iron, and ground down to a thickness which must have been about $\frac{1}{60}$th of an inch, as the prepared disc was found to

Fig. 57.

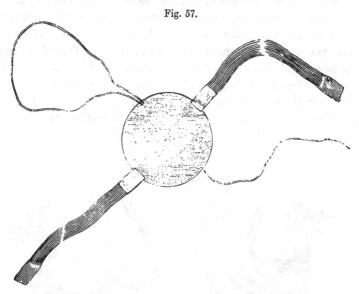

weigh 114 grains. Two stout copper electrodes were soldered to its circumference at opposite points. A point at 90° on the circumference from one of these was taken, and at about $\frac{1}{40}$th of an inch on each side of it were soldered the ends of a piece of No. 18 copper wire two yards long, to serve as a multiplying branch. The disc was put on a copper cradle covered with paper, supported between the flat poles of the Ruhmkorff electro-magnet (Fig. 58), with the line joining its principal electrodes at an angle of 45° to the magnetic axes of the field, and a current from a large single element of Daniell's was sent through it by these electrodes. One electrode of the galvanometer was applied to the middle of the multiplying branch, and the other was moved about on the opposite parts of the circumference of the disc till a position giving no current was found, where it was then soldered. The moveable electrode applied to different points of the multiplying branch, was then found to give sensible galvanometer indications with a motion of a quarter of an inch, and after a very short time the zero-point became tolerably steady. The electro-magnet was then made, with the galva-nometer circuit broken, and when it was closed again a decided indication of a current was observed in the galvanometer. This current was checked by sliding the moveable electrode towards the end of the multiplying branch next the equatoreal part of the magnetic field; and the conclusion was, that the conducting power of the plate, when magnetized, became greater across than along the lines of magnetization, which was confirmed by every repetition and variation of the experiment. Now it is obvious that the intensity of magnetization must have been on the whole greater in the parts of the disc next the poles: hence a diminution of conductivity *across* the lines of magnetization, to the same

Fig. 58.

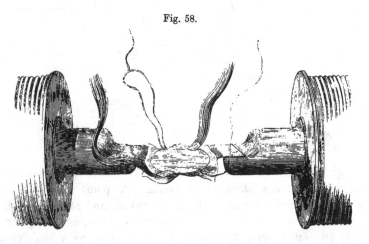

extent as that which we know from Experiment (1) exists along them, would give a contrary effect to that now observed; and it follows that the electric conductivity is in reality greater across than along the lines of magnetization in magnetized iron.

177. This experiment was witnessed by Mr Joule, and afforded a full confirmation of the conclusion (§ 172) which had been established by Experiment 5 above, and which follows from Experiment 1 and Experiment 6, considered together. The effects of applying pieces of hot wood equatoreally or axially to the disc were very clearly observed, and were always similar to those described above (§ 166), indicating a greater resistance to the parts of the current crossing the hot region, than to those passing through the comparatively cool parts, of the iron.

[From the *Proceedings of the Royal Society*, June 15, 1857.]

EFFECTS OF MAGNETIZATION ON THE ELECTRIC CONDUCTIVITY OF NICKEL AND OF IRON.

$177_1'$. I have already communicated to the Royal Society a description of experiments by which I found that iron, when subjected to magnetic force, acquires an increase of resistance to the conduction of electricity along, and a diminution of resistance to the conduction of electricity across, the lines of magnetization. By experiments more recently made, I have ascertained that the electric conductivity of nickel is similarly influenced by magnetism, but to a greater degree, and with a curious difference from iron in the relative magnitudes of the transverse and longitudinal effects.

$177_2'$. In these experiments the effect of transverse magnetization was first tested on a little rectangular piece of nickel 1·2 inch long, ·52 of an inch broad, and ·12 of an inch thick, being the "keeper" of the nickel horse-shoe (§ 143) belonging to the Industrial Museum of Edinburgh, and put at my disposal for experimental purposes through the kindness of Dr George Wilson.

Exactly the method described in § 175 above was followed, and the result, readily found on the first trial, was as stated.

177$_3$'. The effect of longitudinal magnetization on nickel was first found with some difficulty, by an arrangement with the horse-shoe itself, and magnetizing helix (§ 143), the former furnished with suitable electrodes for a powerful current through itself, and the system treated in all respects (including cooling by streams of cold water) as described in § 156, for a corresponding experiment on iron. The result, determined by but a very slight indication, was, as stated above, that longitudinal magnetization augmented the resistance.

177$_4$'. The magnetization of the small piece of metal between the poles of the Ruhmkorff electro-magnet being obviously much more intense than that of the larger piece under the influence merely of the smaller helix, I recurred to the plan of experiment 7 (§ 175) above; by which the effect of transverse magnetization on the little rectangular piece of nickel was first tested, and I had an equal and similar piece of iron, and another of brass, all prepared to be tested, as well as the nickel, with either longitudinal or transverse magnetic force.

177$_5$'. To each of the little rectangles of metal to be tested, a thin slip of copper (instead of lead, as in the experiment of § 175), of the same breadth ('52 of an inch), to serve as a reference conductor, was soldered longitudinally, and to the other end of the metal tested, a piece of copper to serve as an electrode, for the principal current, was soldered. The ends of a testing conductor, (six feet of No. 18 copper wire,) were soldered respectively to the last-mentioned end of the tested metal, and to a point in the reference-conductor found, so that the resistance between it and the junction of the reference-conductor with the tested conductor, should be about equal to the resistance in the latter.

177$_6$'. A single element, consisting of four large double cells of Daniell's (§ 63), exposing in all 10 square feet of zinc surface to 17 square feet of copper, was used to send the testing current through the conducting system thus composed, by electrodes clamped to the ends of the principal conducting channel, just outside the points of attachment of the testing conductor.

$177_7'$. The electro-magnet was excited by various battery arrangements, in different experiments, at best by 52 cells of Daniell's, each exposing 54 square inches of zinc surface to 90 square inches of copper, and arranged in a double battery* equivalent to one battery of 26 elements each of double surface. By accident, only a single battery of 26 elements was used in obtaining the numerical results stated below.

$177_8'$. The nickel was first placed between the flat poles of the electro-magnet, with its length across the lines of force, and, one galvanometer electrode being kept soldered to the junction of the nickel and the copper reference-conductor, the other galvanometer electrode was applied to the testing conductor, till the point (equipotential with that point of junction) which could be touched without giving any deflection of the needle, was found. A multiplying branch, (three feet of No. 18 wire,) was then soldered with its ends $\frac{3}{8}$ths of an inch on each side of this point, and, as soon as the solderings were cool, the corresponding point on this multiplying branch was found. The magnetizing current was after that sent in either direction through the coils of the electro-magnet, and it was found that the moveable galvanometer electrode had to be shifted over about $4\frac{1}{2}$ inches on the multiplying branch towards the end of the testing conductor connected with the nickel, that is to say, in such a direction as to indicate a *diminished resistance* in the nickel. When the same operations were gone through with the nickel placed longitudinally between the poles of the electro-magnet, the zero-point on the multiplying branch was shifted about six inches in the direction which indicated an *increased resistance* in the nickel.

$177_9'$. The piece of iron similarly tested, gave effects in the same direction in each case, and the results originally obtained for iron (§§ 146, 155, 161—177) were thus verified.

$177_{10}'$. No effect whatever could be discovered when the piece of brass was similarly tried. It is much to be desired that experiments with highly increased power, and with a better kind of

---

* This arrangement was found to give about the same strength of current through the coils of the electro-magnet, as a single battery of 52 of the same cells in series, and was therefore preferred as involving only half the amount of chemical action in each cell, and consequently maintaining its effect more constantly during many successive hours of use.

galvanometer, should be made, to discover whatever very small influence is really produced by magnetic force on the comparatively non-magnetic metals.

$177_{11}'$. The shifting of the neutral point on the multiplying branch required to balance the effect produced by the longitudinal magnetization in the iron, was only from $1\frac{1}{2}$ to 2 inches. Three inches were required to balance the opposite effect of the transverse magnetization.

$177_{12}'$. Hence, with the same magnetizing force, the effect of longitudinal magnetization in increasing the resistance, is from three to four times as great in nickel as in iron; but the contrary effect of transverse magnetization is nearly the same in the two metals with the same magnetizing force. It may be remarked, in connexion with this comparison, that nickel was found by Faraday to lose its magnetic inductive capacity much more rapidly with elevation of temperature, and that it must consequently, as I have shown, experience a greater cooling effect with demagnetization* than iron, at the temperature of the metals in the experiment. It will be very important to test the new property for each metal at those higher temperatures at which it is very rapidly losing its magnetic property, and to test it at atmospheric temperature for cobalt, which, as Faraday discovered, actually gains magnetic inductive capacity as its temperature is raised from ordinary atmospheric temperatures, and which, consequently, must experience a heating effect with demagnetization and a cooling effect with magnetization.

$177_{13}'$. The actual amount of the effects of magnetization on conductivity demonstrated by the experiments which have been described, may be estimated with some approach to accuracy from the preceding data. Thus the value of an inch on the multiplying branch would be the same as that of $\frac{1}{36} \times \frac{3}{4}$, or $\frac{1}{48}$ of an inch on the portion of the main testing conductor between its ends. The whole resistance of this $\frac{3}{4}$ of an inch of the main testing conductor, assisted by the attached multiplying branch of 36 inches, is of course less in the ratio of 48 to 49, than that of any simple $\frac{3}{4}$ of an inch of the testing conductor; but in the actual circumstances there will be no loss of accuracy in neglecting

---

* See Nichol's *Cyclopædia of Physical Science*, second edition (1860), article "Thermo-magnetism," [Art. XLVIII. § 207.   Vol. I. above].

so small a difference. Hence the effect of the transverse magnetization of the nickel was to diminish its resistance in the ratio
of half the length of the testing conductor diminished by $\frac{4\frac{1}{2}}{48}$ of an
inch, to that of the same increased by the same, that is to say, in
the ratio of $11\frac{31}{32}$ to $12\frac{1}{32}$, or of 383 to 385. Hence it appears
that the resistance of the nickel, when under the transverse magnetizing force, was less by $\frac{1}{192}$; and similarly, that the resistance,
when under the longitudinal magnetizing force, was greater by
$\frac{1}{144}$, than when freed from magnetic influence; and that the effects
of the transverse and of the longitudinal magnetizing forces on the
iron were to diminish its resistance and to increase its resistance
by $\frac{1}{288}$ and $\frac{1}{500}$ respectively. The first effect which I succeeded
in estimating (§ 155) amounted to only $\frac{1}{3000}$, being the increase
of resistance in an iron wire when longitudinally magnetized
by a not very powerfully excited helix surrounding it. In the
recent experiments the magnetizing force was (we may infer) far
greater.

$177_{14}{}'$. It is to be remarked that the results now brought
forward do not afford ground for a quantitative comparison between the effects of the same degree of magnetization, on the
resistance to electric conduction along and across the lines of
magnetization, in either one metal or the other, in consequence
of the oblong form of the specimens used in the experiment.
It is probable that in each metal, but especially in nickel, of
which the specific inductive capacity is less than that of iron,
the transverse magnetization was more intense than the longitudinal magnetization, since the poles of the electro magnet were
brought closer for the former than for the latter. I hope before
long to be able to make a strict comparison between the two
effects for iron at least, if not for nickel also; and to find for
each metal something of the law of variation of the conductivity
with magnetizing forces of different strengths.

[From *Trans. of the Royal Society*, May 27, 1875.]

PART VI. EFFECTS OF STRESS ON MAGNETIZATION.

178. In Parts III. and IV. of my first series of papers under this title (Trans. of the Royal Society, February 1856 [§§ 104 to 135 above]), I described experiments discovering effects of stress on the thermo-electric quality and the electric resistances of metals. About the time those experiments were made I also made several nugatory attempts to discover the effects of stress on magnetization; and eighteen years have passed before I have been able to resume the investigation. Early in the year 1874 I made arrangements to experiment on the magnetization of iron and steel wires in two different ways—one by observing the deflections of a suspended magnetic needle produced by the magnetization to be tested, the other by observing the throw of a galvanometer-needle, due to the momentary current induced by each sudden change of magnetism. The second method, which for brevity I shall call the ballistic method, was invented by Weber, and has been used with excellent effect by Thalén, Roland, and others. It has great advantages in respect of convenience, and the ease with which accurate results may be obtained by it; but it is not adapted to show slow changes of magnetism, and is therefore not fit for certain important parts of the investigation. On this account I am continuing arrangements for carrying out the first method, although hitherto I have obtained no good results by it.

179. On the other hand, I have found the ballistic method very easy and perfectly satisfactory in every respect, except that it does not show the slow changes of magnetization. It was by it that all the results which I am now going to describe were obtained. The apparatus, which is very simple, is represented in the accompanying sketch (fig. 1).

AA' is the wire whose magnetism is experimented on. In my first experiments it was a piece of steel pianoforte-wire, No. 22*, Birmingham wire-gauge, that is, weighing about 3·54 grammes per metre, and therefore of ·7644 of a millimetre in diameter. It is about 5 metres long, and its upper end is firmly fixed to a

---

* This is the wire used in the American Navy and in British cable-ships for deep-sea soundings. Its strength to resist pull is such that it bears about 230 lbs. (104 kilogrammes), or the weight in air of 29·4 kilometres (or 15·9 nautical miles) of its own length.

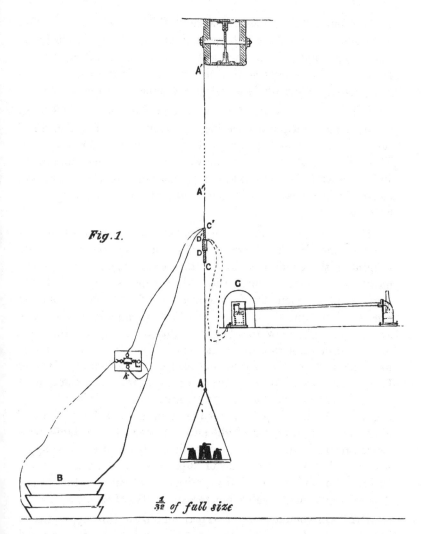

Fig.1.

$\frac{1}{32}$ *of full size*

beam in the ceiling of the Physical Laboratory of the University of Glasgow, where all the experiments have been made. To the lower end is attached a pan bearing weights, by means of which different amounts of pull may be rapidly applied and removed from the wire when desired. Over a portion CC' of this wire 28·7 centimetres long there is wrapped a piece of thin sheet copper, and on the outside of that there is coiled, in two layers, 719·7 centimetres of silk-covered copper wire, the copper weighing 2·502 grammes per metre. The inner layer contains 326 turns and the outer 321. The resistance of this coil when cool is ·511 of an ohm. Its ends are put in communication by thick electrodes with a reversing-key, k, and a battery of three of my "tray" Daniells. The resistance of each of these cells is about ·06 of an ohm, giving for the whole battery a resistance of ·18 of an ohm.

180. Over the coil CC' is wound another coil DD' 9·8 centimetres long, which contains 538 centimetres of wire, No. 26 of the Birmingham wire-gauge, with 38 centimetres for electrodes. This wire is also wound on in two layers, the inner containing 147 turns and the outer 146. The resistance of this induction-coil is 1·432 ohms: weight per metre of the wire being 1·189 grammes.

181. The deflections of the galvanometer are read in my usual manner by the image of a fine wire fixed vertically, close in front of the edge of a flat paraffin- or gas-flame. The screen on which the image is thrown is a white paper scale, divided into fortieths of an inch, fixed at a distance of 126·5 cms. from the mirror. The lamp is placed close behind the middle of the scale, and just enough below to allow its light to pass under the scale to the mirror through a small blackened tube. The galvanometer used is represented in the accompanying sketch (fig. 2). Having been extemporized from a large lecture-room instrument, it is not so well adapted for this investigation as I could wish. It consists of an astatic pair of needles mounted on a light frame of aluminium, and carrying a light mirror placed on the frame, with its centre in the line of the suspension a little above the upper needle, and attached by means of clips, so as to admit of its being turned into and fixed in any position relatively to the frame. The resistance of the galvanometer-coil is ·634 of an ohm.

182. On commencing to experiment with this apparatus in March, 1874, I immediately obtained some very startling and interesting results. I found that when a current was kept flowing through the magnetizing coil, and weights were alternately placed on the pan at the lower end of the wire and taken off, the effect of the pull always diminished the magnetization, and the effect of removing the pull increased it*. The magnetizing current

---

* [Added May 1876. Since the communication of this paper to the Royal Society I have found in Wiedemann's *Galvanismus*, § 499, that similar results had been obtained by Matteucci (*Annales de Chimie et de Physique*, 1858), and by Villari (Pogg. *Annalen*, 1868).]

Fig. 2.

$1/_3$ RD SIZE

being then stopped, and the same operation with the weights repeated, I found similar effects—that is, the application of a pull diminished the residual magnetization, and the removal of the pull increased it. But, to my surprise, the effect was greater in this second case when merely residual magnetism was concerned, than in the first case when the original magnetizing force was still in action. This greater effect in the second case was surprising, because the whole magnetization concerned in the first case was greater than the magnetization concerned in the second case, by the amount of the quasi-elastic magnetization, which goes and comes again every time the magnetizing force is removed and reapplied.

183. The amount of the magnetizing force used may be roughly estimated by taking the electromotive force of each of the three cells as one volt, or 10^8 C. G. S. units. Thus, as the resistance in the circuit was about ·69 of an ohm, the strength of the current must have been about $\dfrac{3 \times 10^8}{·69 \times 10^9}$ or $\dfrac{1}{2·3}$. This was distributed in 647 turns of a solenoid whose axis was 28·7 centimetres long. The whole strength of current circulating round a length δx of the solenoid was therefore $\dfrac{647}{28·7} \times \dfrac{1}{2·3} \delta x$, and the magnetizing force to which the steel wire was subjected $4\pi \dfrac{647}{28·7} \times \dfrac{1}{2·3}$, or 123. Hence, as remarked by Professor Maxwell, who, in reporting on this paper, first made the preceding estimate, we may call the magnetizing force a large one, so large, in fact, that it probably magnetizes the wire nearly to saturation at once. For the sake of comparison it may be remarked that the horizontal and vertical components of the earth's magnetic force at Glasgow are about ·16 and ·43; and in first experimenting with the apparatus now described, I made sure that the magnetization of the wire by the vertical component of the earth's magnetizing force was not essentially concerned in any of the results, by reversing the magnetizing current, then applying and removing weights repeatedly, then stopping the current, and again putting weights on and off repeatedly. The deflections of the galvanometer observed in this succession of operations were not sensibly different from those observed previously, but in reverse directions; that is to say, the results still fulfilled the preceding statements.

The fact that the effect of pull to diminish magnetization, and of taking off pull to increase it, was found to produce a greater difference when the magnetization was solely residual than when it was sustained by the continued influence of the magnetizing force, led me to expect that the effect of making and breaking the circuit of the magnetizing coil and battery should be greater for the wire when pulled than when unpulled. Subsequent experiments proved this to be the case.

184. But, lastly, I found between the quasi-elastic part of the magnetization, produced by alternately applying and removing

the magnetizing force, and the initial reverse magnetization produced by the application of a reverse magnetizing force, a very
surprising difference. The former, as stated above, is greater
when the wire is pulled than when unpulled: the latter is less
in the wire when pulled than when free from pull, but not by
so great a difference; and the whole magnetizational effect of
reversing the current suddenly must therefore be greater when
the wire is pulled than when unpulled, and is found to be so.

185. The following series of experiments, I—XXXI., performed in November and December 1874, all on one and the
same piece of steel wire, first confirmed the conclusions inferred
(§ 182) from the preliminary investigation of the previous March,
then reproduced with more regularity the immediate experimental
results of that preliminary investigation, and lastly discovered
the very remarkable phenomena described in § 184.

186. The general order of procedure followed was this. The
image on the scale of the ballistic galvanometer is watched by
one observer, while a second stands by to make or break circuit of
the magnetizing current, or to put on and off weights, on word of
command from the first. When the image is seen to be steady on
the scale, the number at which it stands is read and recorded as
" z " (zero). The order " make " or " break " or " on " or " off " is
given by the first observer and executed suddenly by the second.
The first observer reads and records the greatest or least number
on the scale reached by the image in consequence of the electro-
magnetic impulse produced by the operation. Finally, the excess
(positive or negative) of this reading above the immediately preceding zero is written down and marked " M " or " B," or " On "
or " Off " as the case may be.

187. The connexions chanced to be so arranged that, with
the direction of current invariably used in the " M's " of Series
I—XXIX., the effect of M was to throw the image to the left,
or in the direction of decreasing numbers: thus until Series XXX.
a negative number always shows increase of magnetization in the
direction of that produced by M. The numbers actually written
down by the observers during the experiments are shown by the
following extracts from their day-book, for a few of the series,
chosen as examples to precede the abridged Tables of results
given below for all the Series I—XXXI.

December 2, 1874.

| VIII. | | | IX. | |
112 lbs. On.			112 lbs. Off.	
Z 376			362	
M.................... 328	− 48		339	− 23
Z 373			359	
B 413	+ 40		380	+ 21
Z 374			360	
M.................... 332	− 42		335	− 25
Z 373			358	
B 413	+ 40		381	+ 23
Z 375			358	
M.................... 333	− 42		336	− 22
Z 371			358	
B 410	+ 39		379	+ 21
Z 372			358	
M.................... 330	− 42		333	− 25
Z 373			357	
B 412	+ 39		380	+ 23
Z 376			359	
M.................... 335	− 41		335	− 24
Z 374			357	
B 413	+ 39		380	+ 23
Z 377			360	
M.................... 335	− 42		335	− 25
Z 377			357	
B 415	+ 38		378	+ 21
Z 377			354	
M.................... 335	− 42		330	− 24
Z 375			353	
B 414	+ 39		376	+ 23
Z 378			355	
M.................... 335	− 43		331	− 24
Z 377			353	
B 416	+ 39		376	+ 23
Z 374			354	
M.................... 332	− 42		330	− 24
Z 372			353	
B 411	+ 39		375	+ 22
Z 374			356	
M.................... 333	− 41		331	− 25
Z 369			355	
B 407	+ 38		376	+ 21

December 7, 1874.

112 lbs. On and Off.

XVI.		XVII.	
No current *.		Current flowing.	
Z.................... 362		357	
On 404	+ 42	377	+ 20
Z.................... 365		354	
Off 342	− 23	335	− 19
Z.................... 360		352	
On .... 386·5	+ 26·5	372	+ 20
Z.................... 360		352	
Off .... 336·5	− 23·5	334	− 18
Z.................... 365		353	
On 393	+ 28	374	+ 21
Z.................... 364		353	
Off 339	− 25	335·5	− 17·5
Z.................... 362		362	
On 388	+ 26	380·5	+ 18·5
Z.................... 365		353	
Off .... 340·5	− 24·5	334·5	− 18·5
Z.................... 367		353	
On 391	+ 24	372·5	+ 19·5
Z.................... 367		350	
Off 345	− 22	331·5	− 18·5
Z.................... 367		357	
On .... 390·5	+ 23·5	376·5	+ 19·5
Z.................... 369		360	
Off .... 347·5	− 21·5	341	− 19
Z.................... 369		359	
On 394	+ 25	379	+ 20
Z.................... 369		357	
Off 346	− 23	331·5	− 25·5
Z.................... 370		354	
On 393	+ 23	373	+ 19
Z.................... 370		353	
Off .... 347·5	− 22·5	333	− 20
Z.................... 370		349	
On 392	+ 22	368	+ 19
Z.................... 368		346	
Off .... 345·5	− 22·5	326	− 20
Z.................... 365		346	
On 392	+ 27	366	+ 20
Z.................... 364		342	
Off .... 341·5	− 22·5	322	− 20

* The effects here observed were diminutions and augmentations of residual · magnetism from previous operations.

188.   Each series from I. to XXIX. was conducted with perfect regularity on one or other of several plans, of which the details are sufficiently exemplified in the preceding unabridged quotations. Omitting now the actual readings and taking merely the differences from the zeros, we have the following full statement of results showing the amount of the electro-magnetic impulse produced by each operation.

[Addition, May 1876.—It must not be assumed that the apparent accumulations of magnetization shown in the last columns of the Tables I.—IX., or that the differences of demagnetizations and magnetizations shown at the foot of each of the Tables down to XVII., express correctly the whole changes of magnetization of the wire; for, as remarked above (§ 178), the ballistic method does not show slow changes of magnetization; and there certainly were slow changes of magnetization not shown in my ballistic experiments; because the algebraic sum of all the deflections observed day after day went on sometimes continually increasing and sometimes continually diminishing, when it was certain that there was no corresponding progressive accumulation of magnetization or of demagnetization.   Some of the experiments by the method of deflection promised in § 178, were performed in last July and August, soon after the communication of this paper, and gave very remarkable results, which were free from the objection of leaving out of account slow changes of magnetization. They include effects of torsion as well as of pull.   I intend to include a statement of them in a paper [Part VII., below] which I hope soon to be able to offer to the Royal Society.]

Series I. to IX.—Magnetizing Current made and broken.

I.—November 26, 1874. Weights off.

Magnetization by Make.	Decrease of Magnetization by Break.	Excess of Magnetization by Make above Demagnetization by Break.	Apparent accumulation of Magnetization.
26	23	3	3
24	21	3	6
25	23	2	8
24	21	3	11
24	21	3	14
23	21	2	16
24	22	2	18
24	21	3	21
24	21	3	24
23	22	1	25

II.—November 26, 1874. 28 lbs. on.

Magnetization by Make.	Decrease of Magnetization by Break.	Excess of Magnetization by Make above Demagnetization by Break.	Apparent accumulation of Magnetization.
26	22	4	4
24	22	2	6
24	26	2	4
24	22	2	6
25	22	3	9
26	23	3	12
24	22	2	14
25	22	3	17
25	23	2	19
24	20	4	23

III.—November 27, 1874. Weights off.

Magnetization by Make.	Decrease of Magnetization by Break.	Excess of Magnetization by Make above Demagnetization by Break.	Apparent accumulation of Magnetization.
24	21	3	3
24	22	2	5
24	21	3	8
24	21	3	11
22	20	2	13
24	22	2	15
24	21	3	18
23	21	2	20
24	20	4	24
23	20	3	27

Series I.—IX.—(continued).

IV.—November 30, 1874. 56 lbs. on.

Magnetization by Make.	Decrease of Magnetization by Break.	Excess of Magnetization by Make above Demagnetization by Break.	Apparent accumulation of Magnetization.
30	26	4	4
29	27	2	6
28	26	2	8
28	26	2	10
28	24	4	14
28	25	3	17
28	25	3	20
29	26	3	23
27	25	2	25
28	26	2	27

V.—November 30, 1874. Weights off.

Magnetization by Make.	Decrease of Magnetization by Break.	Excess of Magnetization by Make above Demagnetization by Break.	Apparent accumulation of Magnetization.
23	21	2	2
24	22	2	4
23	22	1	5
24	22	2	7
24	22	2	9
23	20	3	12
23	21	2	14
24	21	3	17
24	21	3	20
24	21	3	23

VI.—December 1, 1874. 84 lbs. on.

Magnetization by Make.	Decrease of Magnetization by Break.	Excess of Magnetization by Make above Demagnetization by Break.	Apparent accumulation of Magnetization.
35	33	2	2
33	31	2	4
33	30	3	7
36	32	4	11
33	31	2	13
35	32	3	16
35	33	2	18
34	31	3	21
35	31	4	25
35	31	4	29

Series I.—IX.—(continued).

VII.—December 1, 1874. Weights off.

Magnetization by Make.	Decrease of Magnetization by Break.	Excess of Magnetization by Make above Demagnetization by Break.	Apparent accumulation of Magnetization.
26	23	3	3
26	23	3	6
25	22	3	9
26	23	3	12
26	23	3	15
26	23	3	18
25	24	1	19
26	23	3	22
26	23	3	25
26	24	2	27
25	24	1	28

VIII.—December 2, 1874. 112 lbs. on.

Magnetization by Make.	Decrease of Magnetization by Break.	Excess of Magnetization by Make above Demagnetization by Break.	Apparent accumulation of Magnetization.
48	40	8	8
42	40	2	10
42	39	3	13
42	39	3	16
41	39	2	18
42	38	4	22
42	39	3	25
43	39	4	29
42	39	3	32
41	38	3	35

IX.—December 2, 1874. 112 lbs. on.

Magnetization by Make.	Decrease of Magnetization by Break.	Excess of Magnetization by Make above Demagnetization by Break.	Apparent accumulation of Magnetization.
− 23	+ 21	− 2	2
− 25	+ 23	− 2	4
− 22	+ 21	− 1	5
− 25	+ 23	− 2	7
− 24	+ 23	− 1	8
− 25	+ 21	− 4	12
− 24	+ 23	− 1	13
− 24	+ 23	− 1	14
− 24	+ 22	− 2	16
− 25	+ 21	− 4	20

Series X.—XVII.—Weights on and off.

	28 lbs.		56 lbs.	
	X. December 3. No current.	XI. December 3. Current flowing.	XII. December 4. No current.	XIII. December 5. Current flowing.
On	+5·5	+3·5	+15	+10
Off	-5	-4	-10·5	-10
On	+5	+4	+11·5	+9
Off	-4·5	-5	-12	+9
On	+5·5	+5	+11	+10
Off	-5	-3·5	-11·5	-8
On	+5	+4	+9·5	+9·5
Off	-5	-4·5	-11	-8
On	+5	+5	+11	+9
Off	-5	+3·5	-10·5	+7
On	+5·5	-5	+12	+10
Off	-5	+3·5	-10	-7
On	+5	-5·5	+10·5	+10
Off	-4	+4	-10	+9
On	+5	-3·5	+10	+10
Off	-5	+4	-10·5	+9
On	+4·5	-5	+13·5	+9·5
Off	-4·5	+3·5	-10	+10·5
On	+5	-4·5	+12	+9
Off	-5·5		-11	—

Means of the ten 'ons' and ten 'offs.'

Ons. demagnetization.	Offs. magnetization.	Ons. demagnetization.	Offs. magnetization.	Ons. demagnetization.	Offs. magnetization.	Ons. demagnetization.	Offs. magnetization.
5·10	4·85	4	4·4	11·6	10·7	9·7	8·55

Excess of demagnetizations above magnetizations.		Excess of magnetizations above demagnetizations.		Excess of demagnetizations above magnetizations.		Excess of demagnetizations above magnetizations.	
2·5		4		9		11·5	

Series X.—XVII.—Weights on and off (continued).

	84 lbs.		112 lbs.	
	XIV. December 5. No current.	XV. December 7. Current flowing.	XVI. December 7. No current.	XVII. December 7. Current flowing.
On	+ 28	+ 16	+ 42	+ 20
Off	− 16	− 15·5	− 23	− 19
On	+ 18	+ 13·5	+ 26·5	+ 20
Off	− 16	− 14·5	− 23·5	− 18
On	+ 18	+ 15·5	+ 28	+ 21
Off	− 16·5	− 13	− 25	− 17·5
On	+ 18	+ 14	+ 26	+ 18·5
Off	− 18	− 14	− 24·5	− 18·5
On	+ 18	+ 13·5	+ 24	+ 18·5
Off	− 19	− 15	− 22	− 18·5
On	+ 17·5	+ 17	+ 23·5	+ 19·5
Off	− 18·5	− 13·5	− 21·5	− 19
On	+ 18·5	+ 16	+ 25	+ 20
Off	− 17	− 14·5	− 23	− 25·5
On	+ 16·5	+ 17·5	+ 23	+ 19
Off	− 18·5	− 15·5	− 22·5	− 20
On	+ 18	+ 16	+ 22	+ 19
Off	− 17·5	− 14·5	− 22·5	− 20
On	+ 19	+ 14·5	+ 27	+ 20
Off	− 16	− 15·5	− 22·5	− 20

Means of the ten 'ons' and ten 'offs.'

	XIV	XV	XVI	XVII
Ons.	18·95 demagnetization.	15·35 demagnetization.	26·7 demagnetization.	19·55 demagnetization.
Offs.	17·3 magnetization.	14·55 magnetization.	23 magnetization.	19·6 magnetization.
	Excess of demagnetizations above magnetizations.	Excess of demagnetizations above magnetizations.	Excess of demagnetizations above magnetizations.	Excess of magnetizations above demagnetizations.
	16·5	8	37	5

Series XVIII., XIX., XX.—Cycles M, On, B, Off.

	XVIII. December 8.	XIX. December 8.	XX. December 9.
	56 lbs.	84 lbs.	112 lbs.
M....................	− 28	− 27	− 41
On	+ 6	+ 12	+ 20
B	+ 26	+ 31	+ 38
Off	− 7	− 15	− 21
M....................	− 33	− 36	− 40·5
On	+ 9·5	+ 13	+ 18·5
B	+ 25	+ 29·5	+ 38·5
Off	− 7	− 14	− 22
M....................	− 30	− 34	− 42·5
On	+ 7·5	+ 14·5	+ 18
B	+ 27	+ 32·5	+ 38·5
Off	− 8	− 14	− 22·5
M....................	− 34	− 35·5	− 45
On	+ 9·5	+ 13·5	+ 19
B	+ 22	+ 29	+ 39·5
Off..................	− 8·5	− 16	− 21·5
M....................	− 31	− 36	− 42·5
On	+ 10·5	+ 13·5	+ 20
B	+ 28	+ 30·5	+ 38·5
Off	− 8	− 15	− 21

Series XXI., XXII., XXIII.—Cycles On, M, Off, B.

	XXI. December 9.	XXII. December 9.	XXIII. December 10.
	56 lbs.	84 lbs.	112 lbs.
On	+ 15·5	+ 26	+ 37
M....................	− 31	− 39	− 43
Off	− 9	− 15	− 22
B	+ 23	+ 24	+ 23·5
On	+ 16·5	+ 25·5	+ 37
M....................	− 29·5	− 39	− 41
Off	− 10	− 16	− 22
B	+ 22	+ 24	+ 24
On	+ 15·5	+ 26	+ 37·5
M....................	− 31	− 38	− 42
Off	− 10	− 16	− 21
B	+ 22	+ 24	+ 24
On	+ 15	+ 27	+ 36
M....................	− 29	− 39	− 41
Off	− 10	− 16	− 20·5
B	+ 21·5	+ 24	+ 23·5
On	+ 15·5	+ 26	+ 38
M....................	− 28·5	− 38	− 42
Off	− 10	− 16	− 21
B	+ 22	+ 25	+ 24

Series XXIV., XXV., XXVI.—Cycles M, On, Off, B.

	XXIV. December 10.	XXV. December 10.	XXVI. December 10.
	56 lbs.	84 lbs.	112 lbs.
M	− 27	− 25	− 26
On	+ 10	+ 15	+ 20
Off	− 9·5	− 14·5	− 19·5
B	+ 22	+ 24	+ 24
M	− 26	− 24	− 25·5
On	+ 10·5	+ 14	+ 20
Off	− 9	− 14·5	− 19
B	+ 21	+ 22	+ 22
M	− 26	− 24	− 26
On	+ 9·5	+ 15	+ 20
Off	− 10	− 15	− 20
B	+ 22	+ 23	+ 21
M	− 25	− 26	− 24
On	+ 9	+ 15	+ 20
Off	− 9	− 14·5	− 22
B	+ 23	+ 24	+ 24
M	− 25	− 26	− 25·5
On	+ 8	+ 14	+ 21
Off	− 9	− 14·5	− 23
B	+ 23	+ 24	+ 25

Series XXVII., XXVIII., XXIX.

Cycles (the same as those of XVIII., XIX., XX.) M, On, B, Off.

	XXVII. December 14.	XXVIII. December 15.	XXIX. December 16.
	56 lbs.	84 lbs.	112 lbs.
M	− 30	− 30	− 31
On	+ 9·5	+ 14·5	+ 18·5
B	+ 28·5	+ 34	+ 40
Off	− 8	− 14·5	− 21·5
M	− 33	− 39	− 44
On	+ 8·5	+ 13·5	+ 18·5
B	+ 29	+ 34·5	+ 40
Off	− 7	− 15	− 22
M	− 34	− 37·5	− 42
On	+ 8·5	+ 16	+ 18
B	+ 29·5	+ 34·5	+ 40
Off	− 8	− 14	− 20
M	− 34	− 38	− 43
On	+ 9	+ 14	+ 20
B	+ 28	+ 35	+ 39·5
Off	− 6·5	− 14·5	− 23
M	− 34	− 37	− 42
On	+ 9	+ 12·5	+ 20
B	+ 28	+ 34	+ 38
Off	− 7	− 14	− 19

189. Series XXX. and XXXI. were irregular. After putting on and off weights several times, and finding as before diminutions and augmentations of residual magnetism, a weight of 28 lbs. was left on the wire, and the magnetizing circuit was made with the current in the same direction as before; then an additional 28 lbs. on and off: the current once more made and broken, still in same direction; then circuit made in reverse direction, and broken; lastly, 28 lbs. additional put on and taken off. The results were as follows, M (−) being now used to denote institution of current in one final direction, and M (+) in the opposite direction.

Latter part of Series XXX. December 22, 1874.—28 lbs. permanently hung on wire.

"On" and "off" means another weight of 28 lbs. put on and taken off.

On	+ 5·5	B	− 25	On	− 6	On	+ 4·5
Off	− 4·5	On	− 10	Off	+ 6·5	Off	− 7
M (−)	− 42	Off	+ 5·5	On	− 6	On	+ 5
On	+ 5·5	On	− 6	Off	+ 6	Off	− 5
Off	− 6	Off	+ 6	M (−)	− 180	On	+ 4·5
B	+ 24·5	On	− 7	On	+ 5	Off	− 5
M (−)	− 28	Off	+ 6·5	Off	− 6·5	B	+ 26
B	+ 26	On	− 6	On	+ 5		
M (+)	+ 181	Off	+ 5	Off	− 4·5		

Series XXXI. December 22, 1874.—Weight of 28 lbs. hung permanently on the wire.

"On" and "off" means a weight of 56 lbs. hung on in addition and taken off.

On	+ 20	Off	− 9½	M (−) and B many times quickly.		Then immediately	
Off	− 12	B	+ 23½			Off	
On	+ 13½	M (+)	+ 180	M (+)	+ 184	M (−)	− 199
Off	− 12	B	− 24	On. Then M (+) and B many times quickly.		M (+)	+ 200
M (−)	− 40	M (−)	− 181			M (−)	− 200
On	+ 9¼	On	+ 9	M (−)	155	M (+)	+ 200
Off	− 10	B	+ 35	M (+) reading lost.		Then immediately	
B	+ 25½	M (+)	+ 150			On	
M (−)	− 29	B	− 35	M (−)	− 180	M (−)	− 180
Then immediately, without waiting for image coming to rest,		M (−)	− 150	M (+)	+ 179½	M (+)	+ 179
B		Off	− 11	Then immediately, without waiting for image coming to rest,		M (−)	− 179
On	+ 19	B	+ 25			Steel wire removed from core, to test the direct induction from coil to coil.	
Off	− 12	M (+)	+ 182	Off			
On	+ 12	On and M (−) and B many times quickly.		M (−)	− 199		
M (−)	− 40			M (+)	+ 198	M (+)	+ 20
B	+ 36	M (−)	− 34	Then immediately		B	− 19½
M (−)	− 35½	B immediately after.		On			
Off	− 10	M (+)	+ 155	M (−)	− 181		
On	+ 11½	Off, and B and M (+) and B many times quickly.		M (+)	+ 176		
		M (−)	− 185				

190. Interpretation of Series XXX. with help from Series XXXI.—From each M (∓) and following B subtract and add respectively ∓ 19¾, or, say ∓ 20, for induction of coil on coil, measured in the last two experiments of Series XXXI. Therefore the first M (−) of Series XXX. gave 22 of increase to the magnetization of the steel wire, which had remained for six days undisturbed as left at the end of Series XXIX. After an "on" and "off" of 28 lbs.* the current was stopped, and the magnetism of the wire instantly fell 4·5. A second M (−) gave 8 of increase to the magnetization, while the current was continued, of which 6 was instantly lost when the current was stopped a minute or two after. Then M (+) without other disturbance produced 161 of demagnetization and reverse magnetization, of which 5 was lost when the current was stopped a minute or two after, and 5·5 more in the set of six "ons" and "offs" which followed. The residual reverse magnetization must (as we may judge from subsequent results of Series XXX. and XXXI.) have been approximately equal to the immediately previous residual magnetization remaining from all the previous M (−) and other operations to which the wire had been subjected, from the time in November when it was set up for the experiments. Subtracting 10½ from 161 we find 150½ for the number measuring the change from the previous (−) magnetization to the equal (+) magnetization in the wire at the present stage of its history. The amount of this magnetization is therefore 75¼, say 75. With this amount of magnetization in the wire, and 28 lbs. hanging constantly on it, the effect of putting on and off another 28 lbs. is to diminish and increase the magnetism by 6; that is to say, to diminish it and increase it by about $\frac{1}{12}$ of its mean value (which, according to these rough estimates, is 73 of our arbitrary scale). The remainder of Series XXX. speaks intelligibly for itself.

191. Interpretation of Series XXXI.—The first "on" gave a diminution 20 in the magnetization, the second 13½; the first on and the "off," "on," "off" following, gave on the whole a diminution of 9½; that is to say, shook out 9½ of the 75 of the residual magnetism remaining from Series XXX. The effect of repeated "ons" and "offs" on the 65½ of remaining magnetism would no doubt have shaken but very little more out, and would have caused alternate

* Another weight of 28 lbs. was hanging constantly on the wire.

diminutions and augmentations of about 12; that is to say, with 56 lbs. hung on, in addition to the constant 28 lbs., the magnetization of the wire would have been 53, and with only the 28 lbs. it would have been 65.

The M (−), on, off, and B which actually followed, added $15\frac{1}{4}$ to the magnetization, and so brought it up to about $80\frac{3}{4}$, or $5\frac{3}{4}$ more than it had at the beginning of Series XXXI. The subsequent M (−) and B probably produced little, if any, further change in the residual magnetization; and the "on," "off," "on," which followed, confirmed the previous result of augmentation 12 and diminution 12, alternately by off, and on, after something considerable had been shaken out permanently by the first "on." The "off," "on," "off" after M (−) with current still flowing gave about 10 instead of the 12 found previously when the current was not flowing. These results agree in kind and amount with what was to be expected from Series XIV. and XV., considering that "on" and "off" in those series was 84 lbs. on and off (with nothing, or only a very slight steadying weight kept always on), whereas now the "on" and "off" means change from 28 lbs. to 84 lbs. and back.

192. Interpretation of Series XXXI. continued. Taking now the effects of the Ms and Bs, and (§ 191) subtracting 20 from each number, we see that the effect on the magnetization of the wire producible by repeating M (−) and B over and over again without other disturbance, would be about 16 each way with 84 lbs. on, and probably between $5\frac{1}{2}$ and 9 with 28 lbs. on. (This agrees in kind with the conclusion deducible from the comparison between the previous Series II. and VI.; but the absolute magnitudes of the results seem smaller, probably because of less battery-power in those series than in Series XXXI.) The whole effects of the M (+) after M (−) and B, and of the M (−) after M (+) and B, were still, as in Series XXX., each equal to 180 or 181, with only the 28 lbs. on, and therefore (subtracting 20 for the induction from coil to coil) we had 160 or 161, say $160\frac{1}{2}$, for the sum of the demagnetization and reverse magnetization produced by a M following a B and a previous M of the opposite direction. Now came, in the course of a few minutes, a most startling discovery. With the 84 lbs. on, the sum of magnetization and demagnetization produced by the M after B from previous opposite M was 130, or less by $30\frac{1}{2}$ with those than with the 28 lbs. Thus we see that

while the magnetic effect of stopping the current is *greater*, the effect of subsequently instituting the current in the reverse direction is *less*, with the heavy than with the light weight; and less by about three times as great a difference.

193. Consider, lastly, the effect of a sudden reversal of the current. This may be regarded as the sum of the effects of stopping it, and starting it in the reverse direction; and therefore may be expected to be less with the 84 lbs. than with the 28 lbs. by about $\frac{2}{3}$ of the difference between the effects of starting the reverse current with the heavy and with the light weights hung on the wire. This inference is verified in the concluding thirteen results of the series before removal of the steel wire. Thus (§ 190) subtracting 40 for the induction of coil on coil in the reversal, we find 140, 139½, 141, 136, 140, 139, 139 for magnetic effects of reversals with 84 lbs. hung on the wire, of which the mean is 139·2; and 159, 158, 159, 160, 160, 160, for the magnetic effects of reversals with 28 lbs. on, of which the mean is 159·3.

194. After the conclusion of these experiments on the steel wire, I made many experiments of the same kind on soft-iron wires of various qualities substituted for it in the same apparatus, and I have obtained results of the same kind, as to the effects of hanging on and taking off weights, while the magnetizing current is kept flowing. I have also obtained some very remarkable and perplexing results by putting weights on and off with the current not flowing. In one of the iron wires the effect found was opposite to that in steel; that is to say, putting on weight augmented and taking off weight diminished the residual magnetism; in another the same effect as in steel was found, that is, putting on diminished and taking off augmented the residual magnetism. Neither of these was as soft as some of the other wires tried, and the one ("bright soft iron wire," Johnson's) that agreed with steel was remarked on at the time as *much harder* than another that had been previously experimented on ("black soft iron wire," Johnson's). This latter seemed utterly destitute of retentive power under the influence of putting the weights on and off. Like all the others it always experienced a diminution of magnetism by weights on and increase by weights off, when the magnetizing current was flowing. But when the current was stopped large effects (larger than those when the current was

flowing) were produced by putting on and taking off the weights; and these effects were always of the same kind, whichever had been the direction of the current.

195. I have not yet been able to explain these effects by terrestrial magnetic force*, nor to even guess any other possible cause; and have in fact, since the 23rd of December last, been exceedingly perplexed by seeming anomalies which the various soft-iron wires tried have presented, commencing that day with a reversal of the electro-magnetic effect of the "off" and "on" which the first of the wires experimented upon showed, after the magnetizing current had been made for the first time and broken, and a weight of 14 lbs. put on and off several times had first shaken out nearly one-third of the residual magnetism, and then given alternate augmentation by "on" and diminution by "off" of the magnetism that remained. A weight of 28 lbs. was then hung on, and it stretched the wire permanently by about 8 per cent. of its length. Then immediately I found reverse effects by putting off and on and off the 28 lbs., and on and off smaller weights—the "on" giving diminution and the "off" augmentation of what would have been the residual magnetism, if residual magnetism there was, from the first magnetization by the current. This quality remained until an hour or two later, when the current was once more made in the same direction as the first, and· broken again. Then 14 lbs. put on actually gave augmentation of the residual magnetism (instead of shaking out a considerable quantity as the first "on" after the first " B " had done), and the "offs," "ons," and "offs" following gave alternate diminution (by the "offs") and augmentation (by the "ons"); that is to say, the effects were the *same in kind* as the effects which had been observed before the stretching by the 28 lbs.; but they were nearly three times as great in amount as they had been then, with the same weight of 14 lbs. on and off. Thenceforth the same piece of wire (experimented on several more days up till Jan. 12, 1875) and other pieces of similar wire tried after it, with the current made sometimes in one direction and sometimes in the other, and broken,

---

* [Note added January 1877.—Nearly six months later I ascertained that these startlingly great effects actually were due to the vertical component of the earth's force, though this was only about $\frac{1}{300}$ of the magnetizing force of the currents used.]

and different amounts of weight up to 28 lbs. put on and off, always showed increase of the residual magnetism by the "on" and diminution by the "off." Even the first "on" after the stoppage of the current gave always an increase of the magnetism; but when the weight was as much as 28 lbs. the "shaking out" tendency was remarkably shown in the increase by the first "on" being much less than the diminution by the first "off."

196. The soft-iron wire experimented upon, on the 23rd of December gave, with 28 lbs. hanging on it, smaller effects of successive "makes," in one direction, and "breaks" of the current; a greater effect when it was made in the reverse direction; and a smaller sum of these two effects (that is to say, a smaller effect) on the reversal of the current than when it was un-pulled.

197. The investigation is being continued with special arrangements to discover the explanation of the seeming anomalies described above, and with the further object of determining in absolute measure the amounts of all the ascertained effects, at different temperatures up to 100° Cent. It is needless to give in the mean time any minute details of the experiments already made on the soft-iron wires by which the results now described were obtained.

PRELIMINARY NOTICE TO PART VII.

[From the *Proceedings of the Royal Society*, June 10, 1875.]

EFFECTS OF STRESS ON INDUCTIVE MAGNETISM IN SOFT IRON.

197₁'. At the last ordinary meeting of the Royal Society (May 27, 1875), after fully describing experiments (Part VI. above) by which I had found certain remarkable effects of stress on inductive and retained magnetism in steel and soft iron, I briefly referred to seeming anomalies presented by soft iron which had much perplexed me since the 23rd of December, 1874. Differences presented by the different specimens of soft-iron wire which I tried complicated the question very much; but one of them, the softest of all, a wire specially made by Messrs Richard Johnson and Nephew, of Manchester, for this investigation, through the kindness of Mr William H. Johnson, gave a result standing clearly out from the general confusion, and pointing the way to further experiments, by which, within the fortnight which has intervened since my former communication, I have arrived at a complete explanation of all that had formerly seemed anomalous. These experiments have been performed in the Physical Laboratory of the University of Glasgow by Mr Andrew Gray and Mr Thomas Gray, according to instructions which, in my absence, I have sent them from day to day by post and telegraph.

197$_2'$. The guiding result (described near the end of my former paper, and referred to in the last paragraph but one of the Abstract [II., Appendix below] in *Proceedings of the Royal Society* for May 27, 1875) was, that the softest wire, tried with weights on and off repeatedly, after it had been magnetized in either direction by making the current, in the positive or negative direction, and stopping it, gave effects on the ballistic galvanometer which proved a shaking out of residual magnetism by the first two or three ons and offs, and a gradual settlement into a condition in which the effect of "on" was an *augmentation*, and the effect of "off" a *diminution*, of the inductive magnetization due to the vertical component of the earth's magnetizing force. When a fresh piece of the same wire was put into the apparatus and tested with weights on and off it gave this same effect. If the wire had been turned upper end down and tried again in the course of any of the experiments, still the same effect would have been shown. It seemed perfectly clear that in these experiments there was no other efficient dipolar quality of the apparatus by which the positive throw of the ballistic galvanometer could be given by putting on the weight, and the negative throw by taking it off, than the vertical component of the earth's magnetic force.

197$_3'$. Yet I did not consider that I had *explained* the result by the terrestrial influence, because, for *all* the specimens of steel and soft iron, the effect of weights on had been uniformly to *diminish*, and of weights off to *augment* the magnetism when the magnetizing current was kept flowing. And I was, moreover, perplexed by the magnitude of the result—the effect of weights on and off shown by the very soft iron wire, under only the feeble magnetizing influence of the earth, being many times (from three times to nine or ten times) as great as the effects which the same weights on and off produced in the same wires when under vastly greater magnetizing forces of the currents through the helix.

197$_4'$. But by reducing the strength of the magnetizing current gradually, it was clear that the small positive effect of the "on" with the positive current flowing, and the small negative effect with the negative current, must be gradually brought to approximate more and more nearly to the large positive effect of the "on" when there is no current at all. Immediately after my former communication I therefore arranged to have experiments made with different measured strengths of current, feebler and feebler, until the law of the continuity thus pointed out should be ascertained; and so speedily arrived at the following astonishing conclusions:—

197$_5'$. (1) When the magnetizing force does not exceed a certain critical value the alternate effects of *pull* and *relaxation* are respectively to augment and diminish the induced magnetization.

(2) When the magnetizing force exceeds the critical value the effects are—pull diminishes, relaxation augments, the induced magnetization.

(3) The critical value of the magnetizing force for the annealed Johnson soft-iron wire, with 14 lbs. on and off, is about 17 or 18, if (for a moment) we take as unity the vertical component of the terrestrial magnetic force at Glasgow.

(4) The maximum positive effect of the pull on the inductive magnetism is obtained when the magnetizing force is about 4.

(5) The positive effect of the pull when the magnetizing force is 3 is about eight or nine times the amount of the negative effect when the magnetizing force is 25.

197$_6'$. The actual results of the experiments which proved these conclusions are exhibited graphically in the accompanying diagram. The horizontal scale (abscissas) shows the numbers of divisions of the scale of the steady-current galvanometer (called for brevity the "battery-galvanometer") used to measure the strengths of the current through the helix. The scale of ordinates shows the numbers of divisions of the scale of the ballistic galvanometer by which the sudden changes of the magnetism of the

wire produced by 14 lbs. "on" and 14 lbs. "off" were measured.
The ordinates are drawn in the positive direction when the effect

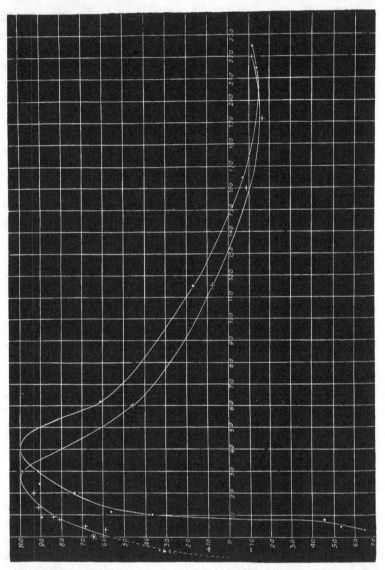

of "on" is to increase and of "off" to diminish the magnetism.
The simple round dots show the results of observations with
currents in the direction called negative (being those which gave
negative deflections of the battery-galvanometer). The spots in

the centre of signs (.+.) show results obtained with currents in
the direction called positive. The star (✳) at the position (64) on
the line of ordinates through the zero of abscissas, shows the mean
effect of many ons and offs with no current flowing—that is to
say, when the sole magnetizing force is the vertical component
of the earth's magnetic force. The curves are drawn as smoothly
as may be by hand, one of them to pass as nearly as it can
(without intolerable roughness) through all the signs (.+.) and the
star at (64), the other through all the plane dots. The latter curve
cuts the line of abscissas at + 8, this being the result (telegraphed
to me this evening) of special experiments made to-day for the
purpose of finding accurately the amount of the negative current
which, by neutralizing the vertical force of the earth or the wire,
gives an accurate zero effect for the "off" and "on." The dotted
prolongation of the curve through the signs (.+.), to cut the line of
abscissas on its negative side, is ideal, and is inserted to illustrate
the relation of this curve to the other. By the two curves cutting
the line of abscissas at + 8 and − 8, we see that 8 is the
strength of the current, measured on the scale of the battery-
galvanometer, which gives a magnetic force in the axis of the helix
equal to the vertical component of the terrestrial magnetic force.

197,′. Next a series of experiments were made to test the
inductive effects of repeatedly making the current always in one
direction, and stopping it, with the weight of 14 lbs. always on, and
again with the weight off, and this with various degrees of current,
feebler than those used in the earlier experiments. The results
with all the different intensities of magnetizing force thus applied,
were the same in kind as that which I found on the 23rd of
December, 1874, operating with a much stronger magnetizing force
on the first soft-iron wire tried; that is to say (contrarily to what
I had found in the steel wires), *the change of magnetization produced by repeated applications and annullings of the magnetizing
force of the helix was greater with the weight off than on.*

[*Note on Diagram, added July* 2, 1875.—A continuation of
the experiments with higher and higher magnetizing powers,
since the communication of this paper, disproves the negative
maximum indicated by the curves on the diagram, and proves an
asymptotic approach to a value approximately − 12, of ordinates,
for infinitely great positive values of the abscissas.]

[From the *Transactions of the Royal Society*, May 23, 1878.]

PART VII. EFFECTS OF STRESS ON THE MAGNET-IZATION OF IRON, NICKEL, AND COBALT.

198. In a preliminary notice (§§ 197$_1$'—197'$_7$ above) of investigations regarding the effects of stress on inductive magnetization in soft iron, communicated to the Royal Society on the 10th of June, 1875, I described experiments which afforded a complete explanation of the seeming anomalies referred to in §§ 194 and 195 above, which had at first been so perplexing. These experiments showed that the diminution of magnetism in a soft-iron wire, which I had found to be produced by pull, while the wire was under the influence of a constant magnetizing force, was to be observed only when the magnetizing force exceeded a certain critical value, and that when the magnetizing force was below that critical value the effect of pull was to increase the magnetism—a result which I afterwards found had been previously obtained by Villari*. The critical value of the magnetizing force I found to be about twenty-four times the vertical component of the terrestrial magnetic force at Glasgow. Hence the magnetizing force which I had used in my first experiment, which (§ 183 above) was nearly 300 times the vertical component of the terrestrial force, must have been about twelve times as great as the critical value. Further (which was most puzzling), I found the absolute amount of the effects of pull to be actually greater with the small magnetizing force of the earth than that of the opposite effects of the 300-fold greater magnetizing force of my early experiments. Thus the effect of the terrestrial force was not only in the right direction, but was of amply sufficient amount to account for the seeming anomalies which had at first been so perplexing; and in going over the details of the old observations I find all the anomalies quite explained. One of them, that particularly referred to in § 195 above, is still interesting. The alternate augmentation of the residual magnetism by "on" and diminution of it by "off," with the weight of 14 lbs., corresponded to the normal effect on residual magnetism in soft iron. The elongation of 8 per cent. produced when the 28 lbs. was hung on, was no

---

* Poggendorf's *Annalen*, 1868; also Wiedemann's *Galvanismus*, Vol. II., § 499.

doubt accompanied by a shaking out of nearly all the residual magnetism, and an inductive magnetization in the opposite direction by the vertical component of the earth's magnetic force. The reversed effects of the " ons" and " offs," observed after this change, were really augmentations and diminutions of magnetism induced by the earth's vertical force, and were therefore the proper effects for soft iron when subject to a magnetizing force of less than the Villari critical value. Further experimental investigation is necessary to explain the *greater amount* of effect, the same in kind as those observed before the stretching by 28 lbs., which the wire showed after it had been stretched by this weight.

199. The experiments indicated in my preliminary notice of June 10, 1875 (§§ 197$_1$'—197'$_7$, above), were the commencement of an elaborate series of investigations by Mr Andrew Gray and Mr Thomas Gray, which have been continued with little intermission from that time until now, and which are still in progress, with the general object of investigating the effects of longitudinal and transverse stress upon the magnetization of different qualities of iron and steel, and of nickel and cobalt. A separate series of investigations made nearly two years ago by Mr Donald Macfarlane on the effects of torsion on the magnetization of soft iron, bringing out some very remarkable results, are also included in this paper (§§ 223—229, below).

§§ 200—212. *Investigation by the Ballistic* method, of the change of Magnetization produced in a specimen of exceedingly soft Iron Wire, by the application and removal of pulling force.*

200. The wire used in these experiments was specially prepared for this investigation by Messrs Richard Johnson and Nephew, Manchester. It was of No. 22 Birmingham wire gauge; its weight per metre was 3·47 grammes, and its diameter was ·075 of a centimetre. A steel pianoforte wire of the same gauge would bear about 230 lbs. on and off, without experiencing in consequence any permanent change of quality. This iron wire was so soft that, after it was stretched by a scale-pan weighing 1 lb., which thenceforth was kept always hanging on it, an additional weight of 14 lbs. on and off gave a permanent elongation of ·4 per cent., and 4 lbs. more gave it a further permanent elongation of 1·6 per cent.

---

* Compare §§ 178, 179 above.

(making in all 2 per cent. of permanent elongation produced by 18 lbs. on and off, the permanent weight of 1 lb. being always on). The weight of 18 lbs. was applied and removed several times on the 14th of May, without producing any more permanent elongation than the 2 per cent. which was observed after the first on and off. During three-quarters of a year after that day no weight of more than 14 lbs. (in addition to the permanent 1 lb.) was ever applied to the wire, and electro-magnetic experiments were made upon it from day to day, with little intermission, with " ons" and "offs" of 14 lbs., but sometimes with 7 lbs. During all this time the length of the wire (about 4 metres) remained sensibly constant for the same weight, and the wire experienced regular elastic elongations and contractions when the weights were applied and removed. In some of the experiments with " ons" and "offs" of 14 lbs., about 90 centims. of the wire were heated to 100° C. by a stream of hot water (as described in § 201 below), but this left no permanent change in the wire.

201. The diagram (Fig. 1, Plate I.) shows the arrangement of the apparatus by means of which the results were obtained. The wire, W, experimented on was attached to a fixed support near the ceiling of the laboratory, and hung vertically downward, passing along the common axis of the magnetizing and induction coils, F, and was kept stretched by the scale-pan, D, which, as stated above, weighed exactly 1 lb. The coils, F, were supported on a clamp attached to the wire at their lower end, as shown in the drawing; and thus, as the length of the induction coil was small compared with the total length of the wire, motion of the wire relatively to the induction coil due to the stretching produced by the applied stress, was in great measure avoided. The magnetizing coil was 86 centims. long, and was composed of two layers of silk-covered copper wire. The inner layer contained 26·7 metres of wire, of No. 23 B.W.G., arranged in a solenoid* of 960 turns, the

* The common use of the word "helix" in this sense is utterly illogical. The idea of helix is not essential, but accidental, and in no practical case is it of any consequence whether it is a right-handed or a left-handed helix. There is nothing of helical quality in a cylindrical tube composed of two metals in two parts of its circumference, with the junctions of these metals kept at unequal temperatures. The thermo-electric current round the circumference of the solenoid produces the kind of magnetizing influence which is commonly produced by a helix, and constitutes precisely the arrangement which Ampère called a solenoid. It is only because the ordinary helix, with electric current flowing through it, produces more

outer layer, 24·3 metres of No. 20 wire, arranged in a solenoid of 728 turns. The resistance per metre of these wires was ·0673 and ·0522 ohm respectively, and the total resistance of the coil after it was wound was 3·134 ohms. The total length of the induction coil, which was contained within the magnetizing coil, was 31·5 centimetres. It was made up of two layers of silk-covered copper wire, of No. 29 B.W.G., laid on in 1439 turns. The wire thus coiled on was 10 metres long, and had a resistance of ·184 ohm per metre, and the total resistance of the coil, including 1 metre of electrodes, was 2·204 ohms.

The magnetizing coil was wound on the outside of a compound tube, made up of two tubes of thin brass, of different diameters, placed one within the other, with their axes coincident; the external diameter of the inner tube being less than the internal diameter of the outer by about 3 millimetres. The induction coil, which was wound on a thin copper tube just fitting the wire experimented on, was enclosed within the inner brass tube in such a position that its ends were at equal distances from the extremities of the magnetizing coil. The space between the two brass tubes formed a channel through which water could be made to flow from the cistern B, and thus, by regulation of the temperature of the water in B, the part of the wire within the magnetizing coil could be kept at any temperature from about 10° to nearly 100° C.

202. In order that the elongations and contractions produced in the wire by the applications and removals of the pulling stress might be observed, a second wire was hung from the same support, and kept stretched by two 28 lbs. weights, hung from the ends of the cross-bar, E. To this wire a scale of half-millimetres was vertically attached, in such a position that a pointer fixed to the magnetizing coil moved along it as the coil moved downward or upward with the application or removal of the weight.

203. The electrodes of the magnetizing coil were connected with the studs 3 and 4 of the commutator, K. One of the other pair of studs was connected with the zinc pole of a battery of my

or less approximately (very approximately indeed in the case of a helix with many turns) the same effect, that it is available for the electro-magnetic uses. It seems desirable, therefore, to take advantage of Ampère's original word "solenoid," and, except in cases in which the helical quality is taken into account, to give up the name "helix." The electro-magnetic solenoid may also be called a bar electro-magnet without soft iron core.

tray Daniells, the other stud with the sliding piece of a resistance-slide, R. This slide was designed for the purpose of allowing the battery strength to be raised continuously from 0 to nearly 1 cell, and from 1 cell to nearly 2 cells, and so on. It consisted of a contact-making slider, S, movable along a bare copper wire connecting the two poles of the cell to be sub-divided. This wire, which was 64 metres in length, and had a resistance of ·67 ohm, was stretched for convenience alternately from one side to the other of a large board, in the manner represented in the diagram (Fig. 1, Plate I.). Thus, with the number of cells and arrangement of connexions figured in the diagram, when the slider was brought up as nearly as possible to C, the current flowing was very nearly that due to 3 cells, and when the slider made contact at any other point of the wire, the current flowing through the magnetizing coil was less than that due to 3 cells by an amount depending on the distance along the wire of the slider from C.

204. A galvanometer, the resistance of which was only a very small fraction of an ohm, was used to measure the strength of the magnetizing current, and was so placed in the circuit that the current always flowed through it in the same direction: thus the whole range of the galvanometer scale was available for measuring the deflections produced by the stronger currents used, without the necessity for shifting the zero of the scale by means of a magnet. In these experiments at first the battery galvanometer was placed for convenience in the circuit between the commutator K and the coils F, and a reversing key used to keep the current through it always in the same direction, but it was afterwards transferred to the position in the circuit shown in the diagram and the reversing key dispensed with. The deflections of the needle of this galvanometer were read on a scale of half-millimetres placed at a distance of 75 centimetres from the mirror, and were used as the values of the magnetizing forces for the abscissas of the curves below. The total resistance in the circuit of the magnetizing coil, which was measured from time to time in the course of the experiments, and showed little or no variation, was 3·828 ohms. The induction coil was placed in circuit with an astatic galvanometer (described in Part VI., § 181 above, and called in that paper the ballistic galvanometer), the "throw" of which, as observed on a scale placed 120 centims., or 2400 of its own divisions, from the mirror,

measured the strength of the induced current. This galvanometer was placed within a case of thin sheet copper, and the whole enclosed within a glass bell jar to guard it from the effects of currents of air.

205. The procedure in experimenting was similar to that described in Part VI., § 186 above. One observer took the readings of the ballistic galvanometer, and made and broke the circuit of the magnetizing current; while a second, on word of command from the first, applied or removed the weight, and noted the elongations or contractions of the wire, as shown by the pointer and scale described above. The results were entered in the register on the system followed in my previous experiments, according to which + M denoted that the current was made in such a direction as to cause the image on the scale of the ballistic galvanometer to move towards the right or towards increasing numbers; − M, that the current was made in the contrary direction; B, that the current was stopped; Z, the zero of the ballistic galvanometer scale; " On," the application of the weight; and " Off," its removal. The polarity of the magnetization produced in the wire by + M was the same as that shown by the wire when under the influence alone of the vertical component of the earth's magnetic force; and consequently a deflection of the image on the scale of the ballistic galvanometer towards the right, produced by a change in the magnetic condition of the wire after it had been thus magnetized, indicated an increase of its magnetization, and a deflection to the left a diminution of its magnetization. This magnetization produced by + M, I shall call positive magnetization, and that produced by − M negative magnetization. Hence a deflection of the ballistic image towards the right indicates an increase of positive magnetization, and a deflection towards the left, a diminution of positive, or an increase of negative, magnetization.

206. The conclusions of the preliminary notice of June 10, 1875 (§§ 197$_1'$—197$_7'$ above), and the statements of § 198 above, are proved by the following table of results obtained by a very careful repetition of the experiments referred to in that notice. The first column of results in the table gives the operations in the order in which they were performed; the second, the deflections of the image on the scale of the ballistic galvanometer obtained after each operation;

and the third column gives the current strengths. During the whole of the experiments cold water was kept flowing from the cistern B through the channel between the coils to the vessel H, to prevent any heating of the coils and wire by the passage of the magnetizing current. The amount of the elongation and contraction of the wire by the application and removal of the weight of 14 lbs. was constant throughout the experiments. The experiment was begun with zero current in each case, and the weight of 14 lbs. applied and removed until equal and opposite effects were obtained

TABLE I.—Temperature of Cold Water.

Results for + M.			Results for – M.		
Operations.	Deflections.	Current strengths.	Operations.	Deflections.	Current strengths.
14 on	+ 26 }	0	14 on	0 }	11
,, off	− 26 }		,, off	0 }	
,, on	+ 33 }	10	,, on	− 27 }	22
,, off	− 33 }		,, off	+ 27 }	
,, on	+ 36 }	22	,, on	− 36 }	46
,, off	− 36 }		,, off	+ 36 }	
,,' on	+ 36·5}	44	,, on	− 36·5}	66
,, off	− 36·5}		,, off	+ 36·5}	
,, on	+ 33·5}	67	,, on	− 33·5}	96
,, off	− 33·5}		,, off	+ 33·5}	
,, on	+ 19 }	143	,, on	− 25 }	145
,, off	− 19 }		,, off	+ 25 }	
,, on	+ 12 }	180	,, on	− 19 }	169
,, off	− 12 }		,, off	+ 19 }	
,, on	+ 6 }	220	,, on	− 15 }	195
,, off	− 6 }		,, off	+ 15 }	
,, on	+ 2 }	246	,, on	− 11 }	223
,, off	− 2 }		,, off	+ 11 }	
,, on	0 }	267	,, on	− 4 }	267
,, off	0 }		,, off	+ 4 }	
,, on	− 4 }	310	,, on	0 }	290
,, off	+ 4 }		,, off	0 }	
,, on	− 7·5}	352	,, on	+ 5 }	355
,, off	+ 7·5}		,, off	− 5 }	
,, on	− 9 }	403	,, on	+ 7·5}	394
,, off	+ 9 }		,, off	− 7·5}	
,, on	− 11 }	447	,, on	+ 9 }	430
,, off	+ 11 }		,, off	− 9 }	
,, on	− 11·5}	503	,, on	+ 13 }	880
,, off	+ 11·5}		,, off	− 13 }	
,, on	− 12 }	600			
,, off	+ 12 }				
,, on	− 13 }	830			
,, off	+ 13 }				

by on and off; and the same process was followed after each augmentation of the magnetizing current. The table contains for each step only the results obtained after this state had been reached.

207. We see from the above table that the vertical component of the magnetizing force of the earth is balanced by an opposite magnetizing force due to a current measured by 11 divisions of the battery galvanometer, or about $\frac{1}{25}$ of the Villari critical value of the magnetizing force, with 14 lbs. "off" and "on."

If now a Daniell's cell be taken as one volt, or 10^8 in the C. G. S. system of units, we have for the magnetizing force of 1 cell estimated in the manner explained above (Part VI., § 183 above), the value $4\pi \dfrac{1688 \times 10^8}{86 \times 3{\cdot}828 \times 10^9} = 6{\cdot}443$. But the strength of the magnetizing current of 1 cell was measured by a deflection on the scale of the battery galvanometer of about 130 divisions, or a little more than twelve times the amount of the magnetizing force of the earth. Hence we have in absolute measure, ·5 as a rough approximation to the value of the vertical component of the earth's magnetic force at Glasgow. The true value, as said above (Part VI., § 183 above), is nearly ·43. (Compare also § 243 below.)

208. Immediately after the results given in Table I. were obtained, the experiments were repeated with the water in the cistern B kept constantly at the temperature of 100° C. Thus the water flowing through the channel between the coils may be considered as having been nearly at the boiling point. The results are given in Table II.

TABLE II.—Temperature 100° C.

Results for +M.			Results for −M.		
Operations.	Deflections.	Current strengths.	Operations.	Deflections.	Current strengths.
14 on	+26 }	0	14 on	0 }	11
,, off	−26 }		,, off	0 }	
,, on	+33 }	27	,, on	−31 }	34
,, off	−33 }		,, off	+31 }	
,, on	+28·5}	75	,, on	−32·5}	56
,, off	−28·5}		,, off	+32·5}	
,, on	+19 }	127	,, on	−31 }	86
,, off	−19 }		,, off	+31 }	
,, on	+11 }	185	,, on	−26·5}	109
,, off	−11 }		,, off	+26·5}	
,, on	+ 6 }	220	,, on	−22 }	138
,, off	− 6 }		,, off	+22 }	
,, on	0 }	274	,, on	−13 }	186
,, off	0 }		,, off	+13 }	
,, on	− 6 }	357	,, on	− 5 }	242
,, off	+ 6 }		,, off	+ 5 }	
,, on	− 8 }	453	,, on	− 0·75}	283
,, off	+ 8 }		,, off	+ 0·75}	
,, on	+ 9 }	830	,, on	+ 3 }	340
,, off	− 9 }		,, off	− 3 }	
			,, on	+ 6 }	406
			,, off	− 6 }	
			,, on	+ 7 }	478
			,, off	− 7 }	
			,, on	+ 7·5}	508
			,, off	− 7·5}	
			,, on	+ 9 }	830
			,, off	− 9 }	

209. For convenience of comparison, the results given in the above tables are represented graphically by the curves in Plates II. and III. of diagrams, in which the abscissas represent magnetizing forces, and the ordinates augmentations and diminutions of the magnetism of the wire. In diagram Plate II., a separate curve is given for + M and − M, at both temperatures; the curves for the higher temperature being drawn in full lines, and those for the lower temperature in dotted lines. In their general features they are similar to the curves given in the preliminary notice (§§ $197_1'$—$197_7'$) above referred to; but they include a more exact determination of the amount of the magnetizing force which gives maximum effect in each case, and of the points in which the curves cut the line of abscissas, and hence of the relation between the magnetizing force of the earth and the Villari critical value. The

magnetizing force for which 14 lbs. "on" or "off" gave maximum effect was now found for the same wire to be about four or five times, and the Villari critical value about twenty-three times, the vertical component of the earth's magnetic force at Glasgow These results also disprove the negative maximum indicated by the curves of that notice, and show (as there stated in the Note at the end) that for higher magnetizing forces than the Villari critical value, the effect approaches a constant amount and the curves become asymptotic.

210. The curves in diagram Plate III. were drawn by taking for ordinates the mean for each temperature of the effect for + M, and the effect for − M, of 14 lbs. "on" or "off," and for abscissas the corresponding current strengths, and therefore show approximately the effect which would have been produced by "on" or "off," had the wire not been affected by the magnetizing force of the earth.

By comparing the curve for cold water with the curve for hot water, we see that when the wire is at the temperature of 100°C., the average maximum effect of "on" or "off" is less than at the ordinary temperature of cold water by about 8 per cent. of the effect in the latter case, and that also, when the Villari critical value has been exceeded, the constant value to which the effect of "on" or "off" approaches is less for the higher temperature than for the lower, but in this case by about 30 per cent. of the amount of the effect for the lower temperature. The two curves also cross one another at a point above the line of abscissas, thus showing a greater critical value of the magnetizing force for the higher temperature than for the lower.

211. The curves in Plates IV. and V. of diagrams give at both temperatures, the average for each strength of magnetizing current of the effects on + M and − M of applying and removing stresses of 7 lbs. and 21 lbs. respectively.

A comparison of these curves with the average curves for 14 lbs. "on" and "off" (Plate II. above) shows:—

(1) That the effect at both temperatures of the application and removal of the stress is greater with 14 lbs. than with 7 lbs., and much greater with 21 lbs. than with 14 lbs.; the maximums at the ordinary temperature in these three cases being respectively 31, 35, and 54.

(2) That the Villari critical value is much greater for 7 lbs.

" on" and "off" than for 14 lbs.: and, though by a smaller difference, greater for 21 lbs. than for 14 lbs.

(3) The difference between the maximum effects of "on" or "off" for the high and low temperatures is greater for 7 lbs. than for either 14 lbs. or 21 lbs., and seems to be greater for 21 lbs. than for 14 lbs.

212. A series of observations of the effects of alternately making and breaking the circuit of the magnetizing coil and battery, were made at both temperatures and for both + M and − M. The method of procedure was as follows :—

With no current flowing, a weight of 14 lbs. was placed in the scale-pan, and removed ten times in rapid succession, and the wire finally left with only the permanent weight of 1 lb. hanging on it. Then beginning with + M, a magnetizing current of small amount was applied, and the effect measured by the "throw" of the ballistic galvanometer. The weight of 14 lbs. was then applied and removed ten times in succession, the circuit broken with nothing but the permanent weight of 1 lb. hanging on the wire, and the deflection of the ballistic galvanometer again noted. The same cycle of operations was then repeated for higher and higher strengths of current until ten cells were placed in circuit with the magnetizing coil.

The same process was then followed with the − M.

These experiments were repeated also at both temperatures with 21 lbs. as the weight applied and removed ten times before each operation. Curves Figs. 6a and 6b of Plate VI. exhibit the result for 14 lbs., and curves Figs. 6c and 6d of Plate VII. those for 21 lbs.

A very striking feature of these results is the great excess of the deflection produced by − M over the deflection produced by + M. It cannot but be due to the terrestrially-induced magnetism existing in the wire each time before the current is made in either direction.

Comparing the results for the ordinary temperature with those for 100° C., we see that the effect at the higher temperature is always considerably less than at the lower temperature. Thus, taking the 21 lb. curves, Figs. 6c and 6d of Plate VII., the deflection after − M with the greatest magnetizing force is 320 for the lower temperature, and about 250, or 22 per cent. less, for the higher, and for the same magnetizing force the other deflections are less at the higher temperature than at the lower, in nearly the same proportion.

213.   Immediately after the results for the temperature 100°C., shown in the curves Figs. 6c and 6d of Plate VII. had been obtained, an experiment was made to determine the amount by which (as stated in the Preliminary Notice of June 10, 1875, § 197, above) the effects of making and breaking the circuit of the magnetizing coil and battery when the wire is pulled, exceed the effects of the same operations when the wire is free from pull. The process was the same as that described in § 212 above, except that after each ten " ons" and " offs" the weight of 21 lbs. was put on and left in the scale-pan and the circuit made or broken before it was again removed.   The experiment was made at only one temperature, 100° C.   The results are given in the curves Fig. 6e of Plate VIII.

A full examination of the results shown in these curves of Plates VI., VII. and VIII., must be reserved for a later communication; and further experiments will be necessary to elucidate them.   Meantime it is interesting to see by comparing the curves Fig. 6e of Plate VIII. with the curves Figs. 6c and 6d of Plate VII. for the same temperature (100° C.), that the effect of the − M is greater with the pulled than with the unpulled wire for every degree of magnetizing force; while the effect of the + M is greater in the pulled wire for magnetizing forces less than 250, and greater in the unpulled wire for magnetizing forces exceeding 250.   This was to be expected from the previously proved (§§ 209, 210 above) greater magnetic susceptibility in the pulled than in the unpulled wire, when the magnetizing force is less than a critical value of 280 or 290; and greater susceptibility in the unpulled than in the pulled wire when the magnetizing force exceeds the critical value; and from the fact that the difference in one direction of the susceptibilities in the pulled and the unpulled wire when the magnetizing force is the Glasgow vertical force, is about three times as much as the difference in the other direction when the magnetizing force is 80 times the Glasgow vertical force.   The effect of the − M includes a reversal of the natural vertical force.   That of the + M is merely an addition to it.

§§ 214—222. *Preliminary investigation by the direct Magneto-metric Method of the effects of transverse stress on the Magnetization of an Iron Tube.*

214. In order to test qualitatively, in the first instance, the effects of transverse stress on the magnetization of iron, experiments were made on a smooth gun-barrel, said to be made of tolerably soft iron. The barrel was fitted at its muzzle with a piston working watertightly in a Bramah stuffing-box, and served round with a magnetizing coil of silk-covered copper wire, separated from the barrel by a copper tube, and containing within it an induction coil, in order that the ballistic method might be used if this was deemed advisable. The barrel was then fixed rigidly to a stone pier in the laboratory, with its breech end resting on a large block of stone which formed the base of the pier. On a shelf, also attached to the pier, and at a convenient distance due magnetic west of the barrel's axis, a small reflecting magnetometer was placed, with its needle on a level with the top of the barrel. A dead-beat galvanometer, the resistance of which was only a small fraction of an ohm, was used to measure the magnetizing current. The barrel having been filled with water was subjected to hydro-static pressure, applied by means of a lever carrying a weight, the lever itself being counterpoised by a weight attached to a cord passing over a pulley above. The pressure (the friction of the piston in its collar being neglected) was measured by the amount of the applied weight and the multiplication of the lever. The effects of the application and removal of the pressure were measured by the deflections of the galvanometer needle, read on a scale of half-millimetres, placed at a distance of one metre from the mirror.

215. The first experiments after the apparatus had been got into working order brought out the remarkable result that the effects of transverse stress on the magnetization of iron are, as to quality, the opposite of those of longitudinal stress; that is to say, when the magnetizing force is less than a certain critical value, the effect of applying transverse stress is to diminish, and of removing it to increase, the induced magnetization; and when this critical value has been exceeded, the effect of the application of transverse stress is to increase, and of its removal to diminish the induced magnetization.

The curves Fig. 7 of Plate IX. show (after the manner of those of Fig. 2, Plate II.) the effect for both + M and − M on the magnetometer needle, when placed on a level with the top of the barrel, produced by applying a transverse outward pressure of (approximately) 1000 lbs. per square inch to the iron of the barrel. The ordinates are given in divisions of the magnetometer scale, and the abscissas in divisions of the scale of the battery galvanometer. It is interesting to note that the critical value of the magnetism is for this position of the magnetometer nearly 80 times the vertical component of the earth's magnetic force at Glasgow.

216. By placing the magnetometer at lower levels relatively to the barrel, it was found that the values of the current required for maximum and for zero effect, by "on" and "off", were less, the nearer the level of the magnetometer needle was to the middle of the length of the bar. The following table shows the principal effects for − M at four different levels of the magnetometer needle :—

Distance* of top of barrel above level of magnetometer needle.	Maximum deflection before critical value was reached.	Magnetizing current for maximum deflection.	Critical value of magnetizing current.
Centims.			
0	32·8	110	430
10·5	31	130	345
21·0	14	100	265
31·5	3	100	147

The whole of the results for the − M in the last three cases are shewn in the curves Fig. 8 of Plate X.

217. The magnetometer needle was placed on a level with the top of the barrel, and at a distance of two metres from its axis. Experiments were then made to find the total magnetization produced by different strengths of magnetizing current, and the effects on it of ten successive applications and removals, while the current was still flowing, of a hydrostatic pressure of 1000 lbs. per square inch; also to find the total residual magnetism after the removal of the magnetizing current, when the tube was left solely under the influence of the earth's vertical magnetic force, and the effect on it of ten successive applications and removals of the same

---

* The length of the barrel and of the magnetizing coil was about 90 centims.

hydrostatic pressure, while the bar was still under this magnetizing influence. The general character of the results of these experiments is difficult to describe in words, but can be seen by inspecting Fig. 9, Plate XI, in which the curves marked A show the total magnetization, those marked B the residual magnetization. The abscissas of these curves are proportional to the magnetizing forces and the ordinates to the observed magnetization : but they have inadvertently been so drawn that their negative ordinates show magnetization of the same polarity to that produced by the inductive influence of the earth's vertical force, and their positive ordinates magnetization of the opposite polarity. The full lines in each set of curves show the magnetization before, the dotted lines after, ten "ons" and "offs" of and pressure of 1000 lbs. per square inch.

The results in these experiments were obtained by beginning with a *negative* (*i.e.*, opposed to magnetizing force of earth) magnetizing current of about 700 divisions of the battery galvanometer scale, which was gradually diminished to zero and then increased until a *positive* current of 700 (20 cells) divisions was reached. This process was then exactly reversed. The results may be examined in the order in which they were obtained, by beginning at the left hand ends of the curves A and B, and passing to the right along the thinner lines, returning to the left along the thicker lines.

218. The magnetometer was found in this position to be at too great a distance from the barrel to show the residual magnetism with accuracy, and accordingly its distance from the barrel was reduced to one metre, and the effects of residual magnetism alone observed. These are shown by curves marked B' of Fig. 10, Plate XII. The method and order of experimenting were here the same as described in § 217; and the explanation of Fig. 9, curves B, applies also to Fig. 10, except that in the latter the directions of the ordinates are reversed from the former; thus in Fig. 10 positive ordinates indicate positive magnetization, negative ordinates negative magnetization (see § 205 above).

219. Beginning at the extreme left of curves marked B' on Fig. 10, Plate XII., and following the arrows, it will be seen that the residual magnetism remains nearly constant in amount until the magnetizing current has been diminished to about 300 divisions, when it begins to take a greater negative value, and continues to

do so until the current is brought down to 50 divisions, when it begins slowly to diminish. After the reversal of the current the full residual magnetism diminishes with great rapidity, passing through zero at about 15 divisions of positive current. It then becomes positive, preserving nearly the same rapidity of change for some distance beyond zero. After a positive current of 150 divisions is reached, the full residual magnetism increases very slowly, and the curve becomes asymptotic towards a value of about 440. The curve showing the residual magnetism after ten "ons" and "offs" has similar characteristics to those for the full residual magnetism; the increase on the left side of zero at about 300 divisions of negative current is, however, more decided.

The general character of the return curves is similar to that just described. It is to be remarked, however, that the zero of current in each of them is much further passed before the zero of magnetization is reached. This difference between the going and returning curves would be done away with, and the curves from left to right and right to left in the diagram would be perfectly symmetrical about the zero of magnetizing current, if the influence of the earth's magnetic force were eliminated. I intend to return to this subject with a modification of the experimental arrangements, to allow the residual magnetism to be observed unaffected by any influence due to magnetizing force in the direction of the length of the tube; instead of as here with the tube always under the magnetizing action of the vertical component of the earth's magnetic force, when the electro-magnetic current is not flowing.

The curves, except in the positions corresponding to zero of current, show that the ten "ons" and "offs" were not sufficient to shake out the effect of the powerful magnetizing force of the current, and allow the barrel to take the magnetization due to the vertical component of the terrestrial magnetic force. Indeed, they only diminished the residual magnetism in the curves both from left to right and from right to left by about 100 divisions, out of total residuals from 230 to 440. It is interesting to see by contrasting the right-hand ends of the curves with the left-hand ends, how much stronger the residual magnetism is when helped, than when opposed by the vertical magnetic force of the earth.

220.   Comparison between the results of the effects of longitudinal and transverse pull shows that an aeolotropic property of

different magnetic inductive susceptibility in different directions, is temporarily developed in soft iron by aeolotropic stress (that is to say by stress not consisting of positive or negative pressure equal in all directions). The results show that with low magnetizing forces, negative pressure perpendicular to one set of parallel planes of soft iron, produces an augmentation of magnetic susceptibility in the direction of the pressure, and diminution of the susceptibility in all directions at right angles to it. The effects of positive pressure have not yet been tested experimentally, but it is certain that they will be opposite to the effects of negative pressure. Independently of experiment, we may also infer that the effects of infinitely small positive pressure perpendicular to one set of parallel planes, and infinitely small negative pressure of equal amount perpendicular to a set of parallel planes at right angles to them, must be equal and opposite in the directions of these pressures, and therefore must leave the magnetic susceptibility unaltered in the directions inclined 45° to them. This is exactly the stress which is experienced in a twisted wire of circular section ; the amount of the stress being zero in the axis of the wire, and being elsewhere in simple proportion to distance from the axis. The directions of the positive and negative pressures at any point of the substance are two lines in the tangent plane of the cylindric surface through it, co-axal with the boundary of the wire, and inclined at 45° to the normal plane section. Hence, when the torsion is infinitely small, the magnetic susceptibility of the wire in the direction of its length must be unaltered, and if finite amounts of torsion produce any change in the magnetic susceptibility, the amount of this change must ultimately (for very small torsions) vary inversely as the square of the amount of torsion, as we see by remarking that whatever effect is produced must be independent of the direction of the torsion, there being nothing of helicoidal quality in longitudinal magnetization.

221. In Wiedemann's *Galvanismus* (vol. II., §§ 476—498) an abstract is given of researches in this subject by Matteucci, Wertheim and Edmund Becquerel *. One main result of all these investigations is that torsion in either direction diminishes the temporary inductive longitudinal magnetization of soft iron.

222. Nearly two years ago I instituted a series of experiments

* Matteucci, *Comptes Rendus*, t. xxiv., 1847 ; Wertheim, *Comptes Rendus*, t. xxxv., p. 702, 1852.

on the subject, chiefly for the purpose of finding the influence of torsion upon the longitudinal magnetization of soft iron wire subjected to different amounts of pulling force. These experiments, in which the magnetizing influence was simply the vertical component of the earth's magnetic force, were carried out by Mr Donald Macfarlane. The mode of experimenting and the results obtained are described in the following report.

§§ 223—229. *Experiments on the effect of torsion and stretching in altering the induced Magnetism of a very soft Iron Wire, subjected to various amounts of constant pull.*

223. *Description of Apparatus.*—In the diagram of apparatus Fig. 11, Pl. VIII. AB is a soft iron wire, No. 22 B.W.G., 81 centims. long, to the ends of which were soldered pieces of No. 16 copper wire; the upper piece, AD, about 5 metres in length, was attached to the ceiling of the room with an arrangement for raising or lowering it through a small space, the lower piece, BC, about 50 centims. long, had attached to its lower end a scale-pan for holding the stretching weight.

E is a small mirror magnetometer, the mirror being 1 centim. diameter, carrying a magnet 8 millimetres in length and suspended at N by a single silk fibre 10 centims. long; I is a lens close to the mirror; F is a paraffin lamp; and GH a scale (bent into a circular arc of which E is the centre), on which is formed the image of a fine wire placed in front of the lamp flame, at E.

L and M are two edges at right angles to one another, fixed to the stand carrying the magnetometer, and just in contact with the wire: their use is to make sure that the wire is maintained in the same position relatively to the centre of the magnet carried by the mirror.

K is an arm soldered to the copper wire, BC, for applying torsion to the wire, and immediately below it is a circle divided at intervals of 20°, with small holes at each division for inserting pegs to keep the arm twisted at any angle while readings are being taken.

A similar arm was soldered to the copper wire AD at O, the two ends of which were in contact with two vertical guides, thus confining the twist to the portion of wire between K and O.

The distance of the wire from the centre of the magnetometer magnet was 8·2 centims.; distance of scale from mirror, 157 centims.; one division of scale ·5 millim.

The experiments were made in this way:—

Having removed the wire to a distance from the magneto-meter, the zero reading of the latter was taken; the wire was placed in position, a stretching weight put in the scale-pan, and the reading on the torsion-circle noted when the wire was free from torsive stress; torsion was then applied by turning the lower end of the wire 20° at a time up to 320°, and the reading on the magnetometer scale at each step taken. Similarly, readings of the deflection were taken as the torsion was taken out, and then continued in the opposite direction as far as − 200° and back to the original starting point. At the end of each of the first seven series the weight was taken off and put on again.

The results are represented in the diagrams of curves Figs. 12 to 31, Plates XIII. to XIX. In each diagram the numbers at the side indicate the readings on the magnetometer scale, and those at the bottom the readings on the torsion circle, a cross (×) marks the positions where the couple of torsion was zero, *b* the begin-ning, and *e* the end of each experiment.

224. Explanatory remarks.

Fig. 12, Pl. XIII.

Time, July 23, $4^h 12^m$ to $5^h 25^m$.

Zero of magnetometer, 90 divisions.

Stretching weight, 13 lbs.

Weight off and put on, final reading rose from 790 to 895.

Fig. 13, Pl. XIII.

July 24, $11^h 0^m$ to $11^h 55^m$.

Stretching weight, 13 lbs. left on from last.

Zero of magnetometer at end, 85.

Weight off and on, final reading rose from 815 to 896.

Fig. 14, Pl. XIII.

Time, July 24, $12^h 20^m$ to $1^h 5^m$.

Stretching weight, 12 lbs.

Zero of magnetometer, beginning 85, end 85.

Weight off and on, final reading rose from 820 to 874.

Fig. 15, Pl. XIV.

Time, July 24, $1^h 15^m$ to $2^h 0^m$.

Stretching weight, 12 lbs.

Zero of magnetometer, beginning 85, end 90.

Weight off and on, final reading rose from 813 to 876.

Fig. 16, Pl. XIV.
  July 26, 12^h 50^m to 1^h 35^m.
  Stretching weight, 11 lbs. on for 46 hours.
  Zero of magnetometer, at end 95.
  Weight off and on, final reading rose from 800 to 850.

Fig. 17, Pl. XIV.
  July 26, 1^h 45^m to 2^h 25^m.
  Stretching weight, 11 lbs.
  Zero of magnetometer, at beginning 95, end 100.
  Weight off and on, final reading rose from 803 to 835.

Fig. 18, Pl. XV.
  July 26, 4^h 20^m to 5^h 10^m.
  Stretching weight, 14 lbs.
  Zero of magnetometer, at beginning 100, at end 100.
  Weight off and on, final reading rose from 750 to 830.

Fig. 19, Pl. XV.
  July 26, 5^h 20^m to 6^h 10^m
  Stretching weight, 14 lbs.
  Zero of magnetometer, at beginning 100, at end 100.

Fig. 20, Pl. XV.
  July 28, 1^h 25^m to 2^h 15^m.
  Stretching weight, 9 lbs.
  Zero of magnetometer, at beginning 100.

Fig. 21, Pl. XVI.
  July 28, 4^h 45^m to 5^h 35^m.
  Stretching weight, 9 lbs. left on from end of experiment
    Fig. 20.
  Zero of magnetometer, at end 105.

Fig. 22, Pl. XVI.
  July 28, 5^h 20^m to 6^h 0^m.
  Stretching weight, 10 lbs.
  Zero of magnetometer, at beginning 105.

Fig. 23, Pl. XVI.
  July 30, 12^h 40^m to 1^h 20^m.
  Stretching weight, 10 lbs. left on 42 hours from preceding
    experiment (Fig. 23).
  Zero of magnetometer, at end 105.

Fig. 24, Pl. XVII.
　July 30, 4^h 20^m to 5^h 10^m.
　Stretching weight, 16 lbs.
　Zero of magnetometer, at beginning 105, at end 105.
Fig. 25, Pl. XVII.
　July 30, 5^h 20^m to 6^h 10^m.
　Stretching weight, 16 lbs.
　Zero of magnetometer, at beginning 105, at end 105.

Note.—This experiment (Fig. 25) was intended to be a continuation of the preceding. The magnetometer zero at the beginning was found, without taking off the weight, by drawing the wire aside; but the disturbance thus occasioned when the wire was replaced raised the final reading of experiment, Fig. 24, from 690 to 712, the initial reading of experiment, Fig. 25.

The experiments represented in the six diagrams which follow were each repeated without stopping, and the repeat is represented by the dotted lines.

Fig. 26, Pl. XVII.
　August 2, 11^h 5^m to 12^h 40^m
　Stretching weight, 18 lbs.
　Zero of magnetometer, at end 95.
Fig. 27, Pl. XVIII.
　August 2, 4^h 0^m to 5^h 20^m.
　Stretching weight, 20 lbs.
　Zero of magnetometer, at beginning 95, at end 85.
Fig. 28, Pl. XVIII.
　August 3, 10^h 25^m to 11^h 45^m.
　Stretching weight, 22 lbs.
　Zero of magnetometer, at end 90.
Fig. 29, Pl. XVIII.
　August 3, 12^h 0^m to 1^h 50^m.
　Stretching weight, 24 lbs.
　Zero of magnetometer, at beginning 90, at end 100.
Fig. 30, Pl. XIX.*
　August 3, 4^h 0^m to 5^h 30^m.
　Stretching weight, 26 lbs.
　Zero of magnetometer, at end 90.

* In this diagram the two dotted lines between + 40 and – 200 coincide.

Fig. 31, Pl. XIX.

August 4, 12^h 20^m to 1^h 50^m.

Stretching weight, 28 lbs.

Zero of magnetometer, at beginning 90, at end 100.

225. From the curves it will be seen that the amount of the effect of torsion in diminishing the magnetization is not greatly influenced by the differences of pull from 10 lbs. to 20 lbs., but that it is greatly diminished by increase of the pull above 20 lbs. For simplicity and uniformity in the comparison, take the amount of the diminution of magnetization by the first application of torsion in the direction called negative, from the whole range of from + 50° to − 200° on the scale of torsion (abscissas of the curves).

We find with pulls of from 10 lbs. to 20 lbs. various amounts of from 180 to 230 magnetometer scale divisions. The seemingly irregular differences between these amounts showed no regular dependence on the amount of pull, but seemed rather to depend upon previous conditions of the wire. But when the weight exceeded 20 lbs. there seemed a somewhat regular diminution in the effect of torsion with increase of pull, as shown in the following table; thus:—

With pull of 20 lbs. the effect was 193

„	22	„	177
„	24	„	150
„	26	„	130
„	28	„	65

226. One very interesting feature common to all the diagrams, and presented even to some degree by the exceptional ones, Figs. 20—23, and Fig. 25, shows that the effect of twisting the wire first in one direction and then in the other, and leaving it free from torsive force, was in every case to leave it with less magnetization than it had at the beginning.

227. These exceptionable diagrams and later continuation of the operations through a second positive torsion and a second negative torsion, represented in the latter halves of the curves of Figs. 26 to 31, Plates XVII. to XIX., show what would be the general character of the effect of continued periodic applications of positive and negative torsion, through equal angles on the two sides of zero.

In every case there is a lagging of quality, showing a residue of effect from previously acting causes. Thus, beginning with a wire which has been reduced to a normal condition by having had the weight off and on, and having been left to itself for twenty-four hours, we found (on the 24th of August) that the magnetization fell from a normal value of 685 down to 550 as the result of twisting it to $+260°$, then to $-260°$, and then to zero of torsion. The second application of positive and negative torsion reduced the magnetization further to 534. It is curious to find that not merely does torsion diminish the magnetization temporarily, but that it leaves so large a permanent diminution. Whether this permanence is absolute in respect to time or not, is an interesting question to be solved by leaving a wire which has been thus dealt with absolutely quiescent from day to day, month to month, year to year, century to century. It seems, however, that but slight mechanical disturbance suffices to shake out the diminution of magnetization left at the end of each of these experiments.

228. The general lagging of effect is shown by the fact that in every ascending branch the curve is lower than in the immediately previous descending branch; and the dotted latter halves of the curves of Figs. 26 to 31, Plates XVII. to XIX. show, by the intersections of their convex portions near the zero, that if the experiment was continued long enough, the history of the variation of magnetization would in every case be represented by a curve like that in the annexed sketch (Fig. 32).

Fig. 32.

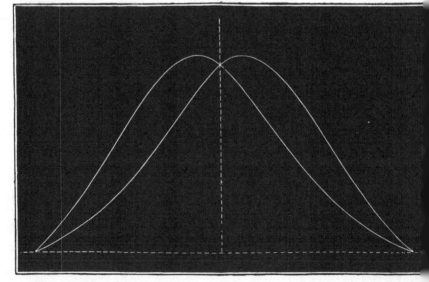

229. A very interesting discovery of Wiedemann's (*Galvanismus*, §§ 491 and 498), at first sight, seemed to find its explanation in the aeolotropic difference of magnetic susceptibility which I have found to be induced by aeolotropic stress in soft iron. The phenomenon consists in the development of logitudinal magnetization by twisting a wire through which a magnetic galvanic current is maintained longitudinally. The annexed double diagram (Figs. 33, 34), copied from Wiedemann's book *, describes the change of ideal

Fig. 33.   Fig. 34.

magnetic molecules which would represent the actually observed effect, which is that the end of the wire by which the current enters becomes a true north pole when the twist given to it is right-handed (or that of an ordinary screw). If this effect were due to greater susceptibility in one direction than in another, the direction of greatest susceptibility would be the direction sloping at 45°, upwards to the right in the front of the right-hand diagram (Fig. 34); that is to say, it would be the direction of positive pull in the stressed material. But the exceedingly intense magnetization by influence of circular lines of force round the cylinder, produced by currents of such strength as Wiedemann may be supposed to have used, must in all probability have been above the critical degree of magnetization at which the effect of pull becomes reversed, and therefore in all probability the direction of least susceptibility in the actual circumstances must have been that of positive pull. Hence, it seems almost impossible to admit the explanation of Wiedemann's result by aeolotropic magnetic

* I have altered the letters *n* and *s* of Wiedemann's book to Gilbert's old wholesome rule of putting *n* to represent true northern polarity, or the polarity of the same name as that of the earth's northern regions, and similarly *s* to represent true southern polarity.

susceptibility in the circumstances*   The true explanation is not easily conjectured: for another cause, also adverse to Wiedemann's result, is operative.   The electric conductivity of the iron is probably least in the direction of the positive pull and greatest in the direction of the negative pull in the stressed material†.   This aeolotropic quality in respect to electric conductivity would cause the electric current, instead of flowing rectilineally along the wire, to flow in left-handedly helical lines in the case represented in Fig. 34, and thus the central parts of the iron cylinder would become really magnetized by, as it were, an ordinary helix, but with very steep thread.   The effect of such a helix is the same as that of a true solenoid superimposed upon a rectilineal current through the wire, and the direction of the current in the supposed circumstances is such that it would give a true south pole at the upper end of the iron rod in Fig. 34.

§§ 230—240.   *On the effects of longitudinal stress on the Magnetization of Nickel and Cobalt.*

§ 230.   Through the kindness of Mr Joseph Wharton, of Philadelphia, U.S., I was enabled to continue my experiments with malleable and cast bars of nickel, and of cast cobalt.

[Note added July 8, 1879.—A qualitative analysis of one of Mr Wharton's nickel bars, performed in the Chemical Laboratory of Glasgow University, by Mr Donald Mackenzie, showed that the bar was not of absolutely pure nickel, but contained some carbonaceous matter and also a trace of iron.   The amount of the latter, however, was very small, and probably could not vitiate to any sensible degree the results of the experiments described below.]

The apparatus used in the preliminary experiments and its arrangement are shown in the diagram Fig. 35, Pl. XX.   Each end of the rod experimented on was inserted into a ferule-shaped clamp, *C* (shown also detached in plan and elevation at Fig. 36, Pl. XX.), the outer surface of which was conical and screw-threaded.

---

* [This experiment has been repeated for me since the communication of this instalment to the Royal Society, by Mr Macfarlane, with currents, not hitherto measured or estimated in absolute measure, but strong enough to greatly heat the iron wire.   The result was always the same as Wiedemann's, and was greatest with the strongest current used.—W. T., May 22, 1878.]

† See §§ 145—153 above.   Also a paper "On the Increase in Resistance to the Passage of an Electric Current, produced in Certain Wires by Stretching" (Tomlinson *Proc. Roy. Soc.*, Dec. 21, 1876).

The clamp, which had in it three longitudinal slits, was then, by means of a conical nut working round it, made to grasp the rod tightly enough to admit of the application of great amounts of longitudinal pull, without much danger of pulling the clamps away from the rod. One of the clamps was then hung from a pin in a strong cross-beam of a frame, so that the bar hung vertically downwards.

A rope, R, made of copper wire, connected the other clamp with a point near the end of a long heavy lever, turning on a fulcrum at that end formed by a knife edge pressing upwards against a brass plate, which formed a bridge between two strong and rigid uprights attached to the floor of the room. A heavy weight of lead was hung on the lever, and could be moved along it to give different amounts of stress. The lever was graduated, and the effect of its own weight was measured, so that the stress applied at any time could be at once read off. When the bar was in position but not under stress, the lever rested on a support high enough to allow the wire rope R to be slack, and was gently removed from this support when the pulling stress was applied. The effects of the various operations were measured by the deflections of the image on the scale of a reflecting magnetometer, M, the needle of which was on a level with the lower end of the bar, and at a distance due magnetic west from its axis, of 12 centims. The scale of the magnetometer was at a distance of 135 centims. from the needle.

231. In these preliminary experiments the bars were under the influence of no magnetizing force except that of the vertical component of the earth's magnetic force at Glasgow. In order that the results obtained with the bars of nickel and cobalt might be readily compared with those obtained from iron in the same circumstances, the experiment was first performed on a bar of tolerably soft iron, of nearly the same dimensions as those of the bar of nickel or cobalt experimented on. The actual results of a set of these experiments are given in the following tables. Table I. contains the results of an experiment on an iron bar 60 centims. long and 8 of a centim. in diameter, and Tables II. and III. the results respectively of two similar experiments performed immediately afterwards, one on a bar of wrought nickel and the other on a bar of cast cobalt. The total magnetization is reckoned

positive when its polarity is the same as that produced by the inductive effect of the earth's magnetic force, and negative when its polarity is of the opposite kind.

TABLE I.—Iron Bar.   Lower end of bar a true South Pole.

Operations.	Magnetometer readings.	Differences.	Total. magnetization.
Bar placed in position	Zero 463 {Image off scale in negative direction.		
Controlling magnet introduced	769		+ 1369
174 lbs. on	675	− 94	+ 1463
,, off	687	+ 12	+ 1451
,, on	660	− 27	+ 1478
,, off	675	+ 15	+ 1463
10 "ons" and "offs"	667	− 8	+ 1471
174 lbs. on	649	− 18	+ 1489
,, off	663	+ 14	+ 1475
259 lbs. on	571	− 92	+ 1567
,, off	597	+ 26	+ 1541
10 "ons" and "offs"	570	− 27	+ 1568
259 lbs. on	540	− 30	+ 1598
,, off	569	+ 29	+ 1569
325 lbs. on	470	− 99	+ 1668
,, off	510	+ 40	+ 1628
10 "ons" and "offs"	474	− 36	+ 1664
325 lbs. on	421	− 53	+ 1717
,, off	457	+ 36	+ 1671

232.   In Table I. it is stated that when the iron bar was placed in position the image on the scale of the magnetometer was driven off the scale in the negative direction.   A controlling magnet, 15 centims. long, placed at right angles to the magnetic meridian in a horizontal line passing through the centre of the needle, was used to bring the image to the division 769 on the scale.   When the nearer end of this magnet was at a distance of 23 centims. from the needle, the image rested at the division 323 on the scale, and when the magnet was brought 6 centims. nearer to the needle, at the division 773 on the scale.   Hence, taking the tangent of the angle of deflection as equal to the angle itself, we may reckon the deflection caused by bringing the iron bar of Table I. into position as 1369 scale divisions, which may be taken as measuring the magnetism of the bar*.   This increased during

* The method of calculation is as follows: Let A denote the deflection produced by placing the bar in position; D and D' the readings with the controlling

twelve successive applications and removals of a pulling stress of
174 lbs. to about 1470 scale divisions, when the bar was found
to have been brought to a nearly permanent condition ; and the
average effect of applying this stress was to increase the magnetism
by 16 divisions, and of removing the stress to diminish the magnetism by the same amount. The magnetism of the bar further
increased during 11 successive applications and removals of a pull of
259 lbs. to 1568 scale divisions, and the average effect of applying
and removing the pull was to increase and diminish the magnetism
by 28 scale divisions. After 11 applications and removals of a
stress of 325 lbs. the magnetism was found to have increased to
1664 divisions; and the average effect of " on " and " off " was
found to be an increase and diminution of 45 divisions.

233. Passing now to the nickel bar of Table II., we see that
when the bar was placed in position it showed 10 divisions of
positive magnetism (or magnetism of the same polarity as that
induced by the earth). The application of 146 lbs. of pull gave
it 29 divisions, and the removal of this pull 10 divisions additional
magnetism, that is, both " on " and " off " increased the magnetism.
A second application of the pull gave a diminution of 5, and
removal an increase of 6, divisions. The remainder of the procedure was similar to that followed in the case of the soft iron,
and with the exception of the first result after the bar was placed
in the clamps, the effect of " on " was always to *diminish* the
magnetism of the bar, and of " off " to *increase* it.

magnet introduced, with its centre at the distances r and r' from the magnet; a the
half length of the magnet; and B a constant depending on the controlling magnet;
then we have,

$$D = A + \frac{B}{(r-a)^2} - \frac{B}{(r+a)^2} = A + \frac{4Bar}{(r^2 - a^2)^2}$$

and similarly

$$D' = A + \frac{4Bar'}{(r'^2 - a^2)^2}.$$

Eliminating B between these two equations, and solving for A we get

$$A = \frac{D+D'}{2} - \frac{D-D'}{2} \frac{r(r'^2 - a^2)^2 + r'(r^2 - a^2)^2}{r(r'^2 - a^2)^2 - r'(r^2 - a^2)^2}.$$

By taking for a the virtual half length of the magnet or the distance of either pole
from the centre of the magnet's length, instead of the actual half-length, a somewhat nearer approximation to the value of A might have been obtained, but, as the
approximation was at best a rough one on account of the size of the angle, it was
not thought necessary to make this refinement.

After the bar was placed in the clamps the effect of the successive operations was on the whole to gradually augment the

TABLE II.—Wrought Nickel Bar. Lower end a true South Pole.

Operations.	Magnetometer readings.	Differences.	Total magnetization.
	Zero 460		
Bar placed in position	450	− 10	+ 10
146 lbs. on	421	− 29	+ 39
,, off	411	− 10	+ 49
,, on	416	+ 5	+ 44
,, off	410	− 6	+ 50
Clamp here slipped.			
Bar reclamped and replaced in	Image at		
position	415	− 45	+ 95
146 lbs. on	412	− 3	+ 98
,, off	393	− 19	+117
,, on	410	+17	+100
,, off	389	− 21	+121
10 "ons" and "offs"	387	− 2	+123
146 lbs. on	410	+23	+100
,, off	388	− 22	+122
174 lbs. on	408	+20	+102
,, off	384	− 24	+126
10 "ons" and "offs"	383	− 1	+127
174 lbs. on	405	+22	+105
,, off	380	− 25	+130
211 lbs. on	405	+25	+105
,, off	376	− 29	+134
10 "ons" and "offs"	375	− 1	+135
211 lbs. on	403	+28	+107
,, off	378	− 25	+132
249 lbs. on	403	+25	+107
,, off	368	− 35	+142
10 "ons" and "offs"	375	+ 7	+135
249 lbs. on	395	+20	+110
,, off	365	− 30	+140
,, on	398	+ 33	+107
,, off	364	− 34	+141
287 lbs. on	397	+33	+108
,, off	357	− 40	+148
10 "ons" and "offs"	353	− 4	+152
287 lbs. on	393	+40	+112
,, off	350	− 43	+155

total magnetization of the bar from the value 10 to the value 155, at which it stood when the experiment was concluded.

234. From Table III. we see that the bar of cast cobalt, when placed in position, had its true north pole down, and gave a deflection of 492 divisions. Ten "ons" and "offs" with 136 lbs. diminished this deflection by 85 divisions. The effect of "on" was then to *increase* the magnetism by 15 divisions, and of "off"

to *diminish* it by the same amount; that is to say, "on" increased the magnetism of the bar, "off" diminished it. A few blows of a mallet reversed this magnetism, and caused the bar to give a deflection of 90 divisions in the opposite direction. The effect of "on" was, as with the nickel bar, to *diminish* the magnetism, and of "off" to *increase* it, the effect of 146 lbs. being 25 divisions

TABLE III.—Cast Cobalt Bar. Lower end at first a true North Pole.

Operations.	Magnetometer readings.	Differences.	Total. magnetization.
	Zero 460		
Cobalt bar placed in position ...	952	+ 492	− 492
146 lbs. on	895	− 57	− 435
„ off	885	− 10	− 425
10 "ons" and "offs"	867	− 18	− 407
146 lbs. on	882	+ 15	− 422
„ off	867	− 15	− 407
Bar struck a few blows with a mallet. Magnetism reversed	370	− 497	+ 90
146 lbs. on	395	+ 25	+ 65
„ off	370	− 25	+ 90
„ on	395	+ 25	+ 65
„ off	373	− 22	+ 87
174 lbs. on	403	+ 30	+ 57
„ off	375	− 28	+ 85
10 "ons" and "offs"	375	0	+ 85
249 lbs. on	Bar here broke, but being held in its place gave a reading of 360.		

of the magnetometer scale, of 174 lbs. 29 divisions. The bar broke before the effect of the application of 249 lbs. could be observed. After the reversal of the magnetism by tapping the bar while under the influence of the vertical component of the earth's magnetic force, there was very little gradual change in the magnetization of the bar.

The seemingly anomalous effect obtained with the cobalt bar, when placed with its true north pole down, according to which the effect of the application of stress was to increase the magnetism of the bar, and of the removal of stress to diminish it, was no doubt due to the magnetizing influence of the earth tending to reverse the retained magnetism of the bar. It will be further investigated in a continuation of experiments on cobalt.

235. The effect of longitudinal stress on the magnetization of nickel when magnetized by a current flowing in a coil surrounding the bar, formed the subject of the next series of investigations. The magnetizing coil was 54 centims. long, and consisted of six layers of silk-covered copper wire of No. 22 B.W.G., each layer forming a solenoid containing 10·7 turns per centim. The resistance of the coil when cool was 7·2 ohms, and the resistance of the electrodes ·3 of an ohm. In this experiment the magnetometer was placed at a distance of 40 centims. from the axis of the bar, and on a level with its lower end; and in order that the deflection due to the total magnetization of the bar might be conveniently measured, the directive force on the magnetometer needle was increased by placing behind it, in the magnetic meridian, a bar magnet, with its true north pole turned towards the north. As in the previous experiments with iron and steel, − M indicates that the electromagnetic field was opposite in polarity to that of the earth, and + M that its polarity was the same as that of the earth. The dead-beat galvanometer used in the experiments on the effects of transverse stress on the magnetization of an iron tube, was again employed to measure the strength of the magnetizing current.

236. The results of a series of these experiments are given in Table IV. It will be seen from that table that the effects of the application and removal of stress were respectively to diminish and to increase the induced magnetization, and that, as in the case of soft iron, this effect reached a maximum with a certain strength of magnetizing current, after which it slowly diminished. In this experiment the critical value of the magnetizing force, corresponding to what has been called above, in the account of experiments in soft iron, the Villari *critical value*, was not reached.

TABLE IV.—Bar of Wrought Nickel.

Operations.	Readings. Zero 554.	Differences.	Total magnetization.	Strength of magnetizing current.
Bar put in position...	352	− 202	+ 202	
− M..............	583	+ 231	− 29	62
10 "ons" and "offs" with 285 lbs.	637	+ 54	− 83	(2 double cells)
285 lbs. on	623	− 14	− 69	
,, off	637	+ 14	− 83	
B..............	568	− 69	− 14	
285 lbs. on	554	− 14	0	
,, off	556	+ 2	− 2	
,, on	553	− 3	+ 1	
,, off	555	+ 2	− 1	
+ M..............	340	− 215	+ 214	62
285 lbs. on	367	+ 27	+ 187	
,, off	332	− 35	+ 222	
,, on	363	+ 31	+ 191	
,, off	329	− 34	+ 225	
B..............	388	+ 59	+ 166	
285 lbs. on	426	+ 38	+ 128	
,, off	403	− 23	+ 151	
,, on	427	+ 24	+ 127	
,, off	404	− 23	+ 150	
− M..............	863	+ 459	− 309	147
285 lbs. on	832	− 31	− 278	
,, off	864	+ 32	− 310	
,, on	833	− 31	− 279	
,, off	864	+ 31	− 310	
B..............	745	− 119	− 191	
285 lbs. on	690	− 55	− 136	
,, off	716	+ 26	− 162	
,, on	689	− 27	− 135	
,, off	716	+ 27	− 162	
+ M..............	226	− 490	+ 328	147
285 lbs. on	260	+ 34	+ 294	
,, off	227	− 33	+ 327	
,, on	260	+ 33	+ 294	
,, off	226	− 34	+ 328	
B..............	343	+ 117	+ 211	
285 lbs. on	400	+ 57	+ 154	
,, off	370	− 30	+ 184	
,, on	400	+ 30	+ 154	
,, off	371	− 29	+ 183	
− M..............	953	+ 582	− 399	282
285 lbs. on	927	− 26	− 373	
,, off	950	+ 23	− 396	
,, on	925	− 25	− 371	
,, off	950	+ 25	− 396	
B..............	777	− 173	− 223	
295 lbs. on	715	− 62	− 161	
,, off	746	+ 31	− 192	
,, on	715	− 31	− 161	
,, off	744	+ 29	− 190	
+ M..............	155	− 589	+ 399	282

TABLE IV.—Bar of Wrought Nickel—continued.

Operations.	Readings. Zero 554.	Differences.	Total magnetization.	Strength of magnetizing current.
285 lbs. on	181	+ 26	+ 373	
,, off	156	− 25	+ 398	
,, on	182	+ 26	+ 372	282
,, off	156	− 26	+ 398	
B	327	+ 171	+ 227	
285 lbs. on	388	+ 61	+ 166	
,, off	357	− 31	+ 197	
,, on	388	+ 31	+ 166	
,, off	358	− 30	+ 196	
Zero changed to	500	Difference) from zero. (		
− M	940	+ 440	− 440	432
285 lbs. on	925	− 15	− 425	
,, off	940	+ 15	− 440	
,, on	923	− 17	− 423	
,, off	939	+ 16	− 439	
B	726	− 213	− 226	
285 lbs. on	665	− 61	− 165	
,, off	697	+ 32	− 197	
,, on	664	− 33	− 164	
,, off	696	+ 32	− 196	
+ M	48	− 648	+ 452	432
285 lbs. on	63	+ 15	+ 437	
,, off	48	− 15	+ 452	
,, on	63	+ 15	+ 437	
,, off	47	− 16	+ 453	
B	268	+ 221	+ 232	
285 lbs. on	329	+ 61	+ 171	
,, off	297	− 32	+ 203	
,, on	330	+ 33	+ 170	
,, off	298	− 32	+ 202	
+ M	13	− 285	+ 487	580
285 lbs. on	25	+ 12	+ 475	
,, off	16	− 9	+ 484	
,, on	28	+ 12	+ 472	
,, off	18	− 10	+ 482	
B	268	+ 250	+ 232	
285 lbs. on	334	+ 66	+ 166	
,, off	299	− 35	+ 201	
,, on	334	+ 35	+ 166	
,, off	299	− 35	+ 201	
− M	963	+ 664	− 463	580
285 lbs. on	954	− 9	− 454	
,, off	963	+ 9	− 463	
,, on	952	− 11	− 452	
,, off	961	+ 9	− 461	
B	724	− 237	− 224	
285 lbs. on	664	− 60	− 164	
,, off	697	+ 33	− 197	
,, on	663	− 34	− 163	
,, off	696	+ 33	− 196	
Zero changed to	617	Difference) from zero. (		
− M	900	+ 283	− 479	708

TABLE IV.—Bar of Wrought Nickel—continued.

Operations.	Readings. Zero 554.	Differences.	Total magnetization.	Strength of magnetizing current.
285 lbs. on	892	− 8	− 471	
,, off	897	+ 5	− 476	708
,, on	889	− 8	− 468	
,, off	897	+ 8	− 476	
,, B	645	− 252	− 224	
285 lbs. on	582	− 63	− 161	
,, off	616	+ 34	− 195	
,, on	582	− 34	− 161	
,, off	615	+ 33	− 194	
Zero changed to	765	Difference from zero.		
+M	74	− 691	+ 497	708
285 lbs. on	81	+ 7	+ 490	
,, off	76	− 5	+ 495	
,, on	83	+ 7	+ 488	
,, off	76	− 7	+ 495	
361 lbs. on	92	+ 16	+ 479	
,, off	80	− 12	+ 491	
,, on	93	+ 13	+ 478	
,, off	80	− 13	+ 491	
285 lbs. on	91	+ 11	+ 480	
,, off	85	− 6	+ 486	
,, on	94	+ 9	+ 477	
,, off	85	− 9	+ 486	
211 lbs. on and off 5 times	90	+ 5	+ 481	
211 lbs. on	96	+ 6	+ 475	
,, off	90	− 6	+ 481	
,, on	96	+ 6	+ 475	
,, off	91	− 5	+ 480	
136 lbs. on	96	+ 5	+ 475	
,, off	93	− 3	+ 478	
,, on	97	+ 4	+ 474	
,, off	93	− 4	+ 478	
,, B	350	+ 257	+ 221	
136 lbs. on	380	+ 30	+ 191	
,, off	363	− 17	+ 208	
,, on	384	+ 21	+ 187	
,, off	365	− 19	+ 206	
,, on	384	+ 19	+ 187	
,, off	365	− 19	+ 206	
361 lbs. on	418	+ 53	+ 153	
,, off	377	− 41	+ 194	
,, on	420	+ 43	+ 151	
,, off	379	− 41	+ 192	
285 lbs. on	412	+ 33	+ 169	
,, off	379	− 33	+ 202	
,, on	411	+ 32	+ 170	
,, off	378	− 33	+ 203	

237. The preceding table shows that the effect of the application of pull was to diminish, and of the removal of pull to increase the magnetism of the bar, whether induced or residual; and that a series of these operations *increased* on the whole the magnetism induced by + M or − M, but diminished on the whole the residual magnetism after B. Further, as stated above, the effect of "on" or "off" on the induced magnetism of the bar increases up to a certain point, and then diminishes as the magnetizing force is increased from zero upwards; while, on the other hand, the effect of "on" or "off" on the residual magnetism goes on increasing as the residual magnetism is increased, and, as does also the residual magnetism, approaches more and more to a certain constant value.

238. On account of the thickness of the bar, a large amount of wire was required to make a coil which would give a sufficiently powerful magnetizing force to reach or pass the critical value, to which the magnetizing force seemed in the preceding experiments to approach. It was found more convenient to continue the experiment with a smaller bar of nickel, kindly lent for the purpose by Professor Tait. This (which was a square bar 45·7 centims. long and ·3 centim. thick) was placed within a coil wound on a thin copper tube of just sufficient internal diameter to admit the bar. The coil contained six layers of silk-covered copper wire, of No. 22 B.W.G., each layer forming a solenoid 42·7 centims. long, containing 10·7 turns per centim. The total resistance of the coil when cool was 4·33 ohms. The resistance of the electrodes was, as before, ·3 of an ohm. The magnetometer needle was in this case placed on a level with the upper end of the bar, and at a distance of 25 centims. from the axis of the bar, and of 108 centims. from the scale of half millims. on which the readings were observed. The stress was not applied by means of the lever, but a weight of 14 lbs. was placed on a pan attached to the bar. This pan, which weighed 1 lb., was left hanging on the bar during the whole experiment.

239. In their general character the results are precisely similar to those shown in Table IV. A maximum effect of 20 divisions was obtained when the magnetizing force was 194 divisions of the battery galvanometer scale, or that due to about 4 cells. As the magnetizing force was increased beyond this value the effects

obtained gradually diminished, and seemed to reach zero when the magnetizing force was about 1000 divisions, or that due to about 40 cells. The results at this point could not, on account of variations of the magnetizing force due to heating of the coil, be relied on as being accurate.

[Note added June 4, 1879.—This result has not been confirmed by experiments lately made with improved apparatus in which the effects of the heating of the magnetizing coil, formerly a great source of trouble, were to a great extent prevented. No sign of a neutral point was found, although battery powers of from 5 to 79 tray cells were employed for the purpose of magnetizing the nickel bar, which was the actual bar formerly experimented on.]

240. An experiment was then made to find whether this critical value could, as in the case of the soft iron tube, be obtained with a smaller degree of magnetizing force when the magnetometer was placed on a level with a point between the middle and end of the bar. Accordingly, the magnetometer was lowered 10 centims., and brought to a distance of 10 centims. from the axis of the bar. The scale was left in its former position, and thus the distance of the mirror from it was increased to 123 centims. The directive force on the needle was also increased to 6·08 times that due to the horizontal component of the earth's magnetic force. The maximum effect of the application and removal of stress was obtained with a magnetizing force of 50 scale-divisions, or that due to one cell. And the critical value was found without difficulty to be 428 divisions of the battery galvanometer scale, or the current due to 10 cells. Beyond that point the effect of " on " and " off " were respectively to increase and to diminish the induced magnetization of the bar, the effect with 707 divisions, or 20 cells, being about $3\frac{1}{2}$ divisions of the scale.

§§ 241—244. *Experiment by the direct Magnetometric Method on the effects of longitudinal stress on the Magnetization of Iron Wire.*

241. In this experiment the magnetizing coil described above (§ 201) was employed. The total length of the wire (which was a piece of Messrs Johnson and Nephew's very soft iron wire, and cut from the same hank as that used in the former experiments) was 97 centims. As only the soldered fastenings of the wire projected

beyond the ends of the coil, the magnetometer needle was placed on a level with the top of the coil, and at a distance from its axis of 25 centims., and the distance of the scale of the magnetometer from the mirror was 108 centims.

242. This experiment confirmed in all essential points the results obtained by the ballistic method in the experiments described above (§§ 200—212). The effects of applying and removing a weight of 14 lbs. were respectively to *increase* and to *diminish* by 31 divisions the magnetism induced in the wire by the vertical component of the earth's magnetic force. The amount of this induced magnetism, when only the pan was hanging on the wire, was also 31 divisions. Hence the application and the removal of the 14 lbs. alternately doubled, and reduced to its previous amount, the magnetization induced by the earth's force.

When the magnetizing current was 4·25 divisions of the battery-galvanometer scale, the application and removal of pull produced no effect, thus showing that the influence of the earth's magnetizing force was exactly counterbalanced by that due to the current. The maximum effect of applying and removing stress was obtained when the magnetizing force was about 50 divisions (one cell gave 70 divisions). The Villari critical value was obtained with 215 divisions of magnetizing current, or 50 times the magnetizing force which balanced the influence of the earth's magnetic force. This is a much greater number than that obtained by the ballistic method; that it is so is no doubt due to the fact that the induction-coil then used was much shorter than the magnetizing coil, and was placed with the centre of its length coincident with that of the magnetizing coil. With the highest strength of current (567 divisions) the effect of " on " and " off " was 8·5 divisions, and the effects were, as in the former experiments, increasing very slowly.

243. The value in absolute units of the vertical component of the earth's magnetic force was calculated from the value of the magnetizing current which balanced it, and the total resistance of coil electrodes and battery, which were all measured for the purpose with great exactness. As before, the electromotive force of one tray cell was taken as one volt, or 10^8 on the C. G. S. system of units. The total resistance with one cell in circuit was 3·503 ohms, the number of turns per centimetre in the coil 19·628, and the

difference of potentials which gave a magnetizing force which just balanced the earth ·0607 of that due to one cell. Hence we have

$$\text{vertical component} = \frac{4\pi \times \text{·0607} \times 10^8 \times 19\text{·}628}{3\text{·}503 \times 10^9} = \text{·429},$$ which must

be very nearly its true value at Glasgow. (Compare with § 207 above.)

244. It is stated above (§ 211) that the Villari critical value was higher for the smaller weights. This result was also verified by the magnetometric method.

The following are the numbers obtained with various amounts of pull:—

SOFT IRON WIRE.

Weights "on" and "off."	Villari critical value.
6 lbs.	248
10 „	227
14 „	215
18 „	190
26 „	185

APPENDIX.　(ABSTRACTS I., II., III.)

ABSTRACT I.　PARTS I.—V.

[From the *Proc. Roy. Soc.* VIII. Feb. 1856; *Phil. Mag.* Nov. 1856.]

The Lecturer gave an exposition of the substance of a paper presented by him to the Society under the above title.

The paper consists of five parts, namely:—(1) On the Electric Convection of Heat; (2) On Thermo-electric Inversions; (3) On the Effects of Mechanical Strain and of Magnetization on the Thermo-electric Qualities of Metals; (4) On Methods for comparing and testing Galvanic Resistances, illustrated by Preliminary Experiments on the Effects of Tension and Magnetization on the Electric Conductivity of metal; (5) On the Effects of Magnetization on the Electric Conductivity of Iron.

(1) In the first part a full account of the experiments, of which the results were communicated to the Royal Society in April 1854*, is preceded by a short statement of the reasoning, founded on incontrovertible principles regarding the source of energy drawn upon by a thermo-electric current, which led the author to commence the experimental investigation with the certainty that the property looked for really existed whether he could find it or not. In confirmation of the extraordinary conclusion then announced,— that an electric current in an unequally heated conductor, if its *nominal direction* be from hot to cold through the metal, causes a cooling effect in iron, and a heating effect in copper,—the author describes new experiments which he has recently made, and which are as decisive in leading to the same conclusion as those by which he had first established it. He also describes experiments by which he had recently given an independent demonstration that brass has the same property as copper, and platinum the same quality as iron, with reference to electric convection of heat; results anticipated†, one as certain, and the other as highly probable, from the previous results regarding electric convection in copper and iron, and from the known thermo-electric relations between these metals and the others.

---

* See *Proc. Roy. Soc.*, May 4, 1854 [Art. LI. Vol. I. above].

† See *Proc. Roy. Soc.*, May 4, 1854 [Art. LI. Vol. I. above]; also "Dynamical Theory of Heat," Part VI. § 135, *Trans. R.S.E.*, May 1, 1854 [Art. XLVIII. Vol. I. above].

(2) The phenomenon of thermo-electric inversion between metals, discovered by Cumming, forms the subject of the second part. A mode of experimenting is described, by which inversions may be readily detected when they exist between any two metals, and, when thermometers are available, the temperature of neutrality determined with precision. Various results of its application are mentioned, of which some are shown in the following Table:—

-14° C.	-12°·2.	-5°·7.	-3°·06.	-1°·5.	8°·2.	33°.	36°.	38°.	44°.	44°.	47°...71°.
P_3	P_1	Silver.	P_1	P_1	P_1	Tin.	P_2	P_2	P_2	Lead.	Different specimens of Silver.
Brass.	Cadmium.	Gold.	Gold.	Silver.	Zinc.	Brass.	Lead.	Brass.	Tin.	Brass.	Different specimens of Zinc.

53°	57*	64°.	71°.	99°.	121°.	130°.	162°·5.	Some temperature between 223° & 253°·5.	237°.	280°.
P_2 Double wire of Palladium 11·31 grs. and Copper 19·41 grs.	Hard steel.	P_1	Gold.	P_1	P_1	P_1	P_2	Iron.	Iron.	Iron.
	Cadmium.	Copper.	Zinc.	Brass.	Lead.	Tin.	Cadmium.	Gold.	Gold. — Silver. Copper.	Iron. — Copper.

The number at the head of each column expresses the temperature Centigrade by mercurial thermometers, at which the two metals written below it are thermo-electrically neutral to one another; and the lower metal in each column is that which passes the other from *bismuth towards antimony as the temperature rises.* P_1, P_2, P_3 denote three particular specimens of platinum wire, used by the author as standards.

* This determination has been added in consequence of information given by Mr JOULE (December 1856), that hardened steel at ordinary temperatures differs thermo-electrically from copper by about one-tenth of the thermo-electric difference of iron from copper.

It was also found that Aluminium must be neutral to either
P_3, or Brass, or P_2, at some temperature between -14° C. and
38° C.; that Brass becomes neutral to Copper at some high temperature, probably between 800° and 1400° C.; Copper to Silver, a
little below the melting-point of Silver; Nickel to Palladium, at
some high temperature, perhaps about a low red heat; and P_3
to impure mercury (that had been used for amalgamating zinc
plates), at a temperature between -10° C. and 0° C. P_3 appears to
become neutral to pure mercury at some temperature not much
below -25° C.

(3) In the third part, effects of mechanical strain, and of
magnetization on the thermo-electric qualities of metals, are investigated. The author had previously communicated to the
Royal Society* results he had obtained regarding the thermo-
electric qualities of copper and of iron wires under longitudinal
stress, namely, that the former exhibits a deviation towards bismuth, and the latter towards antimony, from the same metal in
an unstrained state.

The only kind of stress applicable to a solid which has no
directional attributes, is uniform pressure or traction in all directions. Hence it appeared probable to the author that a simple
longitudinal stress would induce different thermo-electric qualities
in different directions, in any homogeneous non-crystalline metal
subjected to it. But he had found (see *Proc. Roy. Soc.*, May 4,
1854)* that the thermo-electric effect of longitudinal traction on a
wire, either of iron or of copper, is sensible to tests he could readily
command, and more so in the case of the former than in that of
the latter. He therefore made experiments to test the difference
of thermo-electric quality in different directions in a mass of iron
under stress, and fully established the conclusion that the thermo-
electric quality across lines of traction differs from the thermo-
electric quality along lines of traction, as bars of bismuth differ
from bars of antimony. The experiments he has already made
nearly establish the conclusion, that unstrained iron has intermediate thermo-electric quality between those of the two critical
directions in iron under distorting stress.

The experiments of Magnus show that wires hardened by wire-
drawing have different thermo-electric qualities lengthwise from

* April 1854. See *Proc. Roy. Soc.*, May 4, 1854 [Art. LI. Vol. I. above].

wires of the same substance softened by annealing. The author
has verified, that in copper, iron, and tin, simple traction, leaving
permanent elongation, leaves also a thermo-electric effect, the same
as Magnus had found by wire-drawing, which is a composite application of longitudinal traction and lateral compression; and that
in a variety of metals, namely, iron, copper, brass, tin, platinum,
permanent lateral compression (by hammering) leaves still the
same thermo-electric effect, as Magnus had found by wire-drawing.
In cadmium, not examined by Magnus, and lead, which had not
a given result, the experiments now adduced show a thermo-
electric effect of hammering, the same as in all the other metals
except iron. Zinc wire was also tested, and found to exhibit the
same effect as copper, though Magnus had found a reverse quality
as due to wire-drawing. The discrepance in this case is probably
due to the peculiar effect of annealing on zinc wire, making it
brittle and crystalline, which might give a different condition, as
the "annealed" in Magnus's experiment, and the "unhammered"
in the experiment now adduced. Setting aside this case, the
author concludes that generally the effect of permanent lateral
compression is the same as that of permanent longitudinal extension, or of hardening by wire-drawing, upon the thermo-electric
quality of a wire placed longitudinally in an electric circuit; that
in iron it is a deviation from the constrained metal towards
bismuth, and that in all the other metals mentioned it is a
deviation towards antimony; and that in copper and iron it is
the reverse of the effect experienced by the same metal while
under the stress that caused the strain. Since no kind of strain,
except uniform condensation or dilatation in all directions, is free
from the directional attribute, it appeared probable to the author
that the thermo-electric effects remaining in a metal left with a
longitudinal strain, retained after the stress that caused it is
removed, must be different in different directions. He therefore
experimented on iron hardened by longitudinal compression, and
found that it deviates from soft iron towards antimony, or in the
contrary way to iron hardened by longitudinal traction. From
this, and from the results quoted above, it follows that in iron
hardened by compression in one direction, the thermo-electric
qualities in this direction differ from those in lines perpendicular
to it, as antimony differs from bismuth; that the reverse statement applies to iron hardened by traction in one direction; and

that these differing thermo-electric qualities have in each case
the thermo-electric quality of soft iron intermediate between them.

These various results show that the character of the effect in
each case is decided by *distorting stress* or by *distortion*, and leave
entirely open, and only to be answered by further experiments,
the questions: what is the thermo-electric effect of pressure or
traction, applied uniformly in all directions to a metal? and what
is the thermo-electric effect of a permanent condensation or dilatation remaining in the metal, when freed from the force by
which that condensation or dilatation was produced?

Experiments are also described, by which the author found
that in soft iron under magnetic force, and in that retaining
magnetism when removed from the magnetizing force, directions
along the lines of magnetization deviate thermo-electrically towards antimony; and that directions perpendicularly across the
lines of magnetization in soft iron, deviate towards bismuth, from
the unmagnetized metal.   He illustrates this conclusion by an
experiment on a ribbon of iron, magnetized nearly at an angle
of 45° to its length, and heated along one edge while the other
is kept cool.   When the two ends, kept at the same temperature,
are put in communication with the electrodes of a galvanometer,
a powerful current is indicated, in such a direction, that if pur-
sued along a rectangular zigzag from edge to edge through the
band, the course is always *from across to along the lines of magnetization through the hot edge, and from along to across the lines
of magnetization through the cold edge.*

(4) In this part of the communication, attempts made by the
author to find the effects of various influences on electric conductivities of metals are described.   One of these, with a very
unsatisfactory method for testing resistances, led to the conclusion
that longitudinal magnetization diminishes the conducting quality
of iron wire.   The general plan for testing resistances, which he
subsequently adopted as the best he could find, and which has
proved very satisfactory, is next explained ; and as an illustration,
a single experiment on the relative effect of an equal longitudinal
extension on the resistances of iron and copper wires is described.
The conclusion established by this experiment is, that both by
extension with the tractive force still in operation, and by permanent extension retained after a cessation of stress, the conductivity of the substance is more diminished in iron than in

copper; or else that it is more increased in copper than in iron,
or increased in copper while diminished in iron, if it is not in
each metal diminished, as the author is led by a partial investigation of the absolute effect in each metal to believe.

(5)   The result previously arrived at regarding the effect of
longitudinal magnetization on the conductivity of iron is confirmed; and an experiment that would have been found impracti-
cable by the less satisfactory method, proves the same conclusion
for magnetized steel wire, with the magnetizing influence away.
Two very different experiments show further, that the electric
conductivity of magnetized iron is greater across than along the
lines of magnetization.   A last experiment, showing that iron
gains in conducting power by magnetization across the lines of
the electric current, leads to the conclusion that there is a direction inclined obliquely to the lines of magnetization, along which
the conductivity of magnetized iron would remain unchanged on
a cessation of the magnetizing force.

ABSTRACT II. PART VI.

[From the *Proc. Royal Soc.*, May, 1875.]

Effects of Stress on Magnetization.

Weber's method, by aid of electromagnetic induction and a
" ballistic galvanometer" to measure it, which has been practised
with so much success by Thalén, Roland, and others, has been
used in the investigation of which the results are at present
communicated; but partial trials have been made by the direct
magnetometric method (deflections of a needle), and this method
is kept in view for testing slow changes of magnetization which
the electromagnetic method fails to detect.

The metals experimented on have been steel pianoforte-wire,
of the kind used for deep-sea soundings by the American Navy
and British cable-ships; and soft-iron wires of about the same
gauge, but of several different qualities.

I.   *Steel.*

The steel wire weighs about $14\frac{1}{2}$ lbs. per nautical mile and
bears 230 lbs.   Weights of from 28 lbs. to 112 lbs. were hung on
it and taken off, and results described shortly as follows were
found :—

(1)   The magnetization is diminished by hanging on weights, and increased by taking the weights off, when the magnetizing current is kept flowing.

(2)   The residual magnetism remaining after the current is stopped is also diminished by hanging on the weights, and increased by taking them off.

(3)   The absolute amount of the difference of magnetization produced by putting on and taking off weights is greater with the mere residual magnetism when the current is stopped, than with the whole magnetism when the magnetizing current is kept flowing.

(4)   The change of magnetization produced by making the magnetizing current always in one direction and stopping it is greater with the weights on than off.

(5)   After the magnetizing current has been made in either direction and stopped, the effect of making it in the reverse direction is less with the weights on than off.

(6)   The difference announced in (5) is a much greater difference than that in the opposite direction between the effects of stopping the current with weights on and weights off, announced in (4).

(7)   When the current is suddenly reversed, the magnetic effect is less with the weights on than with the weights off.

II.   *Soft-Iron Wires.*

Wires of about the same gauge as the steel were used, but, except one of them, bore only about 28 lbs. instead of 230 lbs. All of three or four kinds tried agreed with the steel in (1).

The first tried behaved (excepting a seeming anomaly, hitherto unexplained) in the reverse manner to steel in respect to (2), (4), (5), and (6); it agreed with the steel in respect to (7).   Another iron wire*, which, though called "soft," was much less soft than the first, agreed with steel in respect to (1) and (2), but [differing from steel in respect to (3)] showed greater effects of weights on and off when the magnetizing current was flowing than when it was stopped.

Other soft-iron wires which were very soft, softer even than the first, agreed with all the steel and iron wires in respect to (1),

---

* It was tested magnetically with weights up to 56 lbs., and broke, unfairly however, when 63 lbs. were hung on.

but gave results when tested for (2) which proved an exceedingly transient character of the residual magnetism, and were otherwise seemingly anomalous.

The investigation is being continued with special arrangements to find the explanation of these apparent anomalies, and with the further object of ascertaining in absolute measure the amounts of all the proved effects at different temperatures up to 100° C.

ABSTRACT III.   PART VII.

[From the *Proc. Royal Soc.*, May, 1878.]

Effects of Stress on Magnetization of Iron, Nickel, and Cobalt.

This paper commences with a detailed description of a series of experiments on the effects of stress on the magnetism of soft iron, of which some first results were described in a preliminary notice, communicated to the Royal Society on the 10th of June, 1875 and published in the *Proceedings* (§§ $197_1'$—$197_7'$ above) of that date. A few months later, the author found that he had been anticipated by Villari[*] in the most remarkable of those results—that showing increase or diminution of magnetization by longitudinal pull, according as the magnetizing force is less than, or greater than, a certain critical value.

In the first series of experiments described in this paper, the amount of the magnetizing force is varied through a range of values from zero to 900, on a scale on which about $12\frac{1}{2}$ is the value of the vertical component of the terrestrial magnetic force at Glasgow, and the effects of hanging on and taking off weights of 7 lbs., 14 lbs., and 21 lbs.[†] in changing the induced magnetism, are observed. The experiments were made at ordinary atmospheric temperatures, and at temperature 100° C. The results are shown in curves (Plates II. to XII. below), of which the abscissas represent the magnetizing forces and the ordinates, the change of magnetism of the wire produced by "ons" and "offs" of the weight while the magnetizing force is kept constant. The Villari critical value was found to differ for the two temperatures, and for different weights: thus approximately:—

[*] Poggendorf's "Annalen," 1868.

[†] The wire was of about No. 22 B. W. G. gauge, and weighing therefore about 14 lbs. per nautical mile. It was so soft that it had experienced a considerable permanent stretch by 21 lbs.; it would probably break with 30 or 40 lbs. Steel pianoforte wire of same gauge bears about 230 lbs.

Amount of weight "on" and "off."	Magnetizing force for which the "on" and "off" produce no change of magnetism.	
	At atmospheric temperature (being about 15° C.)	At temperature 100° C.
7 lbs.	266	280 or 290
14 ,,	281	286
21 ,,	288	310

The maximum effect of the "on" and "off" was found in each case with a magnetizing force of from 50 to 60 of the arbitrary scale divisions (or about four times the Glasgow vertical force). Its amount differed notably, though not greatly, with the temperature, and, as was to be expected, greatly with the different amounts of pull; but it was not nearly three times as much with 21 lbs. as with 7 lbs.; thus approximately :—

Amount of weight "on" and "off."	Maximum effect in the way of augmentation of magnetism by "on" and diminution by "off."	
	Temperature about 15° C.	Temperature 100° C.
7 lbs.	31 scale divisions of ballistic galvanometer.	25 scale divisions.
14 ,,	35 do. do.	32·4 do. do.
21 ,,	54 do. do.	50·3 do. do.

The curves all tend to asymptotes parallel to the line of abscissas on its negative side for infinite magnetizing forces; and they indicate the following ultimate values for the two temperatures, and the different amounts of pull:—

Amount of weight "on" and "off."	Effect in the way of diminution of magnetism by "on" and augmentation by "off" when the magnetizing force is very great.	
	Temperature 15° C.	Temperature 100° C.
7 lbs.	6 scale divisions of ballistic galvanometer.	3 scale divisions.
14 ,,	13·5 do. do.	9·2 do. do.
21 ,,	21 do. do.	15·2 do. do.

For other features the curves themselves as given in the paper
may be looked to.

Later experiments on the effects of pull transverse to the
direction of magnetization showed correspondingly *opposite* effects
to those of longitudinal pull, but with a "critical value" of magnetizing force nearly twice as great. That for longitudinal pull,
according to the preceding figures, was about 23 times the Glasgow vertical force; for the transverse pull the critical value found
was about 60 times the Glasgow vertical force. The transverse
pull was produced by water pressure in the interior of a gun-
barrel applied by a piston and lever at one end. Thus a pressure
of about 1,000 lbs. per square inch, applied and removed at pleasure, gave effects on the magnetism induced in the vertical gun-
barrel by the vertical component of the terrestrial magnetic force,
and, again, by an electric current through a coil of insulated
copper wire round the gun-barrel. When the force magnetizing
the gun-barrel was anything less than about 60 times the Glasgow
vertical force, the magnetization was found to be *less* with the
pressure on than off. When the magnetizing force exceeded that
critical value, the magnetization was *greater* with the pressure on
than off. The residual (retained) magnetism was always less with
the pressure on than off (after ten or a dozen "ons" and "offs"
of the pressure to shake out as much of the magnetization as was
so loosely held as to be shaken out by this agitation).

The vertical component of the terrestrial magnetic force at
Glasgow is about ·43 c.g.s. units. Hence the critical values of
the magnetizing force for longitudinal and transverse pull are
approximately 10 and 25 c.g.s. units. With any magnetizing
force between these limits the effect of pull, whether transverse or
longitudinal, must be to diminish the magnetization. Hence it is
to be inferred that equal pull in all directions would diminish, and
equal positive pressure in all directions would increase, the magnetization under the influence of force between these critical
values, and through some range above and below them; and not
improbably for all amounts, however large or small, of the magnetizing force (?); but further experiment is necessary to answer
this question.

The opposite effects of longitudinal and transverse pull, for
magnetizing forces not between the critical range of from 10 to
25 c.g.s. units, show an aeolotropic magnetic susceptibility in

iron under aeolotropic stress [that is, any stress other than pressure (whether positive or negative) equal in all directions]. Consideration of the relation of this result to Wiedemann's remarkable discovery of the induction of longitudinal magnetization by twisting an iron wire through which an electric current is maintained, is important and suggestive. In the present paper a counter-influence is pointed out, in the aeolotropic change of electric conductivity probably produced in the iron by stress*. This influence was illustrated by experiments made a few days ago for the author, by Mr Macfarlane, in Glasgow, and Mr Bottomley, in the Physical Laboratory of King's College, London, by kind permission of Professor Adams, which show in two very different ways that twisting a brass tube through which a current of electricity is maintained gives to the electric stream lines a spirality of opposite name to that which the twist gives to longitudinal filaments of the substance, and so proves that in aeolotropically stressed brass the electric conductivity is greatest and least in the directions of greatest and least pressure. The same law probably holds for iron. Wiedemann's result is that the end of the iron wire by which the current enters, becomes a true north or a true south pole, according as the twist is that of a right-handed or of a left-handed screw. This is the same direction of effect as would result from the aeolotropy of the magnetic susceptibility produced by the stress if the tangential magnetizing force in the outer part of the wire is less than the critical value, for which the effect of the stress is isotropic; but it is opposite to the effect due to the aeolotropy of the electric conductivity. Yet the author in repeating Wiedemann's experiments has found his result—the same in direction, and greatest in amount—with the strongest currents he has hitherto applied—currents strong enough to heat the wire seriously (but not yet measured or estimated in absolute measure). The reconciliation of the Wiedemann result with the conflicting influence of conductive aeolotropy, and with the influence of aeolotropy of magnetic susceptibility, which also is conflicting when the magnetizing force is great enough, is a difficulty which calls for investigation.

The paper includes a series of experiments on the effects of

---

* See § 161 of Part V. above. Also a paper by Tomlinson, Proc. Roy. Soc., Dec. 21, 1876.

twist on magnetization of iron wire under longitudinal magnetizing force (the Glasgow vertical force alone in this first series).
It confirms results of previous experimenters, Matteucci, Wertheim, and Edmund Becquerel, according to which twist in either
direction diminishes the magnetization, and extends them to
wires under different amounts of longitudinal pull.   When the
pull was great—approaching the limit of elasticity of the wire—
the twist, even when well within the limits of elasticity, had much
less effect in diminishing the magnetism than when the pull was
small.   The results are recorded in curves which show a very
remarkable lagging of effect, or residue of influence of previous
conditions.

The paper concludes with a description of experiments, showing,
in bars of nickel and cobalt, effects of longitudinal pull opposite
to those found by Villari for iron, with magnetizing force below
the critical value—that is to say, the magnetization of the nickel
and cobalt was diminished by pull.   But this effect came to a
maximum, and began to diminish markedly as if towards zero,
when the magnetizing force was diminished.   Hitherto the critical
value, if there is one, has not been reached; but the experiments
are being continued to find it, if it is to be found, with attainable
degrees of magnetizing force.

(Addition, May 23, 1878.)

It had been reached, for nickel, in Glasgow, about the day on
which this abstract was written; advantage having been taken of
a kind loan, by Professor Tait, of a much smaller bar of nickel
than those which had been specially made for the investigation,
and which alone had been previously available.   Mr Thomas Gray,
by whom the experiments were made, in the Physical Laboratory
of the University of Glasgow, in the author's absence, found the
critical value of the magnetizing force for Professor Tait's thin
nickel bar to be about 600 times the Glasgow vertical force.

[The author is indebted to the celebrated metallurgical chemist, Mr Joseph
Wharton, of Philadelphia, for a splendid and unique set of bars, globes, and disks,
of pure nickel and cobalt, which he kindly made, at his request, for this and other
proposed investigations of electro-dynamic qualities of those metals.]

CAMBRIDGE: PRINTED BY C. J. CLAY, M.A. & SON, AT THE UNIVERSITY PRESS.

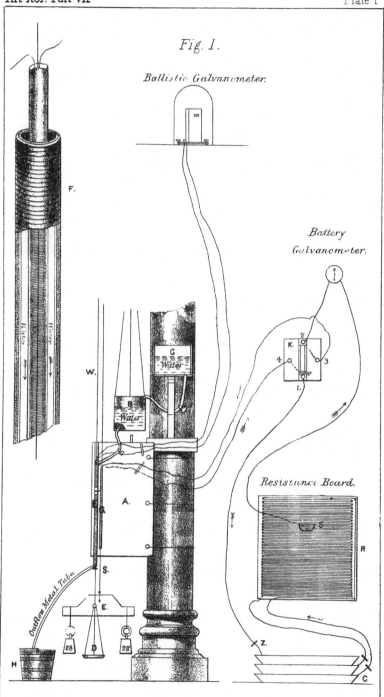

Fig. 1.

Ballistic Galvanometer.

Battery
Galvanometer.

Resistance Board.

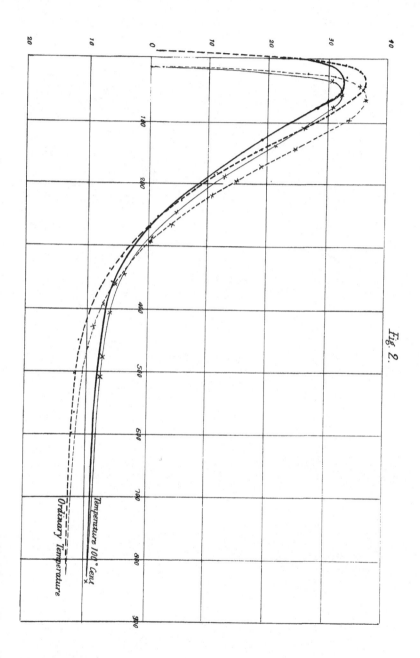

Fig. 2.

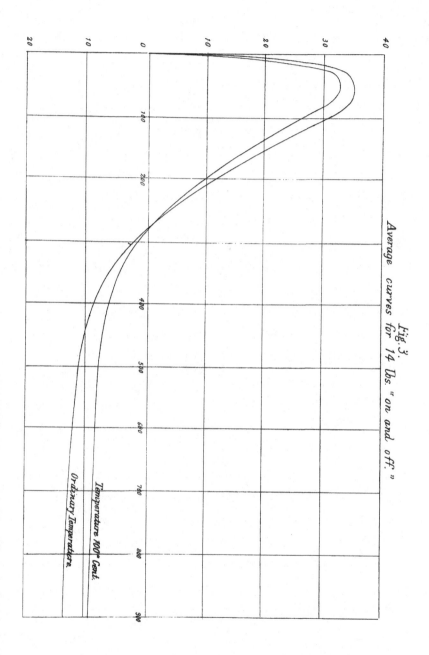

Fig. 3.
Average curves for 14 lbs. "on and off."

Ordinary Temperature.

Temperature 100° Cent.

Plate IV

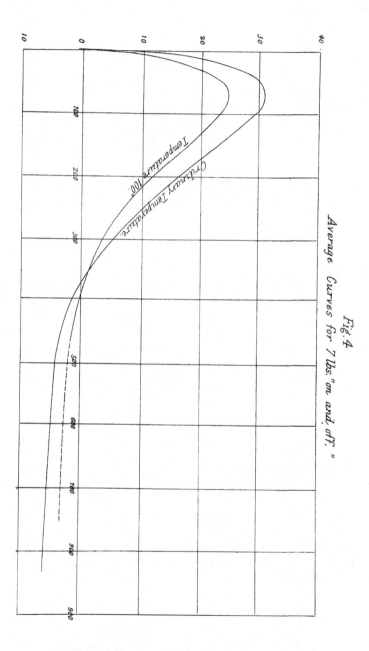

Fig. 4

Average Curves for 7 lbs. "on and off."

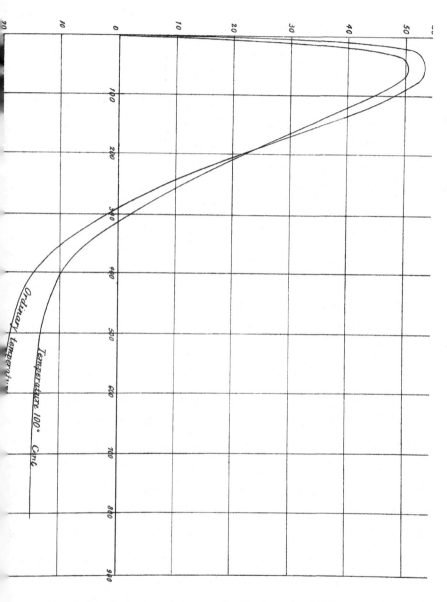

Fig. 5. Average curves for 21 lbs. on and off.

Harrison & Sons. Lith, S! Martins Lane.W.C.

Plate VI

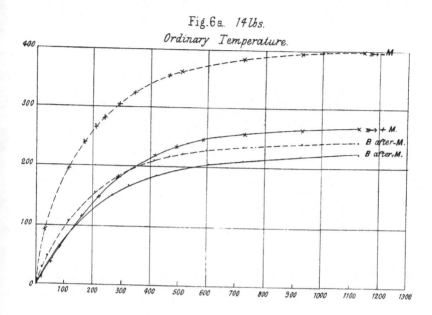

Fig.6a. *14 lbs.*
Ordinary Temperature.

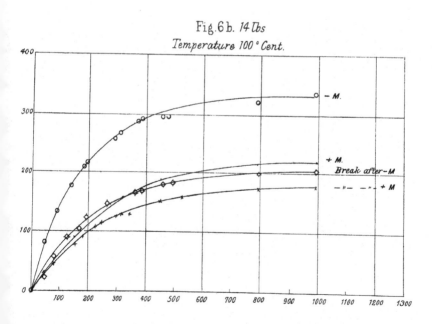

Fig.6b. *14 lbs*
Temperature 100° Cent.

Fig. 6 c. *21 lbs.*
Ordinary Temperature.

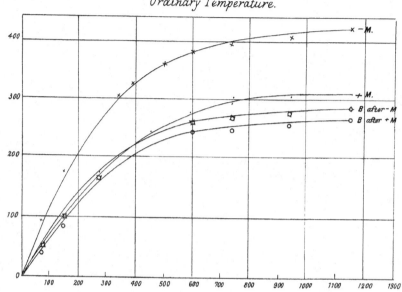

Fig. 6 d. *21 lbs.*
Temperature 100° Cent.

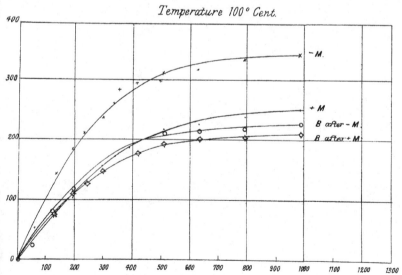

Fig.6e.

Temperature 100° Cent M and B with 21 lbs on.

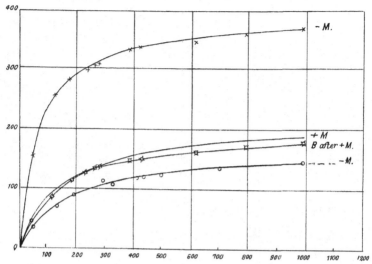

Fig. 11. (§.223).

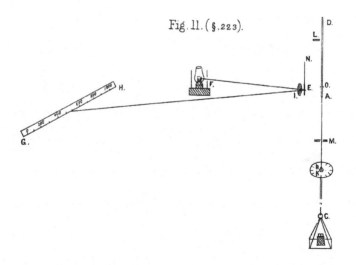

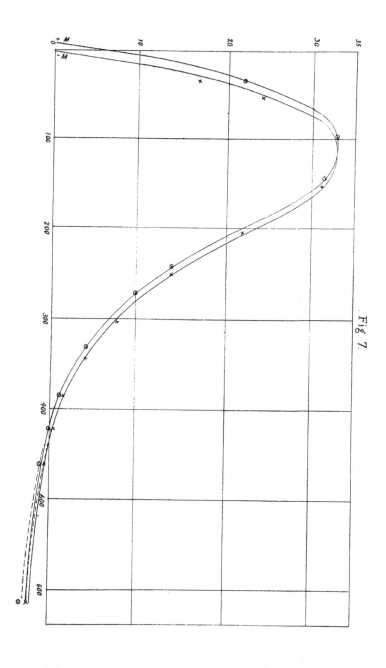

Fig. 7.

Plate X

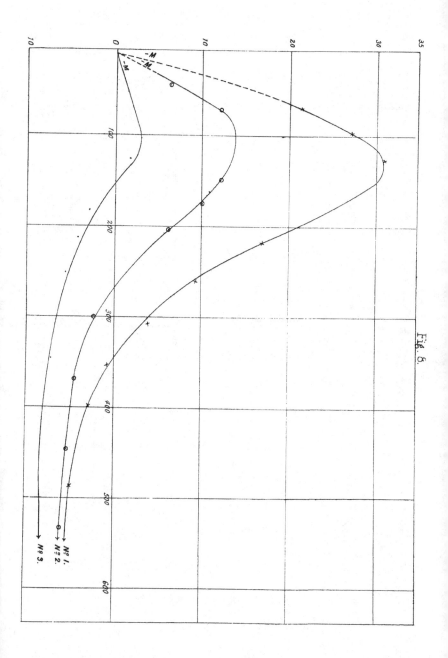

Fig. 8.

B.

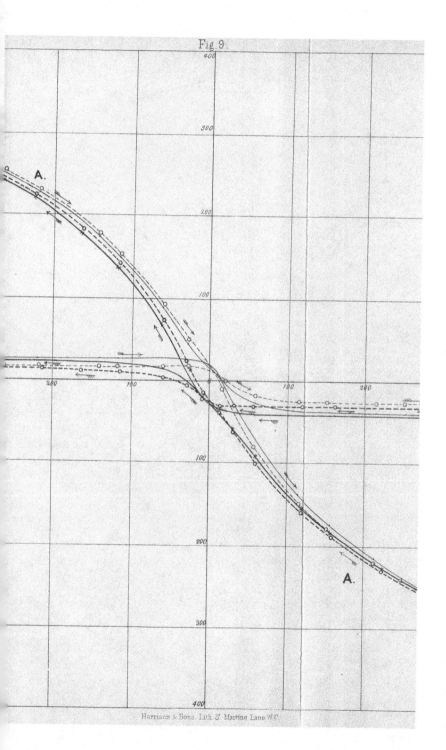

Fig. 9.

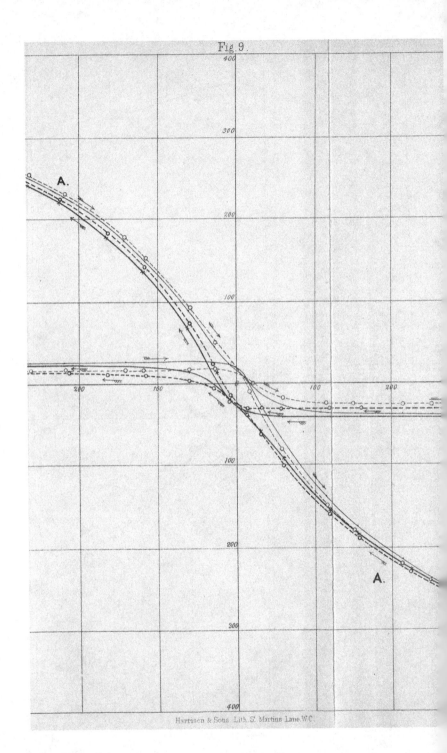

Fig. 9.

Plate XI.

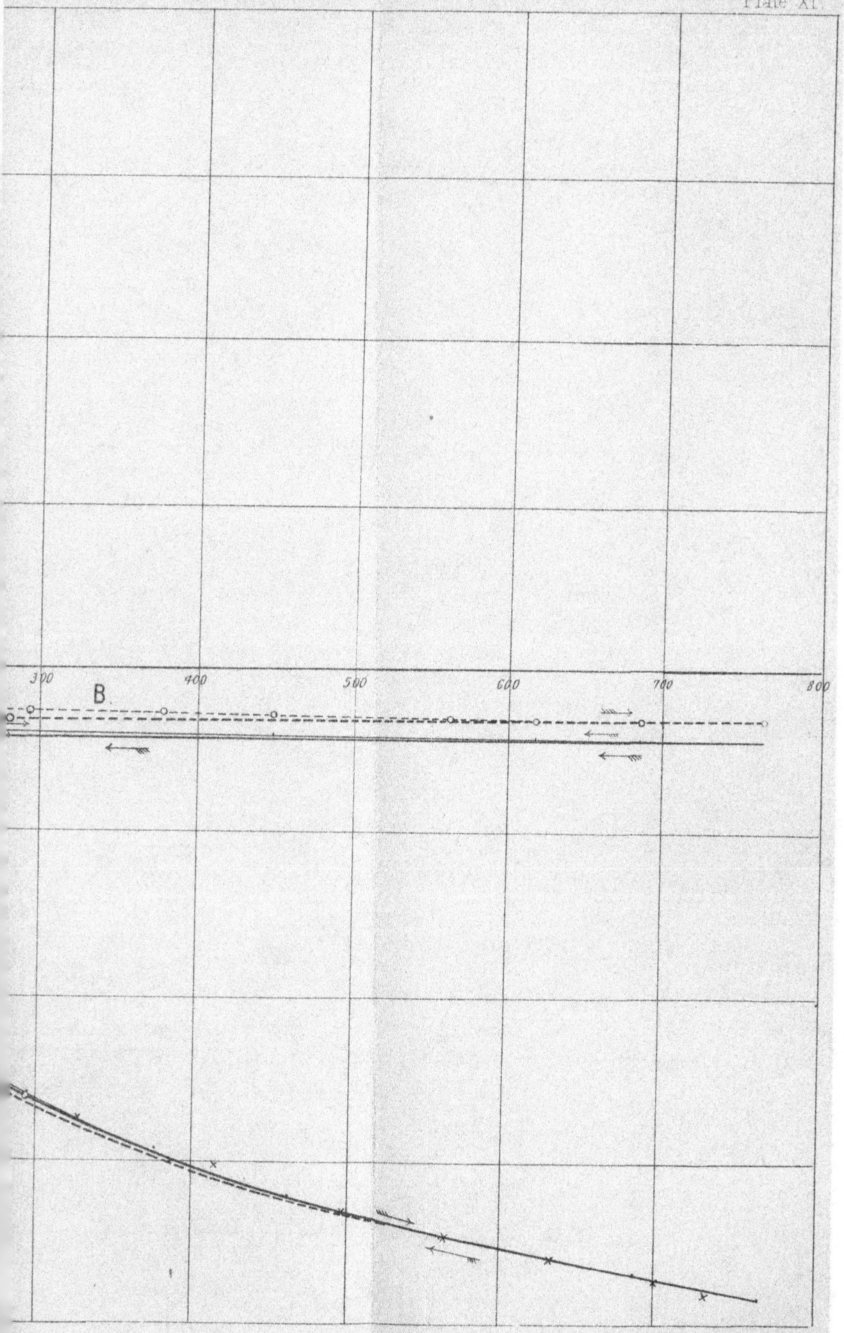

800   700   600   500   400   300

Residue diminished by ten ons an

B.

Full residue.

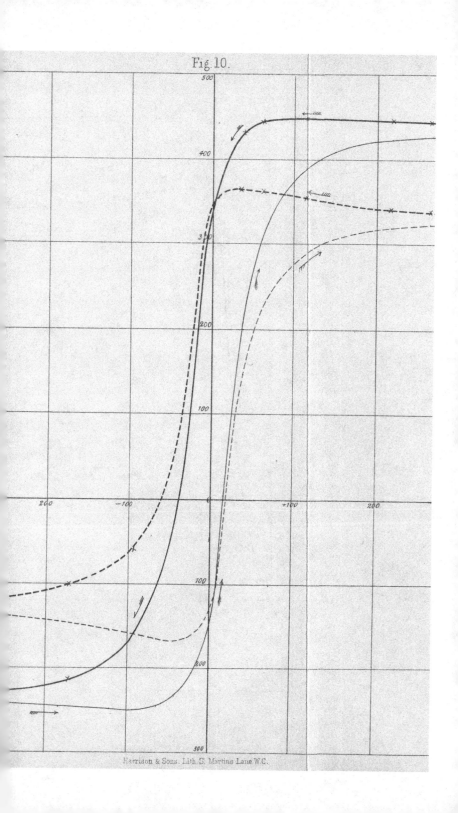

Fig. 10.

Fig. 10.

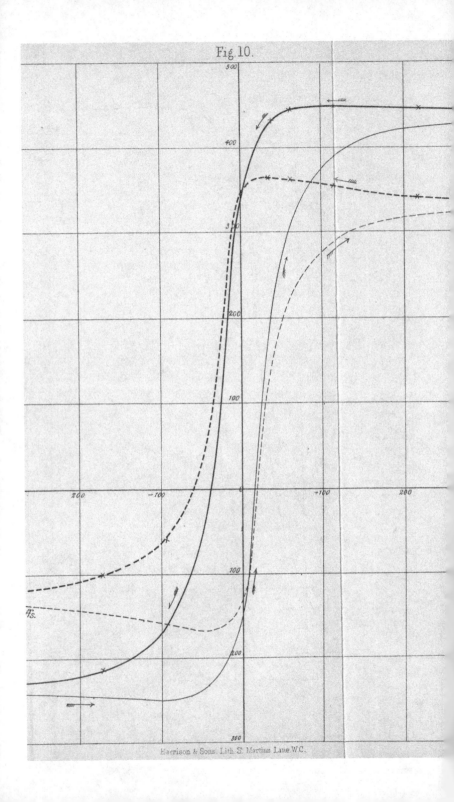

Harrison & Sons. Lith. S. Martins Lane W.C.

Plate XII.

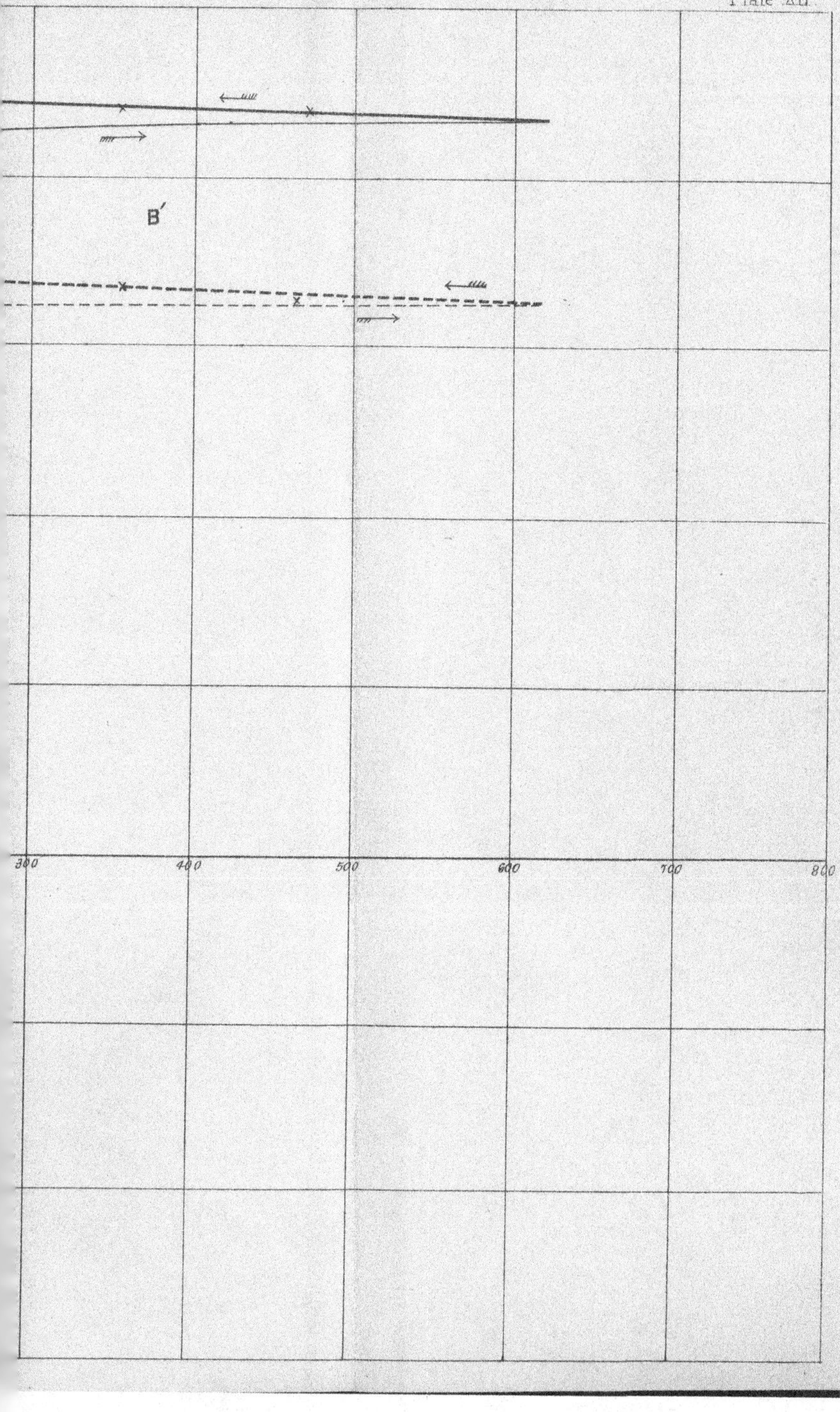

B'

300   400   500   600   700   800

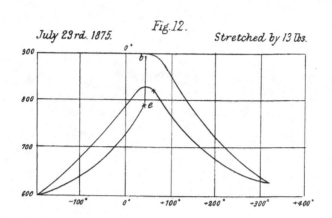

Fig. 12.

July 23rd. 1875.   Stretched by 13 lbs.

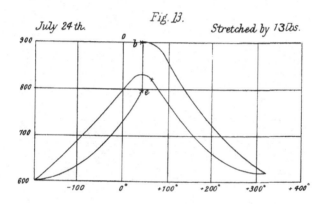

Fig. 13.

July 24th.   Stretched by 13 lbs.

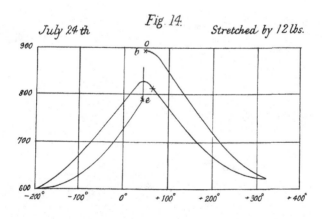

Fig. 14.

July 24th   Stretched by 12 lbs.

Harrison & Sons, Lith, St. Martins Lane, W.C.

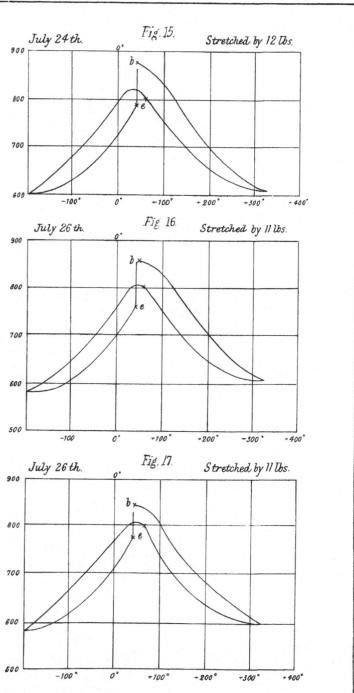

Fig. 15.

July 24th.

Stretched by 12 lbs.

Fig. 16.

July 26th.

Stretched by 11 lbs.

Fig. 17.

July 26th.

Stretched by 11 lbs.

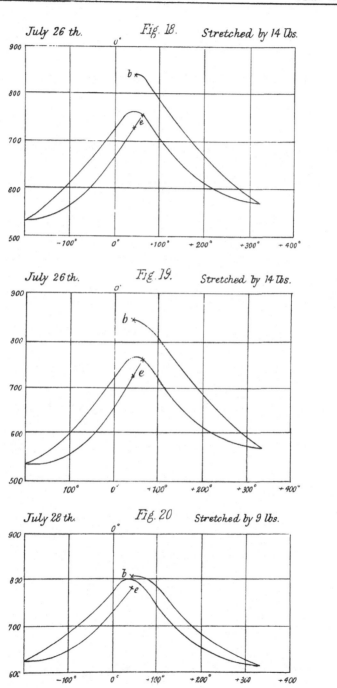

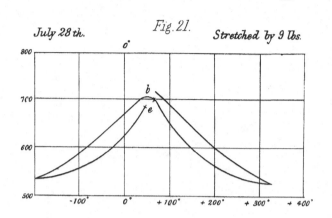

Fig. 21.
July 28th.   Stretched by 9 lbs.

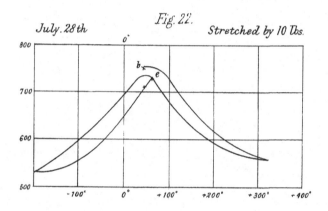

Fig. 22.
July. 28th   Stretched by 10 lbs.

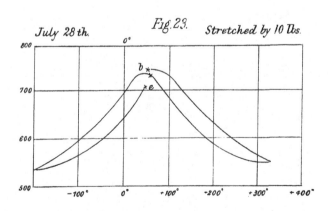

Fig. 23.
July 28th.   Stretched by 10 lbs.

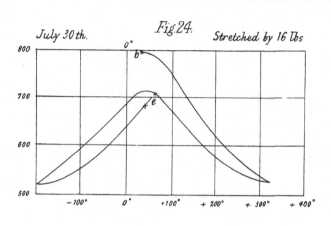

Fig. 24.

July 30th.

Stretched by 16 lbs

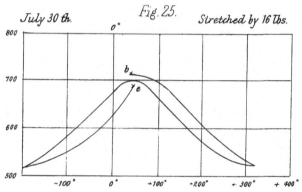

Fig. 25.

July 30th.

Stretched by 16 lbs.

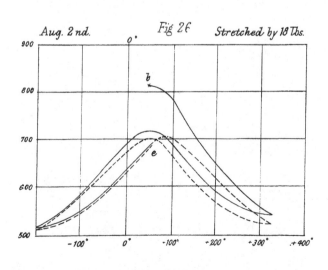

Fig 26

Aug. 2nd.

Stretched by 18 lbs.

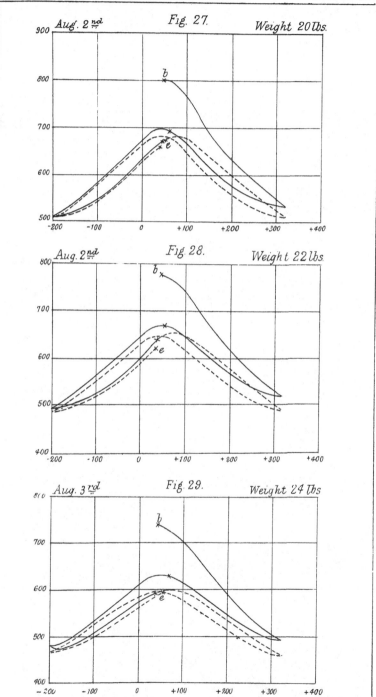

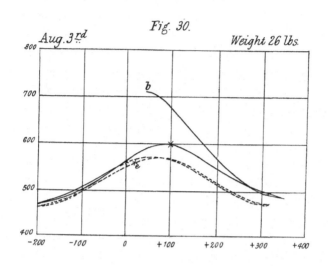

Fig. 30.

Aug. 3rd   Weight 26 lbs.

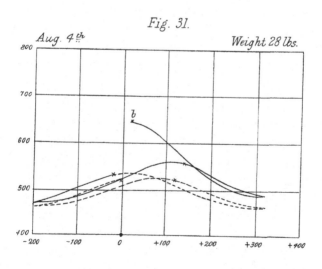

Fig. 31.

Aug. 4th   Weight 28 lbs.

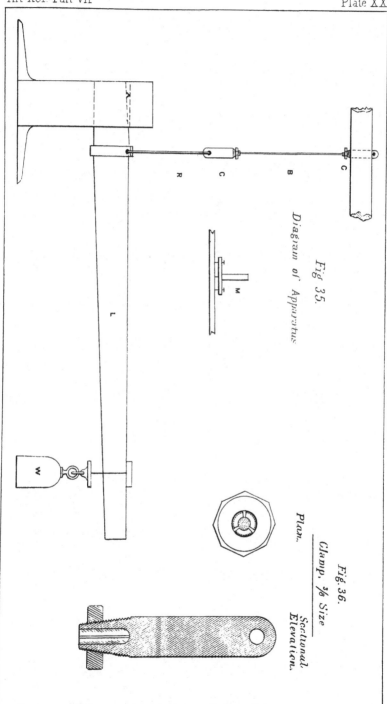

Fig 35.

Diagram of Apparatus

Fig. 36.

Clamp, 3/8 Size

Plan.

Sectional Elevation.

Printed in the United States
By Bookmasters